The Nature of Selection

The Nature of Selection

Evolutionary Theory in Philosophical Focus

Elliott Sober

The University of Chicago Press
Chicago and London

The University of Chicago Press, Chicago 60637
The University of Chicago Press, Ltd., London

Dover Books, Princeton University Press, and Oxford
University Press have granted permission to reprint
material from, respectively, R. A. Fisher, *The Genetical
Theory of Natural Selection*, George C. Williams, *Adaptation
and Natural Selection*, and Richard Dawkins, *The Selfish Gene*.
These are gratefully acknowledged.

Printed in the United States of America

01 00 99 98 97   6 5 4 3 2

ISBN 0-226-76748-5 (pbk.)

Library of Congress Cataloging-in-Publication Data

Sober, Elliott.
   The nature of selection : evolutionary theory in
philosophical focus / Elliott Sober.
      p.   cm.
   Originally published: Cambridge, Mass. : MIT Press, c1984.
   Includes bibliographical references and index.
   1. Science—Philosophy.   2. Evolution—Philosophy.
   3. Evolution (Biology)—Philosophy.   I. Title.
Q175.S6392   1993
501—dc20   93-10367
                                                    CIP

∞ The paper used in this publication meets the minimum
requirements of the American National Standard for
Information Sciences—Permanence of Paper for Printed
Library Materials, ANSI Z39.48-1984.

For the generation before
—Florence and Louis Sober—
and the generation after
—Sam and Aaron Sober—

Contents

Acknowledgments

Until about eight years ago, I had only the most cursory impression of what evolutionary theory is about. As if by osmotic diffusion, I had absorbed from the general scientific culture the fact of evolution as well as a vague notion of Darwin's principle of natural selection. My work in philosophy of science at that time did not concern the details of biological theories. For me, philosophy of science mainly addressed general questions about the nature of science as a whole. Although I was aware that philosophy of psychology and philosophy of physics had identified problems internal to the theories they study, comparable developments in philosophy of biology simply had not impinged on my consciousness.

Then, while browsing through an issue of *Philosophy of Science*, I happened upon Bill Wimsatt's review of George C. Williams' book *Adaptation and Natural Selection*. Wimsatt's enthusiasm suggested that this book might be fun to read. Being well prepared as a philosopher to appreciate the struggle in the sciences between holism and reductionism, I was delighted to find in Williams' work a set of biological problems in which those philosophical questions were live issues of scientific moment. Wimsatt and David Hull encouraged my curiosity, and I am grateful to them for welcoming a newcomer into philosophy of biology. I also am grateful to George Williams for writing a book of such philosophical depth.

My next and largest debt is to Richard Lewontin for inviting me to spend 1980–81 at his laboratory at the Museum of Comparative Zoology at Harvard. He was incredibly generous with his time and ideas; I frequently felt as if I were storing away a fund of issues that would take years to sort out. Stephen Gould was also a wonderful resource, and I thank him, too, for allowing me to pick his brain. The John Simon Guggenheim Foundation and the University of Wisconsin Graduate School made this year away from Madison possible, and so I have them to thank as well.

Once back in Madison, James Crow greatly helped me clarify my ideas about evolutionary theory. The University of Wisconsin's award of a Romnes Faculty Fellowship during 1982–83 made it possible for me to devote more time to this project. The National Science Foundation also has generously assisted in the last two years.

When the manuscript was in its penultimate form, James Crow, Fred Dretske, Ellery Eells, Berent Enc, David Hull, Philip Kitcher, Dennis Stampe, and David Sloan Wilson read parts or the whole and provided copious marginalia. These names and many others pepper the footnotes in what follows. My debts are many; I am embarrassed to have had so much help from both philosophers and biologists.

Norma Sober subjected the prose style of the manuscript to a painstaking critique. Whatever howlers and mixed metaphors survive are there in spite of her protests. Lucy Taylor provided good advice and a deft hand in drawing the figures. Harry and Betty Stanton were all I could have wanted as editors.

In reading the prefaces of other books, I used to wonder about the formulaic closing in which the author shoulders the blame for the faults that remain. What meaning, if any, could this conventional remark possess? Who else could and would be blamed besides the author? Now I think I understand a little. Help received from people of talent incurs a debt. Faults in the product may reflect gifts ill used. An author inevitably hopes that what he writes will have been worth his efforts. I additionally hope that this book is worth the efforts of many, many others.

Elliott Sober
March 1984

The Nature of Selection

Introduction

Clarity is difficult in an evolving science. It is a permanent goal, a possible consequence of struggle. But it is never attained once and for all. As scientific advance brings some issues into focus, others arise that require clarification. Solving and creating problems are not two parts of science, but one.

A science may fall short of perfect clarity in different ways. One is relatively benign. A science may move forward, sideways, and backward as if in a fog that sometimes lifts a little and then resettles. The lack of visible landmarks makes progress difficult to gauge. Theories proliferate; models come and go. Whether theory replacement means theoretical improvement is often hard to determine. But a science enveloped by fog has at least one consolation. A fog does not foster the illusion of clarity; the lack of visibility is patent.

More insidious than the fog is the mirage. Fogs are seen for what they are. Mirages are trickier, engendering the mistaken conviction that things are as they seem. On the other side of the horizon there are some palm trees. Light reflected off a layer of clouds makes them seem to be where they are not. In "seeing" the mirage, one sees something real. One sees the trees, but they are not where they seem to be.[1]

Mirages, unlike hallucinations, are produced by things outside ourselves; they are not sheer fabrications arising from within. This explains why mirages can be jointly experienced by sensible people. Walking together in the desert, we share the same illusion, and this gives further credence to the idea that we are not making a mistake. Vision is a reliable guide to the location of objects; intersubjective agreement is an additional check against error. Mirages can be doubly hard to detect because they exploit this twofold assurance.

This book is about mirages. It is about a set of conceptual problems

---

1. Besides mislocating palm trees, victims of mirages often think they see water. Here it is the sky, reflected in unanticipated ways, that one is really seeing. The composite "oasis"—the image of trees plus water on the horizon—is an illusion whose parts are the misinterpreted reflections of various things, all of them real.

that have impeded the attainment of clarity in evolutionary theory. The mistakes I will discuss are not self-generated. They involve the perception of something real that is subject to a predictable sort of misapprehension. People misinterpret mirages in the pursuit of scientific knowledge just as they do in desert wandering.

Distinguishing reality from illusion is a characteristically philosophical undertaking. It also happens to be part of the ongoing activity of science itself. Indeed, the principal precedent for the kind of analysis I propose to conduct here is a work in evolutionary biology. In 1966, George C. Williams published *Adaptation and Natural Selection*. The field has not been the same since. This landmark in the study of evolution is a philosophical *tour de force*. It is a critique of fallacies and obscurities in the deployment of the concepts of fitness, natural selection, and adaptation. More important even than the specific conclusions that Williams reached is the analytic acuity that he established as a requirement in evolutionary theory.

One of Williams' central insights was to see a mirage for what it is. According to the Darwinian interpretation that Williams defended, natural selection pits organism against organism in the struggle for existence. The characteristics that become common in such a selection process will be those that benefit the organisms that possess them. Yet characteristics that benefit individuals may also benefit the groups in which those individuals live. Some groups avoid extinction more successfully than others because they are composed of fitter organisms. But when this happens, we should not conclude that there is selection for group-beneficial characteristics. Selection works for the good of the organism; a consequence may be that some groups fare better than others. However, it does not follow that selection works for the good of the group. Williams' own expression of this idea cannot be improved upon:

> Benefits to groups can arise as statistical summations of the effects of individual adaptations. When a deer successfully escapes from a bear by running away, we can attribute its success to a long ancestral period of selection for fleetness. Its fleetness is responsible for its having a *low probability* of death from bear attack. The same factor repeated again and again in the herd means not only that it is a herd of fleet deer, but also that it is a fleet herd. The group therefore has a *low rate* of mortality from bear attack. When every individual in the herd flees from a bear, the result is effective protection of the herd (Williams 1966, p. 16).

A faster deer is less likely to be killed by a predator than a slow one. A fast *herd* of deer will therefore be less likely to go extinct than a slow one. The fitness of the group may simply be, as Williams says, a "statistical summation" of the fitness values of the individuals in it. This much is unproblematic. The mistake comes, according to Williams, if we think that selection acts for the good of the group, rather than for the good of the individual.

Just as a fast deer is fitter than a slow one, so the fast herd is fitter than the slow one. But according to Williams, the fitness of the group is a *mirage*. It is a *reflection* of something real—namely, the fitness values of the individuals in the group. It fosters an illusion—namely, that selection works for the good of the group, rather than for the good of the individual. This is an illusion that had tricked a number of evolutionists. Williams set himself the task of exposing the mistake.

In a desert mirage, there really are some palm trees and one is really seeing them. The mistake is to think they are where they seem to be. According to Williams, advocates of group selection made a similar error. Evolution by natural selection does have its causes. But by seeing the group rather than the individual as the unit of selection, these biologists located the causes in the wrong place. The survival and proliferation of groups is merely a reflection of causal processes at work elsewhere.

If one herd is fast and another slow, why not see the extinction of the second and the survival of the first as a case of group selection? In eliminating the slow deer, didn't selection also eliminate the slow herd? One prominent reason that Williams gives, which has remained popular, is *parsimony*. The individual selection hypothesis, so it is said, is *simpler* than the group-level characterization. Several sorts of clarification are needed here. First, what makes the individual selection account the more parsimonious of the two? Second, why is the greater simplicity of one hypothesis a reason to think that it, rather than its competitor, is true? Finally, if we could with equal truth say that selection works against slow deer or that selection works against slow herds, then no methodological canon ought to advise us to think that one hypothesis is true and the other false. Parsimony is a reason for choosing between *non*equivalent hypotheses; the substantive difference between group and individual selection remains to be clarified.

These and other questions must be answered before Williams' critique of the idea of group selection can be accepted. My point here is a preliminary one: Williams identified a distinction of fundamental importance. It is one thing for groups to survive and go extinct because of group selection; it is quite another for this to happen in consequence of a process of individual selection. We will see in Part II that this distinction rests on the difference between group properties being causes and artifacts in a selection process.

Williams stressed the importance of distinguishing the mirage of group selection from the real thing. Few biologists explicitly dissented, but many advanced opinions about group selection that are implicitly at odds with it. For example, it is not uncommon to conceive of the difference between group and individual selection in terms of the analysis of variance. From this point of view, group selection exists when groups differ in fitness and individual selection occurs when the individuals in a group vary in fitness. An immediate consequence of this point of view is that Williams' distinction cannot be drawn. If all the deer in one herd run at one speed and the deer in another run at another, slower, speed, then selection for fleetness will mean that all the variation in fitness is between groups. Williams refuses to conclude that the extinction of the slow herd and the continued existence of the fleet one must be chalked up to group selection. The view that rests on the analysis of variance idea, however, leads to precisely this conclusion.[2]

Another perspective has been influential in the units of selection problem, one at odds both with Williams' distinction between artifact and cause and with the analysis of variance distinction between intra- and intergroup variation in fitness. It is frequently held that if the individuals in a group are relatives, then what might appear to be *group* selection may be redescribed as a case of *kin* selection, which is then claimed to really be a type of *individual* selection. So, for example, if all the individuals in a group were clones of each other, this would mean that the ensuing selection process should be viewed as a case of individual selection. But notice that in this circumstance, all the variation in fitness is between groups. Since the organisms in a group are carbon copies of each other, there is no within-group variation in fitness. The analysis of variance point of view concludes that this case must be an instance of group selection; the point of view that sees kin selection as a form of individual selection reaches precisely the opposite judgment.

It is easy to be confused by this proliferation of selection processes and by the very different definitions of them that biologists exploit. Group, kin, and individual selection need to be disentangled, their differences made clear.

Williams saw the importance of distinguishing a herd of adapted deer from an adapted herd of deer. When a fast herd survives and a

2. The deer herds in this example differ in their chances of extinction but not in their propensities to found new colonies. The same point can be made, however, if we concentrate on group reproduction instead of on group survival. One group may be more successful at founding colonies than another, without this being due to group selection. Williams' idea that a group effect can be an artifact of individual selection holds true here as well.

slow one goes extinct, one can describe the former group as fitter than the latter. But the fact that this process can be "represented" by assigning fitness values to groups hardly shows that it is driven by group selection. Yet, while avoiding the mirage of group selection, Williams is, I think, taken in by the mirage of genic selection. He argues that since a selection process that issues in evolution can always be "represented" in terms of fitness values that attach to single genes, it is correct to think of the single gene as the unit of selection.[3] The idea that the fitness value of a gene may be a mirage—a reflection of selection processes that occur at *higher* levels of organization—is swept aside. The crucial distinction first drawn with respect to the question of group selection is obliterated when a lower level of organization is considered.

There may be a principled reason for this. Parsimony has, after all, been invoked as a reason for casting the unit of selection at as low a level as possible. Other reasons favoring the single gene as the unit of selection have been advanced as well. Whether these show that there really is an oasis on the horizon or merely reinforce one illusion with another, we shall have to see.

Williams realized that the problem of the units of selection is not narrow and technical, but foundational. To think properly about the units of selection requires that we clarify the concepts of fitness, selection, and adaptation. It is for this reason that Williams ends his book with a plea. If we are to avoid the fallacies that have preyed upon evolutionary theory in the past, a precise vocabulary is needed for the scientific study of adaptation. Although the present book is frequently critical of the arguments that Williams advanced, his stress on the importance of logic and clarification has guided my inquiry. I have tried to take up what Williams began.

Because the questions are foundational, there is considerable scope for issues that are philosophical in nature. It is impossible to think about the units of selection controversy unless one thinks about causation, chance, explanation, and reduction. Williams' book is built on a substantial body of doctrine concerning these issues, much of it implicit. I have tried here to contribute what a philosopher can—to uncover presuppositions and make them explicit.

Besides being guided by a set of problems that arose within evolutionary theory, I also have tried to write this book so that it connects with issues that are important within my home discipline, philosophy of science. Biologists are often surprised to learn how little Darwinism has influenced philosophy of science in the last one hundred years.

---

3. Richard Dawkins (1976), in his popularization of this line of thinking called *The Selfish Gene,* repeats this argument.

They frequently think that the philosophical consequences of evolutionary theory must be so profound and unsettling that philosophers have chosen to stick their heads in the sand. Darwin, so it is said, continued the work that Galileo and Newton began of dethroning the human race from its position at the pinnacle of creation. No more can we think of ourselves as the reason that the universe is as it is, our standards of reason and morals having a timeless, unshakable authority. This message, says the biologist, fills philosophers with uncertainty; it even threatens to put philosophy out of business. If reason and morality have evolved, philosophers can no longer simply consult their heartfelt intuitions as a guide to the true, the good, and the beautiful. Perhaps the subjects of epistemology, ethics, and aesthetics ought to be taken over by evolutionary biology, which has managed to shake loose from the illusion that human beings are the measure of all things. No wonder, concludes the biologist I have in mind, that the message of Darwinism is a message that philosophy has chosen not to hear.[4]

To my ear, this explanation of why Darwinism has mattered so little in philosophy of science simply does not ring true. Philosophers now assimilate the fact of evolution with as little difficulty as the fact that the earth is not at the center of the solar system. It is accepted as an unproblematic part of the contemporary scientific world view. But this is scant reason for filling the pages of philosophical monographs with the news! Philosophers have no more reason to trumpet this piece of information than they have for announcing that the earth goes around the sun.

Philosophy of science in this century has been shaped by an interest in physics and mathematics (particularly mathematical logic). The reason is not that these subjects are scientifically important but that they have been connected by philosophers and scientists alike with matters of great philosophical weight. Einstein's theories of special and general relativity have occupied center stage in philosophy of science for a very good reason: as philosophers, we care about issues of a priori knowledge, conventionalism, and about the general principles that permit radically different scientific theories to be compared and evaluated.

---

4. Thus Wilson (1975, p. 3) writes: "The biologist, who is concerned with questions of physiology and evolutionary history, realizes that self-knowledge is constrained and shaped by the emotional control centers in the hypothalamus and limbic system of the brain. These centers flood out consciousness with all the emotions—hate, love, guilt, fear, and others—that are consulted by ethical philosophers who wish to intuit the standards of good and evil. What, we are then compelled to ask, made the hypothalamus and limbic system? They evolved by natural selection. That simple biological statement must be pursued to explain ethics and ethical philosophers, if not epistemology and epistemologists, at all depths."

Evolutionary theory is undoubtedly of great *scientific* importance. But it remains for philosophers of biology to show why it has *philosophical* importance. We must show that by considering evolutionary theory, old problems can be transformed and new problems brought into being. It remains to be seen, I think, how radically the philosophy of science will be reinterpreted. I find the prospect tantalizing but the conclusion far from foregone.

In bringing biology and philosophy together, we must lose sight of neither. Skimming the surface of the biology will hardly do. One does not shift from the philosophy of geology to the philosophy of biology simply by changing one's example of an inductive generalization from "all emeralds are green" to "all swans are white." Nor can philosophers simply plunge into the details of biological debates, thinking that the science somehow matters for its own sake. Of course, it does matter for its own *scientific* sake. But as philosophers, the question of *philosophical* significance must always be paramount.

My remarks about Williams' distinction between an adapted herd and a herd of adapted individuals were meant to give a flavor of the biological material that lies ahead in this book. It was an antipasto, not a full menu. I can provide a comparable foretaste of the philosophical ideas we will encounter by quoting a passage from Bertrand Russell:

> All philosophers, of every school, imagine that causation is one of the fundamental axioms or postulates of science, yet, oddly enough, in advanced sciences such as gravitational astronomy, the word "cause" never occurs. . . . The law of causality, I believe, like much that passes muster among philosophers, is a relic of a bygone age, surviving, like the monarchy, only because it is erroneously supposed to do no harm. . . . No doubt the reason why the old "law of causality" has so long continued to pervade the books of philosophers is simply that the idea of a function is unfamiliar to most of them (Russell 1913).

Russell believed that causality was a mirage that a mature science can see through. We may talk of causes and effects in ordinary life and in a fledgling discipline, but as theory develops these concepts lapse from usage and are replaced by the idea of a mathematical function. Instead of laws of cause and effect, we find only equations.

Russell by no means had the last word on the subject; the role of causal concepts in science has continued to occasion philosophical debate. For example, a philosopher of physics—Patrick Suppes (1970, p. 5)—has commented as follows on Russell's remarks: "Perhaps the most amusing thing about this passage is that its claim about the use of the word 'cause' in physics no longer holds. Contrary to the days when

Russell wrote this essay, the words 'causality' and 'cause' are commonly and widely used by physicists in their most advanced work."

In this book I tell a similar story about the concept of cause in evolutionary theory. From the writings of Darwin to articles in contemporary journals, ideas of causation have been very important in evolutionary biology. Yet they do not surface in any direct way in mathematics; the words "causality" and "cause" are not to be found in the algebra or diffusion equations of quantitative models but in the conceptual framework that motivates them and makes them intelligible.

Philosophers of science generally recognize that the legitimacy of a scientific concept is not to be decided on a priori philosophical grounds but rather by seeing whether science needs that concept to go about its business. The same attitude is appropriate when the concept is one that has been of traditional philosophical interest. If causation is an important issue in evolutionary biology, then philosophers of science must try to understand it, not dismiss it as unintelligible superstition. Much of this book aims at providing a detailed description of the conceptual structure and ontological framework that evolutionary theory deploys. There is much here that can be grist for the philosopher's mill.

Because I have written this book with two audiences in mind, I should provide some guidance concerning its organization. Or, to put the point another way, I should give the reader a hint about what to skip. The chapters in Part I take up general questions about the structure of evolutionary theory, the concept of fitness, the nature of chance, the meaning of adaptation, and the sorts of explanation that the theory of natural selection provides. Some sections (1.1 to 1.5, 3.2, 4.1, 6.1, and 6.2) do not stray very far from discussing evolutionary theory, whereas others (2.1, 2.2, 3.1, 4.3, and 5.1 to 5.3) address more general philosophical issues.

In Part II, I plunge into the details of the units of selection problem. In Chapter 7, I provide a brief historical background and then attempt to identify a set of fallacies that have grown up in this biological problem area. Then, after characterizing in Chapter 8 what I think causation means in the theory of natural selection, I provide a positive account of what the idea of a unit of selection amounts to. In doing this, I think that a number of current research agendas in evolutionary theory can be seen as issuing from a unified cluster of questions about the mechanism of natural selection.

I must apologize to both biologists and philosophers for occasionally boring them with preliminaries. Biologists may roll their eyes when they read my few sentences about what the word "diploid" means. Philosophers will have their chance to yawn when they come to my remark that it is propositions, not concepts, that are tautologies. This may occasionally try the reader's patience, but I hope that the exciting bits will come more frequently than the tedious ones.

If I were a biologist interested mainly in seeing why a philosopher has stuck his nose into the units of selection controversy, I would read only Section 3.2 of Part I and then look at Part II in its entirety. If I were a philosopher mildly curious about evolutionary theory, but more concerned to read about matters of general philosophical interest, I would read through Part I, skimming when the biological details get too boring, and then read Section 7.2 and Chapter 8 of Part II. And finally, if I were a reader of broad learning and wide interests, I would read the entire book, even the index, exclaiming loudly all the while about what a fine time I was having.

PART I

Fitness, Selection, Adaptation

Chapter 1

Evolutionary Theory as a Theory of Forces

What makes mirages and other illusions possible? The answer lies in the concept of causality. A single visual impression can be produced in many ways. The impression of an oasis might be caused by a real oasis on the horizon, or by something else. In choosing to think the former, we select a causal hypothesis that is possible but happens not to be true. When we are taken in by an illusion, we err in an inference from effect to cause.

If scientists merely chronicled sequences of events, without trying to say what caused them, mistaken inferences from effects to causes might not occur. But science is more ambitious, and therefore runs additional risks.

Evolutionary theory aims not only to describe the *patterns* that may be found in the history of life on earth but also to characterize the *processes* that produced them. Its strategy for identifying the *actual* causes that have shaped the tree of life is the usual scientific one: scientists try to characterize a range of *possible* causes of evolution, and then to determine which of these possibilities actually obtained. The actual is understood by first embedding it in the possible.

When R. A. Fisher expressed this opinion in the preface to *The Genetical Theory of Natural Selection* (1930, pp. viii–ix), he quoted the physicist Arthur Stanley Eddington in support. In *The Nature of the Physical World*, Eddington remarked that

> We need scarcely add that the contemplation in natural science of a wider domain than the actual leads to a far better understanding of the actual (Eddington 1928).

It may tell us something about the state of theoretical biology at the time Fisher wrote that he felt obliged to argue that the same idea applies in evolutionary theory:

> For a mathematician this statement is almost a truism. For a biologist, speaking of his own subject, it would suggest an extraordinarily wide outlook. No practical biologist interested in sexual

reproduction would be led to work out the detailed consequences experienced by organisms having three or more sexes; yet what else should he do if he wishes to understand why the sexes are, in fact, always two?

Fisher goes on to remark that even a full characterization of the range of possibilities will at times be too restrictive. Imaginary solutions, he says, also have their place in scientific explanation.

Setting to one side the remark about the usefulness of contemplating *im*possibilities, we can recognize that contemporary evolutionary theory has followed Fisher's advice to the hilt. The theory has assembled a variety of models that describe possible mechanisms that can produce evolution. In assessing the significance of these models, the evolutionist asks which of these possibilities has been actualized and, in particular, which has been at work, not just occasionally, but frequently. Possibilities are articulated not for their own sake but in order to obtain a satisfying theoretical picture of the actual.

Natural selection is one kind of cause. Selection can produce evolution, but there are other mechanisms that are capable of doing the same thing. Furthermore, within the category of selection, we can distinguish a variety of different sorts of selection process. Group selection, individual selection, and genic selection are alternative possibilities; each is a species of the same genus. In this chapter, I will begin the project of characterizing the genus. What is natural selection? How does it differ from other possible causes of evolution? And how are the different possible causes of evolution brought together in a unified theoretical treatment?

In Section 1.1, I provide a bit of history. Darwin's idea of natural selection had a number of anticipations. One useful technique for isolating the biological notion is to locate it in a field of kindred but different ideas. In conceptual analysis as well as in biology, it can help to embed the actual in a range of possibilities.

Section 1.2 shifts from the history of ideas to biology itself. The point is not to relate Darwin's concept of selection to other ideas of selection but to contrast selection as a cause of evolution with other possible causes. In addition, the idea of evolution itself is provided with a preliminary clarification.

Evolutionary theory describes the effects that various causes of evolution may have, both when they act alone and when they act in conjunction. The theory may be thought of in analogy with Newtonian mechanics. Various forces are described, but the theory has at its conceptual center a view of what will happen to the systems it describes when no forces at all impinge. Evolutionary theory's version of a "zero-force law" is examined in Section 1.3.

In Section 1.4, I describe how natural selection is characterized in the very simplest of population genetics models. In Section 1.5, I describe the other side of our theoretical understanding of natural selection. Evolutionary theory describes selection, not just in terms of its consequences but in terms of the ecological conditions that can produce it.

1.1 The Prehistory of a Concept

Charles Darwin was not the first to assert the fact of evolution. Nor was the idea that natural selection may modify the composition of a population entirely without precedent. But Darwin was the first to think of natural selection as the driving force of evolution—as the principal cause of both the adaptedness of organisms to their environments and the existence of organic diversity.

The idea that a selection process may modify the composition of a population had been thought of before, both in and out of biology. The most famous of these anticipations is the one acknowledged by Darwin in his *Autobiography*. Thinking back some thirty-five years to the reflections that led him to the hypothesis of evolution by natural selection, he recalled:

> Fifteen months after I had begun my systematic enquiry, I happened to read for amusement Malthus on Population, and being well prepared to appreciate the struggle for existence which everywhere goes on, from long-continued observation of the habits of animals and plants, it at once struck me that under these circumstances favorable variations would tend to be preserved, and unfavorable ones to be destroyed. The result of this would be the formation of new species. Here, then, I had at last got a theory by which to work (Darwin 1876, p. 120).

Darwin's copious notebooks, in which he worked his way through alternative theories before he reached the idea of natural selection, have survived. Thanks to this rich documentary source, historians have been able to compare this recollection completed in 1876 with the day-to-day trials and errors that brought the Darwin of 1838 to amuse himself with Malthus. Yet, the degree to which Malthus *changed* the direction of Darwin's thought remains controversial: did reading the *Essay on Population* merely crystallize and clarify ideas that Darwin already had at his command, or did it provide him with some fundamentally new insight? This psychological question may never receive an unequivocal answer.[1] Even so, we can recognize in Malthus at least part of the idea that Darwin was to put to such powerful use.[2] Malthus began with the simple idea that a population that expands beyond its own food supplies will be cut back by mortality. Although the case that most concerned him was the number of indigent poor in England, Malthus (1798, pp. 14–16) thought he had reached a principle of universal scope:

> Population, when unchecked, increases in geometrical ratio. Subsistence increases only in an arithmetical ratio. . . . I see no way by which man can escape from the weight of this law which pervades all animated nature.

Death and fecundity are opposing forces that together determine population size. To diminish the former, human beings must reduce the latter. Hence Malthus' advice to the poor that sexual abstinence is the way up from squalor.

Malthus was no evolutionist, and his "law which pervades all animated nature" concerned the regulation of population *numbers*, not the frequency of characteristics found in a population. In fact, Malthus thought his law *prevents* populations from changing much. It was his opinion that an individual deviating from the population norm would be less able to cope with the vicissitudes of life (Mayr 1977). For him, fecundity and death are forces that *prevent* populations from changing. Darwin radically transformed this anticipation of the idea of natural selection. What was for Malthus a process of destruction and stasis became for Darwin an instrument of improvement and change.[3]

Still, it is important to recognize in Malthus' idea an important innovation that survived in Darwin's theory. For Malthus, the natural world is a world of *struggle*; nature is not benignly arranged to bring into existence only that number of organisms that can be supported. Nature is inefficient and cruel, not economical and beneficent. In addition, the Malthusian paradigm pictures competition between organ-

---

1. The reader may wish to compare the emphasis placed on the idea of artificial selection by Ghiselin (1969), Vorzimmer (1970), Young (1971), Schweber (1977), Bowler (1974), and Ruse (1979) with the emphasis placed on Malthus by Limoges (1970), Herbert (1971), and Kohn (1980). Also of interest is Gruber's (1974) psychological study.

2. There is an interesting parallelism in the evolution of the theory of evolution by natural selection: the theory's codiscoverer, Alfred Russel Wallace, reported the importance of Malthus' essay in his own progress toward the hypothesis (Wallace 1905).

3. For discussion of the senses in which natural selection is and is not "progressive," as well as for an account of the place of Malthusian ideas in contemporary evolutionary theory, see Section 6.1.

isms of the same species as an important force. Attention was shifted from the struggle between the lion and the lamb to that between lamb and lamb. Darwin's teacher, Charles Lyell, had thought of competition as an *inter*specific phenomenon; Darwin provided the idea of competition with a Malthusian reorientation. The composition of a population may be modified by differences that obtain among members of that very population (Herbert 1971).

Besides transforming selection from a mechanism of stasis into a mechanism of change, Darwin changed the Malthusian concept in another way. The Malthusian picture was one in which individuals in the same population compete with each other for a resource in short supply. Darwin had to broaden the idea of the "struggle for existence":

> I should premise that I use the term Struggle for Existence in a large and metaphorical sense, including dependence of one being on another, and including (which is more important) not only the life of the individual, but success in leaving progeny. Two canine animals in a time of dearth, may be truly said to struggle with each other which shall get food and live. But a plant on the edge of a desert is said to struggle for life against the drought (Darwin 1859, p. 62).

Consider the difference between tennis and golf. Both games have a winner, but the relationship of the players' scores is different. In tennis, a point for one player necessarily means no point for the other. As each point is played, there is a benefit at stake that one and only one player can obtain. In golf, on the other hand, one player's score does not constrain the other's. In both games, a criterion for winning is describable in relative terms: the player with the better score at the end of a certain time is the winner. The games differ in the way that end state is produced.

Malthusian competition is more like tennis than golf. If one thought that the struggle for existence was only of this sort, one would fail to detect the existence of contests in which one player's score fails to affect the other's. Two plants at the edge of the desert may differ in their abilities to withstand the drought. They thereby struggle against each other, but not in the Malthusian sense. In Darwin's "large and metaphorical sense," all that matters is that the players end up with different scores—that they do unequally well in "the struggle for existence." Whether this relationship takes the form of a "zero-sum game" is entirely incidental (Ghiselin 1974a, p. 51).

A second precedent for the idea of natural selection was artificial selection. Darwin read a great deal about the techniques that plant and animal breeders used to modify living forms (Ruse 1975, 1979). Just as in the case of this debt to Malthus, the precise psychological role of this research in Darwin's development remains controversial. But we may recognize here an idea whose structure approximates the one that Darwin finally crystallized.

Certainly, attention to artificial selection alerted Darwin to the immense store of variability that natural populations contain. This was a premise of the first importance; selection is impossible unless there are variants to select among. Yet the facts of artificial selection did not point unequivocally to the hypothesis of evolution by natural selection. Just as in the case of Malthus, the idea required transformation and a new theoretical context. Three facts blocked an easy extrapolation from artificial to natural selection. One concerned the observed effects of artificial selection, the other two its causes.

Plant and animal breeders had been able to create new varieties within a species. But when Darwin wrote, artificial selection had never managed to produce a new species. Darwin's contemporaries standardly took this to show that there are limits beyond which a species cannot be modified. A breeder may tinker and fine-tune, but nothing more. Darwin looked at the same data, harnessed them to an argument by extrapolation, and reached the opposite conclusion: If *artificial* selection has achieved so much in the comparatively short time of recorded human history, then *natural* selection can be expected to generate even more impressive results on the time scale of natural history.[4]

The second obstacle that Darwin had to overcome in using artificial selection as a conceptual precedent for his notion of natural selection was *consciousness*. Artificial selection is the product of intelligent manipulation. Why think that organisms could be adapted to their environments without this sort of guidance?

Here we must see that Darwin's argument contradicted what was at the time *the* most influential argument for the existence of God—the argument from design. William Paley (1819) (to single out one of the many writers who formulated this line of thought in detail) observed that if we found a watch on a beach, we would unhesitatingly hypothesize that it had been created by some intelligent being. The idea that it might have resulted from the random action of waves on sand strikes us as preposterous. But if the intricacy of the watch can be

---

4. The contemporary situation is much less conjectural. Experiments, for example, ones involving *polyploidy*, have produced reproductive isolation between populations. In polyploidy, the number of chromosomes found in an offspring is some multiple of the number found in the parent. Offspring will thereby be reproductively isolated from parent, and if this event happens several times in the offspring generation, a new self-contained breeding population may be established. It has been found that the chemical colchicine induces polyploidy.

explained only by the design hypothesis, how can we doubt that the far greater intricacy of an organism requires the activity of a powerful Intelligence? In the absence of some alternative plausible mechanism, the argument from design deserved to be taken seriously.[5] Darwin's contribution was to develop a nontheistic alternative.

One of the premises of the design argument—that design requires a designer—was an impediment to using artificial selection as an evidential source for natural selection. Because of this, it is sometimes claimed that "artificial" and "natural" selection are only analogous concepts, and that Darwin's insight was based on a sort of "metaphorical extension." Yet there is room to doubt that "artificial" and "natural" selection are only "analogically" or "metaphorically" related. When physicists do a laboratory experiment on electricity, the result is in part due to their conscious manipulation. But this hardly shows that "artificial electricity" and "natural electricity" are only "analogically related." Darwin himself seems to have thought of artificial selection as an *experiment* that the human race had inadvertently conducted on the question of natural selection (Gruber 1974, p. 166). No stretching of concepts was required for "selection" to apply literally to the world that lies beyond conscious human manipulation. Artificial selection is not selection that takes place *outside* of nature, but selection that occurs within a particular niche found *in* nature.

The third structural difference between artificial and natural selection also cuts very deep. It is the difference between a deterministic process and a probabilistic one. An animal breeder will typically lay down an absolute criterion that dictates which sorts of animals or plants will be the parents of the next generation in the breeding population. A dairy farmer interested in increasing milk production, for example, might select the best yielding cow, or only those cows in the top n percent, or only those cows whose milk yield exceeds a specified value. In any case, nothing is left to chance. A criterion is stipulated that determines, for each cow in the herd, whether or not it will be allowed to breed.

Artificial selection of this sort, in effect, divides the population into two sorts of individuals—those who will reproduce and those who will not. Darwin, in his idea of natural selection, did not suppose that nature typically went to such extremes. Darwin thought that natural selection usually operates on very small differences among organisms, ones that do not divide the population into two categories but rather grade them

---

5. David Hume's criticisms of the design argument in his *Dialogues Concerning Natural Religion* (1779) strike me as entirely unconvincing. If they refute the argument from design, they also show that it isn't rational to postulate the existence of unobserved entities to explain observed phenomena. Hume never demonstrated that the design argument is *bad science*.

according to very modest differences in life prospects. Selection confers on some organisms a slightly better *chance* of surviving and reproducing. Nature is subtle where the plant or animal breeder might prefer to be more heavy-handed. Darwin had to see that small differences in probability, given long enough to do their work, could transform a population at least as decisively as the animal breeder's stricter regimen could on a shorter time scale.

A third anticipation of the idea of natural selection came from Adam Smith and the Scottish economists. We know that Darwin steeped himself in the writings of the political economists (Schweber 1980). Again, we find the logical structure of the idea of natural selection, but this time in a nonbiological setting. In a *laissez-faire* market, individuals or firms compete with each other. As time passes, some go "extinct," not in the sense that the individuals have died, but in the sense that they are no longer engaged in that particular line of business. Classical economists did not believe that individuals prosper and go bankrupt at random. Rather, the idea was that if we follow an initial population of entrepreneurs historically, we will note that the market eliminates the "unfit." Nor is this an intended consequence—the result of some guiding intelligence having an interest in raising the average level of business skill. In the social realm just as in nature, apparent design may exist without a designer.

In contrast with the views of Malthus, the classical economists thought that the selection process they described issued in benign consequences. As if by an "invisible hand," competition between self-interested entrepreneurs results in an increase in public wealth and well-being. In the economic world, populations *change*, and they change for the better.[6]

Each of the three precedents for the idea of natural selection I have mentioned required reworking before Darwin could reach his theory. Malthus pictured selection as an instrument of stasis. The facts of artificial selection showed Darwin what selection could produce through conscious human manipulation. And the Scottish economists offered a nonbiological model in which a selection process improves a population as an unintended consequence of individual optimization. Let us now take up the structure of Darwin's idea of evolution by natural selection directly.

1.2 History and Theory

In a causal chain, the links are events—some proximate to, some more

---

6. We will have occasion to consider a further point of contact between Darwin's ideas on selection and classical economic thought in Section 6.1. Note also that contemporary work on the theory of the firm has brought the biological idea of natural selection back into social theory. See Hirshliefer (1977, 1982).

distant from, the one that we wish to explain. Evolutionary theory, both in Darwin's hands and in the hands of contemporary evolutionists, has a plurality of interests. Biologists divide the explanatory labor—some focusing on one link in a causal chain, others on others.

Darwin's ambition was to establish two hypotheses: that evolution has occurred and that natural selection has been one of its preeminent causes. By "evolution" he subsumed both the modifications that occur within a species and the large-scale changes that are involved in the production of new species.

The Origin of Species was, in Darwin's words, "one long argument" for this twofold proposition (Darwin 1859, p. 459). Although the two hypotheses are separable, the strategy for confirming them was the same. Darwin amassed a large variety of facts that could be explained in a unified manner by his conjectures but which either could not be explained at all or could be explained only in a more ad hoc way by alternative hypotheses (like the idea that God had created an array of immutable species).

The geographical distribution of forms that we observe in living organisms and infer to have existed from the fossil record Darwin sought to explain by the hypothesis of evolution. Why are most marsupials found in Australia? Why are cave insects in Europe so different from those in North America, if both inhabit environments that are so similar? A related group of questions has to do not with the distribution of species in space and time but with the distribution of characteristics among species. How can we explain this pattern of similarity and difference? Such questions as these Darwin tried to answer by appeal to the hypothesis of evolution—of descent with modification. The power of the hypothesis of evolution to explain in a uniform way a staggering variety of phenomena of this sort gains for that hypothesis a significant confirmation.

But the fact of evolution was not for Darwin an unexplained explainer. That evolution occurs in a population, and pursues one path rather than another, could itself be accounted for by the hypothesis of natural selection.[7] In the five hundred pages of the first edition of *On the Origin of Species*, Darwin used the idea of natural selection to explain a wide range of observations. It is remarkable that a hypothesis of such explanatory power could be so utterly simple conceptually: If the organisms in a population differ in their ability to survive and reproduce,

---

7. Kitcher (1984a) emphasizes how Darwin divided his efforts in *The Origin* between showing that the hypothesis of evolution and the hypothesis of natural selection are each confirmed by their abilities to explain.

and if the characteristics that affect these abilities are transmitted from parents to offspring, then the population will evolve.[8]

Darwin's argument demands that natural populations should exhibit *heritable variation in fitness* (to use the contemporary jargon). Darwin devoted great energy to establishing this threefold fact. Less central to the immediate thrust of his reasoning was the *explanation* of why populations should have this property. The origin of variation and the mechanism of inheritance eluded him and evolutionary theory for some time. Darwin's hypothesis of *pangenesis*, developed in the decade following the publication of the *Origin*, was one of many unsuccessful biological attempts to solve this problem. It was only in this century when Mendel's theory of inheritance was rediscovered, systematically confirmed, and then wedded to evolutionary theory in the so-called modern synthesis, that Darwin's explanation of adaptedness and evolution could itself be explained.

In summary, Darwin posited, and evolutionary theory has continued to scrutinize, a causal chain that might be represented as follows. By "distribution of characters" I mean to subsume facts of geographical distribution and of variation within and among species.

Each link in this chain stands for an array of phenomena for which an explanation might be requested. Darwin explained items to the right by appeal to hypotheses concerning items to the left. The leftmost entry, however, stands for a set of facts that Darwin was able to *confirm* but never able to *explain*.

How is it possible to have convincing evidence for a hypothesis without being able to account for why it is true? This is a familiar situation in a science. For example, the correlation of the phases of the moon with the tides was well documented for a long time before Newton; but until Newton developed his theory of gravitation, there was no explanation of why the correlation should obtain. Nor is this situation merely temporary, a defect that attaches only to "immature" sciences. A science may push back the frontier of ignorance. But no matter how far a science succeeds in doing this, there will be a body of propositions, which at the time are *fundamental*, in the sense that they are used to explain other facts, but no explanation is known for them. Darwin's frontier was the fact of heritable variation in fitness. The frontiers of contemporary evolutionary theory lie elsewhere.

---

8. Evolution follows from the satisfaction of these two conditions only when a third requirement is met. This will be stated in a moment.

Although there are fascinating questions to be asked about the connection of the hypothesis of evolution with facts of character distribution,[9] my present concern is with the connection of heritable variation in fitness to the idea of *evolution*. First, evolution does not logically require heritable variation in the fitnesses of organisms. As we will see, evolution can occur by other means. Second, the existence of heritable variation in fitness in a population of organisms does not suffice to ensure that evolution will occur. Other evolutionary forces may intervene and nullify the effects that selection tends to produce. So we should formulate the Darwinian idea first by saying that *evolution by the natural selection of organisms* will occur only if organisms exhibit heritable variation in fitness. And, as a first approximation (to be refined later), we can say that this necessary condition will also be sufficient, if no other evolutionary force intrudes.

We can segment Darwin's reasoning about selection into two components. First, there is the *general characterization* of the conditions required for evolution by natural selection. Second, there is the *historical hypothesis* that those conditions obtained and that the natural selection of organisms has been the principal force governing the evolution of life, including the origin of species.[10]

It is obvious that Darwin's historical hypothesis is empirical. To claim that there is variation in a population, that this variation includes differences in characteristics that affect the capacity to survive and reproduce, that these characteristics are inherited, and that no other potential force materially influences the population is to advance a highly nontrivial set of conjectures! Reflection on the meaning of the expression "natural selection" will never by itself generate any evidence

9. Here is one: Can the hypothesis of evolution—descent by modification—be said to explain facts about geographical distribution and species similarity and difference *apart from particular conceptions of the mechanism that has caused evolution*? That is, does the idea of evolution itself carry any explanatory punch, or does it do so only when connected with a more detailed conception of what has produced evolution? This question is a fundamental one for the systematist's inferences about phylogenetic relationship: how much knowledge about the *processes* that produced observed character distributions is needed before one can reconstruct the *pattern* of historical relationships that obtain among species? This issue is beyond the scope of the present work, but see Sober (1983c, 1984a) for discussion.

10. Selection is sometimes thought of as an "ahistorical" cause of evolution, since in some simple models the starting frequencies of a trait undergoing natural selection do not influence the end result of the selection process. Examples will be provided in Sections 1.4 and 6.1; the latter section also includes an example in which this is not true. In any event, this sense of "historicity" differs from the one at issue here. Roughly, a historical hypothesis says what *did* happen, not what *would* happen if the conditions were right. The claim that selection influenced a particular population is therefore a historical hypothesis.

for or against these various claims. Darwin exerted himself to show that these conditions are satisfied generally, and biologists who have continued to apply the Darwinian paradigm have done the same.

When a characteristic becomes prevalent in a population, one possible explanation is that this happened in virtue of Darwinian selection among individuals. But as with any inference from effect to cause, we must recognize that a given effect may, in theory, be produced in various ways. Gould and Lewontin (1979) have argued that many evolutionists have uncritically assumed that the explanation of a trait's becoming common must be found in its adaptive significance. As an antidote to this uncritical allegiance to the "adaptationist programme," they suggest that alternative hypotheses provided by evolutionary theory deserve a fuller hearing.

Consider the human chin. Apparently, there never was selection for having a chin. Rather, selection for certain other features of jaw structure yielded a chin as an inevitable architectural consequence. If the genes that produce chins are the same as the genes that produce jaws, selection for jaws will bring chins along as a free rider. If jaws and chins were independent characteristics, each could be independently manipulated by natural selection. But if architectural constraints tie traits together in packages, the power of natural selection to produce novelty will be reduced.

So chins may become prevalent because of selection without there being selection for having a chin. Biologists now call this sort of process *pleiotropy*.[11] Although its genetic basis was unknown to Darwin, the idea of free riders on selection was something he considered. Indeed, it was precisely this hypothesis that Darwin favored in his dispute with the codiscoverer of the hypothesis of evolution by natural selection, Alfred Russel Wallace, over the origin of higher cognitive capacities in human evolution (Gould 1980c).

Wallace argued that there could have been no selective advantage for certain sophisticated intellectual abilities in our ancestral past. How could the ability to learn calculus have benefited a cave dweller? Wallace reasoned that natural selection could not have produced these higher mental faculties; he conjectured that the beneficence of some supernatural Intelligence must have been responsible for them. For Darwin, this inference did not ring true. Granted, there was no selection for the ability to do calculus. But that ability could nevertheless emerge via natural selection. Suppose that the ability to do science and math-

---

11. Geneticists sometimes prefer to reserve the term "pleiotropy" for a situation in which a *single gene* has multiple phenotypic effects. I use the term in a somewhat more liberal sense, according to which a *cluster* of genes may have several effects on the phenotype.

ematics were correlated with certain simpler skills of reasoning and communication that were beneficial in our ancestral past. Then the capacity for more abstract reasoning would have evolved as a free rider.

One cannot assume that the correlations of traits found in a population are engraved in stone. If a gene produces one beneficial and one harmful phenotypic effect, there is no reason in principle why a modifier gene should not occur somewhere else in an organism's genetic complement. In theory, the new gene could allow the first gene to continue producing the beneficial effect but prevent it from producing the harmful one. If this lucky mutation occurs, natural selection may favor it, and the correlation of the two characters may thereby be destroyed. To stress the importance of pleiotropy in a particular evolutionary problem is to say not that this *cannot* happen but that it frequently *does not* happen.

Thus pleiotropy can itself evolve; it is in principle vulnerable to the tinkering of natural selection, although the extent to which this has happened is controversial. The same holds true of correlations of characters that may occur in a group of related species. The most famous of these allometric ratios is the ratio of brain size to body weight in major vertebrate groups; a plot of log brain weight to log body weight falls about a straight line, although there is some scatter (Martin 1981). The fit of the data points to this line makes it implausible that each species has independently evolved its own ratio. The fact that some species are above the plotted line and others below, however, suggests that there is some room for manipulation. The suspicion is that the traits are significantly bound together; it would be a mistake to think of two independent selection processes at work on a species, one modifying body weight, the other modifying brain size. As in the case pleiotropy, traits are tied together into packages. In principle, natural selection may undo the ties that bind. But the question that excites biological debate is whether in practice this is the rule or the exception.

A further alternative to individual selection as an evolutionary force is chance. Kimura's (1968, 1979, 1983) and King and Jukes' (1969) "neutralist" hypothesis maintains that a large proportion of the molecular differences found between species may be selectively neutral (i.e., equal in fitness). At least at the molecular level, evolution may to a considerable extent be the result of a random walk.

The idea of a neutral character is not limited in its application to the molecular level. Nor do two characteristics have to be *precisely* the same in the costs and benefits they impose on the organisms that have them, for neutralism to become an issue. As we will see (e.g., in Section 5.1), modest differences in the fitness of characters coupled with small population size provide just the opening that nonselective evolution requires. In cases like this, an advantageous characteristic may fail to become prevalent, and a deleterious one may sweep through the population. In this circumstance, selection plays some role, but it is a decidedly weaker force than many have supposed.

Even when selection accounts for the evolutionary trajectories found in each of several populations, it needn't follow that *differences* in morphology, physiology, and behavior must have adaptive significance. Lewontin (1978) has offered the nice example of the rhinoceros' horn. Rhinoceroses presumably have horns for defense. But why does the Indian rhinoceros have one horn, while the African rhino has two? Conceivably, there may have been some special selective advantage that number of horns represented. But it is entirely possible that the two traits are equally fit. The two ancestral populations of present-day rhinos may have differed in composition, so that selection for nasal armament may have yielded different phenotypic results. Here it isn't that a chance process produced the traits in question (as in the neutralist hypothesis); rather, the idea is that historical differences, rather than differences in selective significance, may account for variation.

In addition, selection at levels of organization other than that of the individual organism—the group above and the gene below—has been proposed as a mechanism of evolution distinct from the Darwinian selection of organisms. Imagine a population in which organisms are either *selfish* or *altruistic*. For an organism to have either of these characteristics does not require consciousness or intentional action. Rather, altruism and selfishness are kinds of *behavior*: selfish individuals act in ways that enhance their own prospects for surviving and reproducing, whereas altruistic individuals aid others at their own expense.[12] The Darwinian idea of individual selection predicts that selfish characteristics will become more common and altruistic ones more rare. Evolutionary biologists, beginning with Darwin himself, have considered the possibility that altruistic characteristics are found in nature and therefore require an explanation other than individual selection. One suggestion has been group selection. Although individual selection may favor selfishness over altruism, group selection might favor groups in which altruists are common over groups in which they are rare. Tantalizing conceptual questions will arise in Part II when we consider what it means for genes or organisms or groups to be units of selection. I mention the idea here just to indicate one more alternative to individual selection as a mechanism of evolution.

Besides these alternatives that are of current theoretical interest, standard evolutionary theory lists a number of forces that may modify the composition of a population. Mutation and migration may cause evo-

12. This definition of altruism will be refined in Section 9.3.

lution, for example. If gene A mutates into gene a more frequently than a mutates into A (or if the immigration and emigration rates are similarly asymmetrical), the frequencies of the two genes may change. Not that biologists now believe that these are *major* causes of evolutionary change. The point is that they are empirical possibilities, and so Darwin's preferred alternative is clearly empirical as well.[13]

All the alternatives just listed, including Darwin's, are *historical hypotheses*. Each makes a claim about what caused the changes that have occurred in life on earth. From Darwin's time to the present, a central undertaking of evolutionary theory has been to determine which of these historical hypotheses is correct. Questions of this sort can differ in their degree of generality. One may ask what caused the evolution of some single characteristic in a single population or species ("Why are polar bears white?"), in a group of related species ("Why has the genetic system known as haplodiploidy evolved several times independently in the social insects?"), or in a much larger range of cases ("Why is sexual reproduction as common as it is among the species found on earth?"). Note that these questions, though they differ in scope, are always about particular populations; these questions are singular in form.

Complementing this avenue of inquiry, evolutionary biology has also developed a theory of *forces*. This describes the *possible causes* of evolution. The various models provided by the theory of forces describe how a population will evolve if it begins in a certain initial state and is subject to certain causal influences along the way. As is usual in science, there is a division of labor between theoretical articulation and empirical application. It is one thing to contribute to the theory of forces by developing a model that describes a possible evolutionary trajectory. It is quite another matter to establish a historical hypothesis by showing that this or that population fills the bill.

As we have seen, Darwin was able to characterize how the force of natural selection works in a remarkably simple way: if the organisms in a population that possess one characteristic (call it F) are better able to survive and reproduce than the organisms with the alternative characteristic (*not-F*), and if F and *not-F* are passed on from parent to offspring, then the proportion of individuals with characteristic F will increase. To be sure, a *ceteris paribus* clause needs to be added here:

13. Creationists are quick to exploit debates within evolutionary biology by claiming that they are signs that evolutionary theory is ill-founded. They ignore the fact that controversy is a sign that a science isn't *dead*. They also confuse the uncontroversial fact that evolution has occurred with highly controversial claims about its causes. The former is an overwhelmingly confirmed matter of history; the latter are food for theoretical thought. See Kitcher (1982a) for discussion.

heritable variation in fitness will result in evolution only on the assumption that no counteracting forces cancel its effects.[14] And, of course, some clarification is required of what it means to say that a trait is "inherited." But as a first approximation, the conditional claim that *if* there is heritable variation in fitness, *then* there will be evolution (if nothing interferes) seems to be remarkably straightforward.

This if-then statement, Darwin's conditional, differs from his historical hypothesis in a fundamental way. Although the latter implies the existence of particular populations undergoing a distinctive causal process, the former statement is entirely silent on this. Darwin's conditional doesn't imply that there was ever life on earth or that a population must automatically experience individual selection as its principal evolutionary force. This is a typical feature of scientific laws; Newton's law of gravitation, for example, does not imply that there ever were any objects with mass or that gravitational attraction is the principal determinant of the trajectory of bodies. Laws simply describe what *would* happen to a system, *if* it were to possess certain specified features.

Two centuries before Darwin, the revolution in physics provided a powerful presumption that the laws of nature are written in the language of mathematics. But Darwin (if not his twentieth-century successors) apparently felt no need of such a hypothesis. Not only are numbers notably unnecessary in his theoretical formulations; in addition, the conditional about the impact of selection on the composition of a population is obvious to the point of triviality. Consider a basket of marbles, half of them green. If I transfer all the green marbles but only 90 percent of the nongreen ones to another basket, what will happen? No experiment is needed to find out. It is obvious that if nothing interferes, the percentage of green will increase.[15]

Physical theories typically require observational confirmation. But the Darwinian *general principle*—as opposed to the historical appli-

14. Geoffrey Joseph has suggested that *ceteris absentibus* might be a more apt expression here. As in the Newtonian case, for a law's prediction of the effect of a force to hold true, what is required is not that all other forces be *equal in magnitude* but that they be *absent*. Although this idea seems to apply straightforwardly to the physics of billiard balls, it needs to be modified somewhat in order to apply to many standard models in population genetics. In these models (examples to be discussed in Sections 1.4 and 6.1), the outcome of an evolutionary process depends on the relative magnitudes of various forces, not on their absolute values. In such cases, what is required for the model's prediction to hold true is not that all other forces be absent but that their net value (if opposite in sign) must be less than the one described in the model.

15. The utter transparency of this little example is anything but representative of what has happened to evolutionary theory in this century. Present evolutionary models still connect assumptions and conclusions by mathematical argument, but the conclusions are often much less obvious—i.e., the mathematics is *hard*!

cations that Darwin and others made of the conditional—has struck many as immune from empirical check. One might expect this difference to be cause for rejoicing. How delightful that the fundamental postulate of a theory can be known to be true without carrying out an experiment! How sublime that it is immune from empirical falsification! But no such reaction took place. Physics-worship ensured that any apparent difference between physics and evolutionary theory would be interpreted as a defect in the biology. Defenders of Darwin reacted by arguing that the theory is, contrary to appearances, empirical; detractors chuckled over the allegation that it is not. I will try to clarify and resolve this problem in Chapter 2.

Darwinian selection of organisms is one possible kind of selection. And selection, whether at the individual level or somewhere else, is only one possible cause of evolution. Unfortunately, "evolution" and "selection" are often used interchangeably (by nonbiologists), as when it is insisted that no evolutionary explanation can be given of a trait because it is implausible to think of the trait as having any selective significance. Natural selection is not just any sort of "biasing effect" in the composition of a population, and evolution is not just any sort of change in composition. This leads to the question: What is evolution?

Biologists nowadays often assert that a population has evolved in just those circumstances in which its gene frequencies have changed. This is not to deny that the evolution of phenotypic adaptations (wings, opposable thumbs, eyes) is what really interests many evolutionary biologists. However, these phenotypic changes are standardly regarded as possible effects or causes of evolution, rather than as providing a criterion for when evolution has occurred.

This genetic criterion for the occurrence of evolution implies that populations can change in composition without evolving. A group of human beings may increase its average height over several generations without a change in gene frequencies. Improved nutrition can augment the proportion of six-footers, gene frequencies holding constant all the while. In a case like this, the characterization of evolution just given will deny that the population has evolved. Similarly, if the census size of a population doubled without affecting gene frequencies, we would again have a case of change without evolution.

In evolutionary theory, gene frequencies are standardly computed per capita. Suppose two land-dwelling organisms live on a rock surrounded by water. They are genetically different—one has copies of gene A but not of B, while the other has copies only of B. The A organism gets fat and the B organism gets thin. The proportion of A-type *cells* in this two-organism population increases. But this isn't evolution in the proprietary sense described above, since the per capita frequency remains 50/50. This head-counting paradigm, as I will call it, has not gone unchallenged (Section 9.4). But for now, note just that it is standard fare in the biologist's criterion for when evolution has occurred.[16]

One should not make too much of the fact that evolution is usually understood in this way. I can see nothing terribly wrong with describing the example just given as a kind of evolution that occurs at the cellular level. Indeed, it is not uncommon for evolutionists to think of the proliferation of cancer cells at the expense of normal cells in an organism's body as a kind of selection process, where selection of this sort can produce a change in the composition of the population of cells.

Other examples that are more standardly discussed in evolution textbooks also are at odds with the "definition" of evolution just given. Population genetics describes various mechanisms that can affect the frequencies of different *combinations* of genes, without changing gene frequencies. For example, certain patterns of inbreeding and population subdivision lead to a predictable decline in heterozygosity.[17] This can be evolution in an interesting sense but needn't involve a change in gene frequencies. The same holds true for the evolutionary consequences of *recombination*. In addition, population change mediated by extrachromosomal mechanisms of inheritance should count as evolution, though *gene* frequencies may remain constant. I conclude that the definition of evolution as change in gene frequencies is a useful rule of thumb, not a hard-and-fast principle.

The idea of evolution just described and even more liberal conceptions that are still genetical in orientation rule out purely *cultural* evolution, in at least one sense. The possibility that cultural characteristics may *cause* evolution is not excluded from theoretical consideration, of course. Beliefs and cultural practices are aspects of the phenotypes of some organisms and may at times affect their chances of survival and reproduction. By the same token, cultural characteristics that have a "genetic basis" may change frequencies and thereby evolve in the proprietary sense of the term. But the criterion does rule out changes in the frequencies of cultural traits that are not reflected in any changes in gene frequencies. The rise of Islam in the Middle Ages and of capitalism in certain Italian city-states are cases in point. Changes of this sort, like the increase in height due to nutrition mentioned above, are not part of the theory's subject matter.[18]

16. I owe this observation about the "head-counting paradigm" to Richard Lewontin.
17. "Heterozygosity" will be defined in Section 1.3; the effect of random genetic drift on heterozygosity is discussed in Section 4.1.
18. However, some evolutionists have attempted to bring genetic evolution and cultural evolution together within a common framework. See, for example, Cavalli-Sforza and Feldman (1981) and Richerson and Boyd (1978).

"Evolution" is not a term that names only the large-scale "important" changes in the tree of life that so justifiably excite our admiration. When a gene in a population goes from 54 to 55 percent on a particular Tuesday, that's also evolution. This is not to say that such minor twitches in a population's composition have the same interest as, say, the evolution of the opposable thumb. But the process of evolution is one thing; what interests us about it may be quite another.

One of the main goals of what follows is to distinguish between different possible causes of evolution. *All* possible causes of evolution may be characterized in terms of their "biasing effects." Selection may transform gene frequencies, but so may mutation and migration. And just as each possible evolutionary force may be described in terms of its impact on gene frequencies, so it is possible for a cause of evolution to be present without producing changes in gene frequencies. The *ceteris paribus* rider must not be forgotten. Selection doesn't imply evolution any more than evolution implies selection.

All this is merely to locate evolutionary theory in familiar territory: it is a theory of forces. The distinctive features of evolutionary theory emerge in the specific sorts of forces that it considers. But evolutionary theory is a rather standard *kind* of theory. This idea will carry us some distance toward understanding the structure of evolutionary theory— how its models are put together and how they make contact with the reality of the living world.

1.3 Zero-Force States

A theory of forces begins with a claim about what will happen to a system when *no* forces act on it. The theory then specifies what effects each possible force will have when it acts alone. Then the theory progresses to a treatment of the pairwise effects of forces, then to triples, and so on, until all possible forces treated by the theory are taken into account. Since most objects in the real world are bombarded by a multiplicity of forces, this increase in complexity brings with it an increase in realism.

A theory must discover how to combine the forces it describes. If I push a billiard ball north and you simultaneously shove it west, science and common sense predict that the ball will move northwest. Newtonian mechanics has made vector addition a familiar paradigm for computing the net effect of forces acting in concert. But it is only one example, and other more complex interactions are certainly possible. If you drink enough acid and nothing else, you will get sick. If you drink enough alkali and nothing else, the same thing will happen. But what will happen if you drink alkali and acid together? Instead of getting twice as sick, you suffer no ill effects at all.[19] Each theory of forces must solve this compositional problem for itself, there being no antecedent recipe that is guaranteed to work for all cases.

However, before this compositional problem can arise, a theory of forces must say what each force will achieve when it acts alone. And prior to this question, the theory must describe what the system will do when no forces impinge. This question may seem easy to answer: When no forces act, the system will not *change*. But this trivial answer must give way to a more substantive one; what counts as change and what as staying the same is a delicate theoretical matter. For example, in Newtonian theory, an object will remain at rest or in uniform motion if it is not acted on by a force. Here "no change" means no change in *velocity*. Yet, in the Aristotelian physics that Newton's superseded, the zero-force state would be one of rest at the center of the earth. For Aristotle, "no change" meant no change in *location* (Lloyd 1968).[20]

The zero-force state in evolutionary theory is specified by the Hardy-Weinberg law of population genetics. It describes how the frequencies of traits in gametes (sperms and eggs) will be related to the frequencies of traits in the organisms that those gametes form. The traits here are single genes. Suppose that each organism in a population is *diploid*; its chromosomes come in pairs. Suppose further that there are only two alternative alleles (genes) *A* and *a* that may occur at a location (a "locus") on each of two homologous chromosomes.[21] Then there are three possible diploid genotypes: *AA*, *aa*, and *Aa*. The first two are called *homozygotes*; the last is the *heterozygote*. When an organism produces gametes, normally the chromosome pairs split and each gamete receives one or the other of the homologues. The gametes are therefore *haploid*. Reproduction then unites two haploid gametes to produce a fertilized egg (the zygote) with the full diploid complement of chromosomes.

19. The example is Cartwright's (1979).
20. Here I gloss over the difficult historical question of saying what "force" meant to Aristotle. It is objectionably Whiggish, I think, to simply assume that Aristotle and Newton were talking about the same thing ("force") but disagreed about its properties.
21. For convenience, one may want to think of each chromosome as a string of beads, and a pair of homologous chromosomes as a pair of such strings lying side by side so that one can compare the beads that are at the same location on the two strands. Although this picture will be adequate for the purposes of discussion here, it would not be if one wanted to think about what a gene is in a reasonably sophisticated molecular way. Here a functional conception of a gene (as something that produces an amino acid) will not coincide with a molecular conception of a gene (as a particular sequence of nucleotides). "Gene" is a fairly flexible term, which can be made as precise as the problem at hand requires. See Kitcher (1982b) for discussion.

Suppose that one generation of a population produces gametes and that gene A occurs in the gamete pool in frequency p and gene a in frequency q, where $p + q = 1$. In addition, what holds for the gametes as a group also holds for the sperms and for the eggs: A has frequency p in the sperm pool and also in the egg pool. There are many such gametes—far more than are "needed" to form the next generation. Suppose that the zygotes are formed at random. Then the frequencies of AA, aa, and Aa will be p^2, q^2, and $2pq$, respectively. Expressed in this way, the Hardy-Weinberg law does not assume that there is no selection, mutation, or migration. It simply exhibits a consequence of random mating, when the frequency of each allele is the same in the male and female gametes. However, another formulation of the principle implies that it has the status of a zero-force law in population genetics.

Instead of comparing the frequency of a gene in the gamete pool with the frequencies of the genotypes at the moment of fertilization, let's look further ahead in the life cycle. The zygotes formed will gradually become adults. There may be differences in the mortality rates that the different genotypes experience. Mutation and migration may similarly affect what the population looks like in the process. Chance events may also have an impact. But if none of these impinge, then the genotypic frequencies will retain the three values mentioned above.[22]

So far, I have described the Hardy-Weinberg law as applying to the transition from gametes to the organisms that are constructed out of them. Evolutionary theory also has something to say about how gene frequencies will behave when subject to no evolutionary forces in the other half of the life cycle—during the transition from organisms to the gamete pool. If no evolutionary forces act, then the frequencies of genes in the gamete pool will be the same as their frequencies in the organisms of the previous generation. This means, for example, that organisms carrying copies of one allele do not produce more gametes than organisms carrying the alternative, and that there is random mating. Also implied is the "fairness' of the Mendelian mechanism of meiosis: Each allele represented in a heterozygote should be represented in 50 percent of that individual's gametes.[23]

Thus a population of diploid organisms that produces haploid gametes engages in an alternating process of splitting and joining. Chromosome pairs in an organism split apart during meiosis, and haploid gametes unite to form a new diploid organism during reproduction. Note that

22. If we assume nonoverlapping generations, the Hardy-Weinberg ratios characterize the whole population. If the generations overlap, the law describes the ratio in a single generation.
23. The evolutionary force called "meiotic drive," to be discussed in Section 9.1, involves a mechanism that "cheats" the Mendelian system (Crow 1979).

in this alternating process the variability of a population is preserved, if no evolutionary forces impinge. If *alleles* are represented at given frequencies at the beginning, they will continue to maintain that representation, as long as nothing interferes. Once *genotypes* have gone to their Hardy-Weinberg equilibrium values, there they stay, in the absence of an evolutionary force.

For natural selection to do its work, there must be variability in the population. Darwin knew this, which is why he was concerned about the possibility of blending inheritance. For the same reason, R. A. Fisher began *The Genetical Theory of Natural Selection* with a consideration of this issue. Blending inheritance destroys variation; if the character state of an offspring is somewhere intermediate between the character states of its parents, the process of reproduction itself exerts a homogenizing influence on the population. The Mendelian mechanism of assemblage and disassemblage, on the other hand, retains the identity of the hereditary components. It is, of course, possible in a Mendelian system that an offspring may have a *phenotype* intermediate between those of its parents; we can imagine a situation in which height obeys this rule. But at the genetic level, the "particles" do not blend. This is of the first importance if evolution by natural selection is to be possible.

In evolutionary theory as in Newtonian physics, the principal use of a zero-force law is to discover when evolutionary forces *have* played a role. If genotype frequencies depart from Hardy-Weinberg equilibrium, some force must have been at work. For example, selection, mutation, or migration may cause the genotypic frequencies at the adult stage to differ from the allelic frequencies in the gamete pool from which those adults were constructed. And "sampling error" (*random genetic drift*) may lead the frequency of a gene among the zygotes to differ from the frequency of that gene in the gamete pool. After all, if there are a billion sperms and a billion eggs, each having exactly 40 percent *A* and 60 percent *a*, there is no guarantee that if you randomly sample from that gamete pool to construct 200 organisms, you will end up with a 40/60 split in the 400 genes found in those zygotes. When the zygotic or postzygotic frequencies depart from the gametic frequencies, an evolutionary force must have impinged.

The Hardy-Weinberg law does not license an inference in the reverse direction. Although a population that departs from Hardy-Weinberg equilibrium must have experienced an evolutionary force, one cannot assume that a population that remains at Hardy-Weinberg equilibrium has not been so affected. Various forces may have been present and canceled out each other. For example, there may be a balance between mutation and selection; if *A* mutates into *a* more readily than *a* mutates into *A*, but *A* is selectively advantageous, the net result may be conformity to the Hardy-Weinberg ratios. Change implies a force, but the absence of change does not imply the absence of force.

The parallel with Newtonian mechanics is straightforward. An accelerating billiard ball must have been acted on by a force. But if we find a ball that has remained at rest or in uniform motion (acceleration = 0), this does not tell us whether any forces have been present. Perhaps no forces impinged, or perhaps you gave a strong push to the north just when I gave an equal and opposite shove to the south.

The Hardy-Weinberg law allows that the distribution of physical characteristics may change from one generation to another without any evolutionary forces acting. Suppose we observe a population that has 50 percent homozygotes *AA*, no heterozygotes, and 50 percent *aa*. Suppose these organisms form gametes, mate at random, and that no evolutionary forces impinge in the process. This generation of parents will then produce offspring in the ratio 1/4 *AA*, 1/2 *Aa*, and 1/4 *aa*. If the three genotypes have different phenotypes—say, *AA* organisms are tall, *aa* organisms are short, and *Aa* individuals are middling—then the frequencies of genotypes *and* phenotypes will have changed in the above transition. But according to the motto "evolution is change in gene frequencies," the population will not have evolved, and no evolutionary forces acted on it. This is a simple case in which the frequencies of gene *combinations* can change even though gene frequencies do not.

One might notice the change in the population's distribution of heights and naively reason as follows: The population changed with respect to the proportion of individuals having various phenotypes. Change requires a cause. So there must be something—a force—that caused this change. But no such force is recorded in evolutionary theory, any more than Newtonian theory requires that an object's change in location or the fact that it stops accelerating must be chalked up to the imposition of a force.

Of course, if we step back from the array of forces covered by evolutionary theory, we *can* explain the frequency changes in phenotypes and genotypes that occurred. They were due to the Mendelian process, wherein organisms produce gametes and gametes produce organisms, and to the laws of developmental genetics that explain why an organism with a given genotype has a given phenotype. But it is important to realize that in modern evolutionary theory, Mendelism is the background against which evolutionary forces are described. It is not itself treated as a force.[24] An analogy comes to mind: Gravitation is treated

24. *Some* aspects of the Mendelian process—e.g., the "linkage" of genes that are located on the same chromosome—are treated as evolutionary forces. The point is that a substantive Mendelian mechanism is assumed to be at work even when all evolutionary forces are said to be absent.

as a force in Newtonian theory, but not in general relativity. In the latter, gravity shows up as a set of paths ("geodesics") along which objects will move if they are not subjected to a force. Gravity is not itself a force, but the background against which forces are described.

Although my treatment of the Hardy-Weinberg law as foundational in population genetics is rather standard, I want to take note of a dissenting opinion. Kimura (1983, p. 7) thinks that the usual formulation of the principle as a zero-force law is "anachronistic" and that the principle is a useful rule of thumb *"even when gene frequencies are changing . . .* if we enumerate genes and genotypes immediately after fertilization." I cannot disagree with these judgments or with Kimura's more direct way of formulating a zero-force law for population genetics:

> The fact that gene frequencies do not change without disturbing factors is obvious without any computation as soon as we note that homologous genes on the chromosomes segregate regularly at meiosis and that genes are self-reproducing entities. We tend to forget that formation of genotypes (zygotes) is simply a grouping by two of homologous genes. Higher organisms including ourselves happen to have a well-developed diploid phase; no one would pay any attention to the Hardy-Weinberg law if we humans were haploid.

Indeed, Kimura's observation *is* more general than the Hardy-Weinberg formulation, even though alternation of haploid and diploid phases is hardly an idiosyncratic feature of our species. What the Hardy-Weinberg formulation of a zero-force law adds to the idea that gene frequencies do not change without cause is the thought that the frequencies of gene *combinations* (genotypes) also reach an equilibrium. This is a substantive addition, since there are causal processes that can affect combination frequencies without altering gene frequencies.

We have seen that evolution is standardly defined in terms of what happens to genes; similarly, the zero-force law (whether we give it the Hardy-Weinberg formulation or merely say that gene frequencies do not change when no evolutionary forces impinge) is stated in the language of gene and genotype frequencies. Yet, as mentioned earlier, fitness, at least in its Darwinian construal, is not so described. It is, in the first instance, a property of *organisms*. In addition, it is a property that organisms possess in virtue of their morphological, behavioral, and other *phenotypic* characteristics. How can a theory of forces represent genotypic and phenotypic phenomena in a common currency?

Organisms may survive and reproduce in virtue of their phenotypes, as Darwin thought. And a nonselective force like migration may bring some organisms into a population and take others out of it. Population genetics represents these processes only insofar as they are reflected in gene frequencies. If green organisms survive better than blue ones, population genetics will notice this difference only if the genes in blue organisms differ from those in green ones. Similarly for migration and every other evolutionary force. If evolution is change in gene frequencies, then the causes of evolution must be described in terms of their effects on those frequencies.

Thus there is a conceptual gap to be bridged. If the *causes* of evolution are described in phenotypic terms, and if the *effects* of evolution are calibrated in the language of gene frequencies, then a "translation manual" must be provided that goes back and forth between these two modes of description. If we look carefully at the connection between the different stages in the life cycle, we can discern the sorts of transformation rules that a fully developed theory of forces will provide. A simplified description of these is provided by Figure 1 (adapted from Lewontin 1974b, p. 14).

Fertilized eggs develop into organisms with different phenotypes. Laws of developmental genetics describe how a fertilized egg with a given genotype will produce different phenotypes, depending on the environment in which it develops. Here genotypic properties are mapped onto phenotypic ones by the laws T_1. Given an array of organisms in a generation, let us say all at the juvenile stage, natural selection and other evolutionary forces may modify the composition of that generation as it matures to the adult stage. This transition is described by laws (T_2) that map phenotypic descriptions onto other phenotypic descriptions. The adults then create gametes that then join to form fertilized eggs. Again selection may play a role, this time in producing differences in reproductive success. Phenotypic properties of parents may thereby affect the genetic composition of the gamete pool from which the next generation of zygotes will be constructed. Here, the requisite laws of transformation T_3 move from the phenotypic to the genotypic level. The composition of gametes to form zygotes is described at the genetic level by a set of principles (T_4) that includes the Hardy-Weinberg law, and we are off and running again with a new generation of fertilized eggs.

When modeling evolutionary processes, knowledge of correlations will sometimes suffice if knowledge of causes is unavailable. Developmental genetics may at present tell us little about why individuals with a given genotype develop a particular phenotype. However, if we know that genotype and phenotype are correlated in a certain way, that information may suffice for us to describe the way a population evolves. This will become clearer when we examine a simple model of a selection process in the next section.

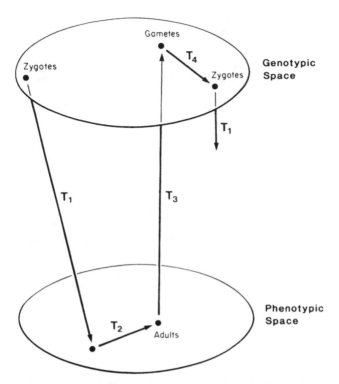

Figure 1. Some of the laws of transformation required in evolutionary theory. T_1 describes how the environment mediates the development of fertilized eggs with given genotypes into organisms with various phenotypes. T_2 describes the way differential mortality can modify the array of phenotypes found within the space of a single generation. T_3 describes how phenotypic differences among adults can influence the composition of the gamete pool, thereby reflecting fertility selection. T_4 describes how the genetic composition of the gamete pool influences the genetic properties of zygotes (adapted from Lewontin 1974b).

1.4 Darwinian Fitness

With a zero-force law in hand, we may proceed to the set of singleton-force laws laid down by evolutionary theory. There are several of these, one for mutation, one for migration, one for random genetic drift, one for linkage, one for inbreeding, and so on. The singleton law that I want to focus on now describes the effects of natural selection when it is unencumbered by other evolutionary forces. Although there are certainly no real populations in which natural selection is the only evolutionary force at work,[25] some natural populations approximately

25. Since natural populations are always finite in number, random genetic drift must always play *some* role, however small, in evolution. Drift will be discussed in Chapter 4.

conform to this idealization, and so there is a use for simple models in which selection is taken to act alone. I will proceed by describing one well-known example to which this sort of model has been applied.

Many human populations in Africa, the Middle East, and the Mediterranean are *polymorphic* for alleles that I'll call S and A. Such populations contain individuals with the three genotypes SS, SA, and AA. The red-blood cells of SS homozygotes are sickle-shaped. These people suffer from a severe anemia that is usually fatal in childhood. A person who is homozygous for the A allele, on the other hand, does not have the sickle-shaped blood cells, and so is not anemic. The heterozygotes, finally, have the sickle-shaped red-blood cells in very low frequencies (usually around 1 percent) and do not experience anemia. So the S allele contributes to sickling in blood cells; when the gene is present in double dose, the sickling is severe enough to produce anemia.[26]

Biologists have observed that the S allele is stably present in the populations in question. Why should this be so? Why doesn't natural selection eliminate the characteristic? Although there are numerous conceivable explanations (for example, the continual introduction of new copies of the allele by mutation or migration), empirical studies have shown that a second sort of selection pressure counteracts the tendency for selection against anemia to wipe out the allele. It was discovered that heterozygotes have a greater resistance to malaria than the AA homozygotes.

Thus, in malarial regions, the heterozygote is the fittest of the three genotypes. People with this genotype do not suffer from anemia and enjoy enhanced resistance to malaria. Next in the fitness ordering are individuals who are homozygous for A; they have no special resistance to malaria, but neither do they have anemia. Last in the fitness ordering, because of their severe anemia, are the SS homozygotes.
It is intuitive that these fitness relationships lead the S and A alleles to be maintained in a balance. Heterozygotes are better able to survive, and when they reproduce, the Mendelian shuffling of genes will produce new homozygotes and heterozygotes in the next generation.

Evolutionary theory accommodates these observations in a simple quantitative model.[27] We begin with a gamete pool in which the two

26. The causal role of the S allele has been pinpointed to a single amino acid substitution in the hemoglobin molecule.
27. Templeton (1982) provides a rather less simplified treatment of this example than the customary one given here, in that he considers a three-allele system. Interesting consequences concerning the optimizing character of natural selection emerge from his discussion.

alleles A and S occur with frequencies p and q, respectively, where $p +$ $q = 1$. It is assumed that the gametes combine at random to form zygotes, so that the initial frequencies of the three genotypes are the Hardy-Weinberg values of p^2, $2pq$, and q^2. This configuration of the population before selection occurs is shown in the following chart.

	Genotypes		
	AA	AS	SS
Frequency before selection	p^2	$2pq$	q^2
Fitnesses	w_1	w_2	w_3
Frequency after selection	$p^2 w_1 / \overline{w}$	$2pq w_2 / \overline{w}$	$q^2 w_3 / \overline{w}$

The numbers w_1, w_2, and w_3 are the fitness values of the three genotypes. That is, they are the average fitnesses of the individuals in the population having the three characteristics. They represent a kind of "mortality tax" that the three classes of individuals pay in their transition from zygote to adult.

The frequency of the three genotypes after selection can't be computed simply by multiplying their frequencies before selection by their fitnesses. Frequencies before and after must each sum to 1. To make the numbers come out right, we "normalize" by dividing each product by the number $\overline{w}$ (pronounced "W-bar"). This is the average fitness of organisms in the population before selection and has the value $w_1 p^2 + w_2 2pq + w_3 q^2$.

The difference that selection makes to the frequencies of the three genotypes within the space of a single generation depends on the ordering of the three fitness values. The genotype with the highest fitness value will increase in frequency. The example we are considering is a case of *heterozygote superiority* (*heterosis*); so w_2 is larger than w_1 and w_3. One usually wants to consider evolutionary processes as they proceed over a number of generations. How does this model, which represents selection only within the space of a single generation, permit this sort of iterated application? Selection occurs in the transition from the egg to the adult stage. Then the surviving adults reproduce and their offspring experience the same regimen of selection in their development into adults. They then reproduce, and the process begins again. In principle, selection may impinge on both survivorship and reproduction; however, this model accords no role to fertility selection. It is assumed, for example, that genotypes do not differ in their chances of being sterile or of securing mates. At the adult stage in each generation, the genotypes divide to form gametes, and these then randomly reassemble to yield the next generation of zygotes. Let me stress that this model assumes *constant* viabilities, i.e., the fitness values of the three genotypes do not themselves evolve. Though the composition of the population may be modified, selection exerts a fixed tax on the genotypes in every generation.[28]

So far, I have described a little of what happens to *genotype* frequencies in this model. But what happens to *gene* frequencies, i.e., the frequencies of the *A* and *S* alleles? The model predicts that the population will eventually come to a gene frequency equilibrium. At this point, the mortality tax will still affect the frequencies of *genotypes* as they pass from the egg to the adult stage. But the frequencies of the *genes* in one generation will be the same as their frequencies in the next. The equilibrium frequency of *A* ($\hat{p}$) will have the value:

$$\hat{p} = (w_3 - w_2) / [(w_1 - w_2) + (w_3 - w_2)]$$

Let's take a simple numerical example to illustrate how this works. We won't consider the fitness values of the sickle-cell system but a more extreme possibility called the "balanced lethal." Suppose the homozygotes have fitnesses equal to 0, and that the heterozygote has a fitness of 1. This means, effectively, that all the homozygotes die in the transition from zygote to adult but none of the heterozygotes do. Easy computation reveals that at the end of the first generation, the genotypes will have frequencies of 0, 1, and 0, respectively. The alleles now each occur at a frequency of 0.5. The surviving heterozygotes then reproduce, and the Mendelian shuffling will generate offspring in which the genotypes occur in the frequency 1/4, 1/2, 1/4. Notice that although *genotype* frequencies have changed from the previous generation, *gene* (allele) frequencies have not. When selection occurs again, genotype frequencies return to the 0-1-0 pattern. If we follow the population over many generations, we will see it zigzagging between 0-1-0 and 1/4 - 1/2 - 1/4, with gene frequencies holding constant at 50/50.

The equilibrium frequency $\hat{p}$ is a function of the fitnesses, not of the starting frequencies of the genes or genotypes. No matter what the initial frequency is in the population, it will evolve to the same equilibrium point. For the case of heterozygote superiority, in which $w_2 > w_1, w_3$, the equation for $\hat{p}$ predicts that the population will evolve to an intermediate frequency for the two alleles. The population will remain polymorphic for the two genes, rather than driving one or the other to fixation. However, if the heterozygote were intermediate in fitness,

28. In Section 6.1, we will consider a model of selection in which the fitness of a characteristic changes as a function of its own frequency.

the equation would predict that the polymorphism would be eliminated.[29]

The model requires that we assign fitness values to traits (genotypes), not to individual organisms one by one. It doesn't matter how fit any particular organism is; what matters is the average fitness of organisms with the same genotype. Similarly, whether a population with both alleles will remain polymorphic or will lose its variation depends only on the *ordering* of the three genotypic fitnesses, not on their exact values. This means that the model allows us to explain why a population is polymorphic without knowing its previous gene frequencies, without knowing how fit any particular organism is, and without knowing the exact values of the selection pressures impinging upon the population. If you say that the locus is *heterotic*, you've said it all.[30]

What does this model predict will happen to the sickle-cell system if malaria is eradicated? In that case, heterozygotes will no longer benefit from their enhanced immunity, and so w_2 will equal w_1, with both still exceeding w_3. In this case $\hat{p}$ equals 1; so the S allele should disappear.[31]

This model provides a very simple example of how the genetic system can, in a manner of speaking, "get in the way of" evolution by natural selection. In the sickle-cell system, anemia is disadvantageous and resistance to malaria is beneficial. Selection works against the former and in favor of the latter. One might therefore reason that if selection of these two sorts is the only evolutionary force at work, then the population should eventually contain 100 percent malaria-resistant and 0 percent anemic individuals. If the genetic system were "favorable," this *would* be the result. But in the present instance, the way the genetic system codes phenotypic characteristics plays an important role in influencing the evolutionary outcome. Anemia and vulnerability to malaria continue to be represented in the population, even though they are always selected against. This illustrates a quite general property of evolutionary reasoning, one that extends far beyond the confines of this simple model: *Whenever a model describes phenotypic characteristics in terms of their relative advantageousness, and then reaches a conclusion about the evolution of those characteristics, there is a genetic presupposition that is needed to underwrite the inference.*

29. The equation gives values outside the interval [0,1] if the heterozygote is intermediate. This means that there is no "feasible" (internal) equilibrium. This and the case of *un-*derdominance ($w_2 < w_1, w_3$) will be discussed more fully in Section 6.1.

30. This is by no means a universal property of selection models or of models in which selection and other forces are treated jointly. A simple example in which starting frequencies affect the evolutionary outcome will be discussed in Section 6.1.

31. This prediction, applied to the sickle-cell case, has not to my knowledge been confirmed. It is quite possible that the selection process is working so slowly that it cannot be measured very well.

The only evolutionary force represented in this model is natural selection. Constant genotypic fitnesses are the only parameters that affect changes in gene frequencies. How are the values for these three quantities obtained? Biologists observe that SS homozygotes commonly die of anemia, that AS heterozygotes commonly enjoy enhanced resistance to malaria, and so on. These facts about individuals are then pooled into summary statements about what happens to sets of individuals sharing the same genotype. Is the fitness of a genotype then simply the percentage of individuals who happen to survive from egg to adult (weighted by $\overline{w}$)? The answer to this question is an emphatic *no*.

Evolutionary theory understands fitness as a probabilistic quantity. An organism's fitness in the above model is its *chance* of surviving. A genotype's fitness is the average of the relevant probabilities attaching to the organisms who have that genotype. The census information of who lives and dies is evidence used for *estimating* fitness; actual survival rates do not *define* fitness values.

An analogy may bring the probabilistic character of the fitness concept into clearer focus. Suppose we had three coins and wanted to know what probability each has of coming up heads if tossed. We might toss each coin a number of times to find out. But the percentage of heads obtained in a series of tosses would not *define* the coin's probability; rather, the actual performance of a coin would be *evidence* concerning how biased the coin is.

A coin's probability of coming up heads, we might assume, is not a function of how often it actually is tossed. A coin that is never tossed, a coin that is tossed an odd number of times, and a coin that is tossed a million times may have the same probability of landing heads-up. Although a coin may land heads-up exactly half the number of times it is tossed if it is tossed a million times, it cannot do this if it is tossed an odd number of times or if it is never tossed at all. If coins differing in the number of times they are tossed may yet have the same probability of coming up heads, we cannot identify the probability with the actual frequency of heads.

The same line of reasoning applies to fitness. In the above model in which genotypes differ in their viabilities, the fitness of a genotype does not depend on how many organisms in the population actually have that genotype. The genotypic fitnesses are survival probabilities, which is to say that they represent the average chance an organism of a given type has of surviving from egg to adult. Populations of different sizes and compositions may be characterized by the same set of fitness values, just as coins that differ in the number of times they are tossed may have the same bias.

There is a limiting case in which the survival percentage can be expected to come very close to the probabilities. In precisely the same way, there is a limiting case in which the percentage of heads can be expected to converge on a coin's probability of landing heads. This is the case of infinite population size. A fair coin (one with a 0.5 chance of landing heads-up) may not come very close to 50 percent heads if it is tossed only 10 times. But Bernoulli's law of large numbers assures us that as we increase the number of tosses (which are assumed to be independent of each other), the probability of coming within some fixed range of 0.5 gets larger and larger. So the odds of getting between 40 and 60 percent heads if the coin is tossed 10 times is not all that great, but the odds of falling in that range if the coin is tossed 1000 times are very good indeed. If we increase the number of tosses indefinitely, the probability of getting arbitrarily close to 0.5 converges on 1.0. This is the meaning behind the idea in population genetics that models in which selection alone are described presuppose infinite population size. It is only in this idealized case that the actual survival frequencies of the various genotypes can be counted on to equal their probabilities.[32]

How, then, can such models apply to the real world of finite population size? In the same way that a model that ignores mutation can nevertheless apply to populations in which there is a nonzero mutation rate. If we are interested only in the qualitative predictions of the model—if one of the genes will go to fixation or the two will be maintained in a balanced polymorphism—other evolutionary forces may be ignored so long as taking them into account would not materially affect this qualitative prediction. If, on the other hand, a more precise sort of prediction were demanded of a model, one might want to take into account a larger number of the evolutionary forces involved. The adequacy of the model depends on the kind of question being asked.

This constant-viability model of selection describes how selection may affect one-half of an organism's life cycle. It shows how differences in the probability of survival from egg to adult may shift the frequencies

32. The nature of this assurance is worth clarifying. As Van Fraassen (1977) and Skyrms (1980) have pointed out, it is logically possible that a fair coin should come up all heads even if it were tossed an infinite number of times. There is no logical guarantee that relative frequencies will match probabilities, even in the limit. Bernoulli's theorem says as much as can be said on the matter. This point has an important bearing on the foundations of probability. Probability is not reducible to the idea of hypothetical relative frequency in the limit. Rather the *probability* of heads may be characterized in terms of the *probability* of having the frequencies fall in a certain range, as the sample size is increased. If the concepts of probability and hypothetical relative frequency are no more intimately connected than this, then it is a mistake to think that the former can be reduced to the latter.

of genes and genotypes in a population. The other half of the life cycle—the process that takes an adult organism to the offspring it creates—corresponds to the other half of the notion of Darwinian fitness, which in its full sense encompasses both survival *and* reproduction.

The component of fitness that reflects differences in fertility, rather than viability, is also understood probabilistically. But unlike survival, which is a yes/no affair, reproductive success comes in degrees. An organism may have one offspring, or two, or three, etc. To each of these possible reproductive outcomes, a probability is assigned. We then may calculate the mathematical *expectation* of the organism's number of offspring, which is the sum of the different possible numbers of offspring (i), weighted by their probabilities (p_i). Assuming as we have that each offspring is produced by two parents, we credit each parent with one half of its expected number of offspring. That is, an organism's reproductive fitness (w) may be expressed as

$$w = \sum_{i=0}^{\infty} \frac{ip_i}{2}$$

The fitness of a genotype is then defined (as in the viability model) as the average fitness of the organisms with that genotype.

In the simple viability model of selection, the only differences between genotypes were differences in the chances of surviving from egg to adult. In a parallel simple fertility model, genotypes would be assumed to be identical in their chances of surviving but would differ in their expected number of offspring. The meiotic formation of gametes, followed by random mating, would then show a larger number of offspring associated with the fitter genotypes. As before, fitness is understood in terms of its probability consequences for the frequencies of traits. Infinite population size assures us that frequencies will reflect probabilities, with random sampling error playing no role.

The viability model with which we began implies that fitter traits are the ones that will most likely increase in frequency.[33] This might suggest that we can define fitness by saying that trait X is fitter than trait Y if and only if trait X will probably increase in frequency at the expense of Y. But, as mentioned earlier, every evolutionary force may be characterized in terms of its "biasing effects" on trait frequencies.

---

33. *In the space of a single generation.* However, when we look across a generational divide, this need no longer be true. When a population is at its gene frequency equilibrium, the frequencies of genotypes at one point in the life cycle—the adult stage, say—will be the same as their frequencies at the same point in the life cycle during the next generation. Note that when the balanced lethal system described above reaches equilibrium, the adult stage is 0–1–0 and the egg stage is 1/4–1/2–1/4 in each subsequent generation.

Mutation and migration may also predict that X will probably become more common and Y more rare.[34] Selection is not distinguished by the fact that it has a biasing effect but by the sources of the biasing it predicts. Evolution predicted by differences in Darwinian fitness is driven by differences in viability and fertility among organisms.

A number of interesting questions remain to be asked about the relation of fitness and chance. A model like the one just sketched for the sickle-cell trait assigns a fitness value to each genotype; this number represents the average probability that an organism of that genotype has of surviving from egg to adult. How are these probabilities to be assigned? At best, we observe organisms living and dying; how are we to understand the idea that they have a *probability* of survival? Scriven (1959) tells a little story that makes this puzzle salient. Identical twins are out for a walk when one is struck by lightning. Intuitively, we might view this as a chance event rather than as a case of selection. The twins, we may want to say, were identical in fitness. But why should we say this? Why not view the lightning strike as a momentary selective event?

One doesn't resolve this puzzle by insisting that fitness is a property of groups of organisms (e.g., organisms sharing a common genotype) and not of organisms taken one at a time. The fitness of a genotype is, after all, a property that is defined in terms of the fitness of organisms. Additionally, what can happen to a twin can also happen to a group of organisms with the same genotype. A population evolving under the selection pressures of malaria and sickle-cell anemia may be perturbed by "chance" events. Lightning may "by chance" kill more copies of one gene than another. Why should we model this event as a nonselective occurrence rather than say that at that particular place and time there was a third selective force at work in addition to the two recognized in the above model?

I will address this problem and related ones in Sections 4.2 and 4.3. The present point is that selection is a force whose effects are calculated in the currency of gene frequencies. Darwinian fitness is a property of an organism, one reflecting its chances of survival and reproductive success. In Part II, we will see that fitness is a property attaching to objects at other levels of organization—to genes, groups, and even species. I have sketched the simple population genetical model of the sickle-cell trait to orient discussion. It would be premature to define fitness with this as the only example to consider. As we will see in the next section and in much of Part II, there is more to fitness than Darwinian fitness.

34. Another defect involved in computing a trait's fitness by comparing its frequencies before and after selection will be discussed in Section 5.2.

Another kind of incompleteness attaches to the discussion so far. Viability and fertility models present only half of the idea of Darwinian fitness. They describe the *consequences* of fitness differences for gene and genotype frequencies; they show what can be inferred from facts about fitness. Evolutionary theory offers a complementary perspective in which fitness differences are themselves conceptualized as consequences—as the conclusions of arguments, rather than as their premises. To this other side of fitness we now turn.

1.5 Source Laws and Supervenience

In likening fitness to the probability that a coin has of coming up heads (Section 1.4), I emphasized the way a dispositional property is often understood in terms of its consequences. To say that a coin is "biased" or that a lump of sugar is "soluble" is to say something about what will happen, either probably or necessarily, if certain "triggering" conditions are satisfied. Not that these conditions must be satisfied—a biased coin may never be tossed and a soluble lump of sugar need never be immersed. But in ascribing dispositional properties, we are saying what the future will bring (either necessarily or probably) if the conditions are right.

When a coin is biased or a lump of sugar is soluble, the question arises: *Why* is the object disposed to behave in the way indicated? In the case of the coin, it may be the physical property of asymmetrical balance that biases it. This is why the chance of heads exceeds the chance of tails, when the coin is tossed in a certain way. For the water-soluble lump of sugar, the answer lies in its chemical structure. Immersing the lump will cause it to dissolve because the molecules are arranged in a certain way.

We must recognize two important facts about dispositional properties. They may be described in terms of an *associated behavior* and in terms of a *physical basis*. If the disposition is deterministic, then the associated behavior must occur if the triggering conditions are satisfied. If it is probabilistic, then satisfying the triggering conditions implies a probability distribution of possible behaviors. The physical basis of an object's dispositional property—of a coin's bias or a sugar lump's solubility—is the property of the object that makes the triggering conditions capable of causing the associated behavior.

I have already noted the associated behavior of the dispositional property known as Darwinian fitness. To ascribe a level of fitness to an organism is to say what its chances of surviving and reproducing are. In the example of the sickle-cell trait, I described the physical basis of the fact that individuals who are heterozygous at the locus in question are on average fitter than individual homozygotes. Or, more precisely, I should say *bases*; it is immunity to malaria *and* susceptibility to anemia that jointly confer on the three genotypes their respective fitness values. And this is just the tip of the iceberg: An organism's chance of surviving from egg to adult and its prospects of reproductive success will usually not have a simple physical basis but will be the upshot of a multiplicity of physical properties.

This same complexity becomes apparent when we compare the physical bases of fitness in one organism with the physical bases of fitness in another. Lumps of sugar display a pleasing simplicity. The ones in my kitchen and the ones in yours have the same dispositional property for the same reason. Chemistry reveals that the water solubility of sugar has a single underlying physical basis.

Organisms are more various and the analysis of fitness is correspondingly more difficult. Zebras differ in their chances of being caught by a lion because some are faster than others. The physical basis of their fitness differences may, to simplify a little, be found in the architecture of their legs. Some are better engines for running than others. Now consider a population of cockroaches that have different chances of being killed by DDT. The physical basis of their fitness differences may, to simplify in the same way, be found in the construction of their digestive systems. There is a physical basis for the fact that one zebra is fitter than another. There is also a physical basis for the fact that one cockroach is fitter than another. However, it would be surprising if the physical bases were the same in both cases. There seems to be no single physical property that varies as fitness varies in all cases.

A property is said to *supervene* on a set of physical properties if it satisfies two conditions.[35] First, the property must not itself be physical, in that different objects may share the property and yet be physically quite different. Second, two systems that are physically identical must both have or both lack the property in question. Fitness appears to be a case in point. Fitness apparently cannot be identified with any single physical property. And if there were two organisms that were physically identical and lived in physically identical environments, they would have to have the same fitness value.[36]

Predating the development of this point about fitness by philosophers of biology (see, e.g., Brandon 1978, Rosenberg 1978, and Mills and Beatty 1979) is a succinct statement of it by R.A. Fisher (1930, p. 39). After comparing his "fundamental theorem of natural selection" (which

---

35. See Kim (1978) for a more formal characterization.
36. Note that fitness is properly viewed as a property of the relation of organism to environment, not as an "intrinsic" property of the organism alone.

I will examine in Section 6.1) with the second law of thermodynamics, he notes this difference: "Fitness, although measured by a uniform method, is qualitatively different for every different organism, whereas entropy, like temperature, is taken to have the same meaning for all physical systems."

One salient fact about much of evolutionary theory, as opposed to other areas in biology like molecular biology and biochemistry, is that the properties investigated are supervenient.[37] Theories of predator/prey interactions, diversity/stability models, and the origins and maintenance of sexuality, to name only a few examples, pay very little attention to the physical mechanisms whereby organisms are able to have the properties in question. Even a race of robots or of organisms from another planet whose mechanism of heredity is based on a structure other than DNA would potentially fall within the scope of these theories.

The simple constant-viability model of selection (Section 1.4) displays one advantage of describing physical systems in terms of a highly abstract, nonphysical property like "fitness." It predicts a balanced polymorphism in a population in which the heterozygote is the fittest of the three genotypes. This model applies to human populations subject to malaria and anemia, but in principle applies to any species, regardless of the physical reason that heterozygotes are fittest. Evolutionary biology frequently arrives at models of great generality by abstracting away from physical details. Fitness is one of the central conceptual tools in this explanatory strategy.

The supervenience of fitness and other evolutionary properties explains why evolutionary theory is opposed to vitalism, without thereby being reducible to any physical theory. Vitalism and reductionism have a thousand meanings apiece. But each is undermined by the fact of supervenience in this sense: Vitalism holds that there is, in addition to all the physical properties (including relational ones) that an organism may possess, an extra added ingredient. This *élan vital* is supposed to take the matter of which an organism is made and transform it into a biological entity. If there really were such an extra added ingredient, it should be possible for two physically identical systems to differ with respect to their biological properties. One system might have this elusive admixture, while the other would lack it. However, vitalism will fail if biological properties supervene on physical ones, since supervenience complies with the physicalistic doctrine of "no difference without a

---

37. In this respect evolutionary theory resembles the picture of cognitive psychology and computer science developed by Putnam (1967), Block and Fodor (1972), and Fodor (1975). These authors argue that psychological properties cannot be identified with physical properties, because there are multiple physical mechanisms that may allow an organism (or a complicated enough computer) to be in a given psychological state.

physical difference." On the other hand, by rejecting vitalism, we are not forced to think that the properties studied in evolutionary biology are, at bottom, physical properties. Although one physicalistic explanation may account for fitness differences among zebras, and a second such story may explain fitness differences among cockroaches, no physicalistic account can be offered of *what fitness is*. The reason is simply that fitness is not a physical property.

These facts about fitness are reflected in the structure of evolutionary theory conceived as a theory of forces. A theory of forces must contain both *source laws* and *consequence laws*. The former describe the circumstances that produce forces; the latter describe how forces, once they exist, produce changes in the systems they impinge upon. The classical law of gravitation is a source law; it says that when two objects of given *mass* are separated by a given *distance*, there will be a gravitational force of a certain magnitude. Coulomb's law is another example: It says that when two objects of given *charge* are separated by a given *distance*, an electrical force of a certain magnitude exists. Source laws describe the physical circumstances that generate particular kinds of forces.[38]

A "consequence law," as the name suggests, describes what follows from the presence or absence of forces. "$F = ma$" is the classical example. It says that objects of fixed mass accelerate when and only when they are subject to a net force. No mention is made here of the physical conditions that produce forces, but only of what forces do, once they exist.

Source laws and consequence laws combine to predict the changes that will occur in a system that is subjected to a set of forces. The source laws allow us to calculate the gravitational force, the electrical force, and the values of the other forces that impinge on an object. Each is a component force. Each may be plugged into "$F = ma$" to calculate the component of acceleration it induces. Then a law for combining component forces (the Newtonian assumption is that of vector addition) is employed to calculate the net force. This predicts the acceleration the object will experience.

The source laws of physical theory have the austere beauty of a

---

38. What is the relation of a force to its source? In classical physics, it is a mistake to think of a pair of massive objects separated by a given distance as *causing* a gravitational force to come into existence. The theory holds that the force comes into existence simultaneously with the appearance of the massive objects; so, if causes must precede their effects, we shouldn't view the relation as causal. In the same way, the physical difference in leg structure in a herd of zebra should not be viewed as *causing* the zebras to differ in fitness, since we may imagine that the fitness difference came into existence as soon as the difference in leg structure did.

desert landscape. Just four types of force are recognized, and some scientists hope to make this list even shorter (Davies 1979). By contrast, the theory of natural selection exhibits the lush foliage of a tropical rain forest. The physical circumstances that can generate fitness differences are many. Perhaps someday these will be regimented and reduced in number. But at present evolutionary theory offers a multiplicity of models suggesting a thousand avenues whereby the morphology, physiology, and behavior of organisms can be related to the environment in such a way that a selection process is set in motion.

The supervenience of fitness—the fact that fitness is not a single physical property—helps explain why *general* source laws are hard to come by in evolutionary theory. We can say with complete generality what physical conditions are necessary and sufficient for the existence of a gravitational force. But the physical basis of fitness varies from case to case, and so a specification of source laws in terms of physical properties of the environment must be more piecemeal. Nevertheless, there is scope for considerable generality in the source laws of natural selection. Although supervenience excludes evolutionary theory from the paradise of a *universal* (physical) source law for natural selection, it does not condemn the theory to a piecemeal and ad hoc treatment of each evolving population as unique (Sober 1984c). There are alternatives intermediate between heaven on the one hand and hell on the other. Evolutionary theory provides general theories about fitness by connecting fitness with other *supervenient* properties.

An instructive example of how this kind of thinking proceeds is provided by Fisher's (1930) explanation of why a 1:1 ratio of males to females so frequently occurs. I will describe Fisher's argument in some detail, since it not only sheds light on the concept of fitness but also provides an example of a source law that generalizes over a wide variety of physically distinct populations.[39]

What mix of males and females would make for the largest group? The intuitive answer is also the right one—the group will maximize its size if it has a female-biased sex ratio in which there are only as many males as it takes to fertilize all the females. So, as Hamilton (1967) notes, if groups were selected in virtue of their productivities, we should standardly observe in nature many populations that have more females than males. Yet, the main (though by no means only) pattern is a 1:1 ratio. How could a 1:1 ratio result from *individuals* (rather than groups) maximizing their own reproductive interests?

In the second edition of *The Descent of Man*, Darwin (1886, pp. 259–260) had noted that the sex ratio problem is a difficult one:

---

39. More recent additions to sex-ratio theory, in particular, Hamilton's (1967) theory of "local mate competition," will be discussed in Section 9.3.

> In no case, as far as we can see, would an inherited tendency to produce sexes in equal numbers or to produce one sex in excess, be a direct advantage or disadvantage to certain individuals more than to others; for instance, an individual with a tendency to produce more males than females would not succeed better in the battle for life than an individual with an opposite tendency; and therefore a tendency of this kind could not be gained through natural selection. . . . I formerly thought that when a tendency to produce the sexes in equal numbers was advantageous to the species, it would follow from natural selection, but I now see that the whole problem is so intricate that it is safer to leave its solution to the future.

If one looks at this problem from the perspective of Darwinian fitness, it is indeed puzzling how the story might proceed. For a parent who produces five daughters and five sons and a parent who produces ten daughters have produced the same number of offspring. The two parents might well have had the same expected number of offspring, in which case they would have been equal in Darwinian fitness. If Darwinian fitness were a fully adequate measure of fitness, natural selection would be unable to discriminate between these two parents; the fact that they differ in the sex ratio of their progeny is a difference that would be invisible to selection. For natural selection to come to grips with the sex-ratio problem, one needs to expand the concept of fitness beyond the simple structure laid down thus far. The key is to look at *three* generations rather than just *two*. A parent manipulating the sex ratio of her offspring may thereby maximize the number of *grand*offspring she ultimately produces.[40]

In saying that "Darwinian fitness" is a tool that is insensitive to the three-generation problem posed by the sex ratio, I do not mean that Darwin never thought of selection in this way. His discussion of why there are sterile castes in the social insects (Section 7.1) appears to have basically the same logical structure that R. A. Fisher exploited in explaining 1:1 sex ratios. And what is perhaps of more direct relevance

40. I describe the female parent as manipulating the sex ratio of her progeny because the example that has been prominent in recent discussion is one in which the sex ratio is apparently under female control. This is the so-called haplodiploid system among the social insects in which a female stores the sperm obtained from a single fertilization and rations them out during her lifetime. Under haplodiploidy, fertilized eggs develop into daughters and unfertilized eggs develop into sons. There is, however, no reason in principle why fathers should not influence the sex ratio of offspring, and the fact that males are the heterogametic sex in so many species provides a conceivable mechanism for this to be so.

is the fact that in the *first* edition of *The Descent of Man*, Darwin came quite close to the equilibration argument that Fisher was later to develop:

> Let us now take the case of a species producing ... an excess of one sex—we will say the males—these being superfluous and useless, or nearly useless. Could the sexes be equalized through natural selection? We may feel sure, from all characters being variable, that certain pairs would produce a somewhat less excess of males over females than other pairs. The former, supposing the actual number of the offspring to remain constant, would necessarily produce more females, and would therefore be more productive. On the doctrine of chances a greater number of the offspring of the more productive pairs would survive; and these would inherit a tendency to procreate fewer males and more females. Thus a tendency towards the equalization of the sexes would be brought about (Darwin 1872, p. 316).

It would be interesting to find out why Darwin withdrew this explanation, substituting in the second edition the skeptical remarks quoted before.

How should a parent manipulate her ratio of sons to daughters, if the point is to maximize the contribution she makes to the generation of grandoffspring? It might seem that sons will enjoy a greater degree of reproductive success than daughters. After all, there is almost no limit on the number of zygotes that a male may father, whereas there is a much lower material limit on the number of fertilized eggs that a female may produce. So it may appear that the best investment is to produce all males. But the problem with this strategy is that although a male may produce many offspring, it also is possible for him to produce none at all. Females, on the other hand, are in a sense more assured of producing some offspring, although no female will be able to produce the number of offspring that a particularly lucky male will be able to father.

One must not be misled by the fact that sons produce more *gametes* than daughters do. What we must consider is whether sons will, on average, be more or less reproductively successful than daughters. The issue is grandoffspring, not gametes. The obvious fact we must bear in mind is that every grandoffspring has to have two parents.

We are to visualize this three-generation process as follows. A population of fertilized females produces its offspring. Those offspring then randomly mate to produce the subsequent generation of grandoffspring. A female in the first generation may adjust the sex ratio of her progeny so as to maximize the number of grandoffspring she ultimately produces. The question is: What mix of sons and daughters should she produce?

The answer depends on the ratio of sons to daughters present in the population at large. Suppose that at the zygote stage there are f females and m males. Some of these individuals then survive to reproductive age, randomly mate, and produce N offspring altogether. This means that, on average, the f females produce N/f offspring apiece, whereas, on average, the m males produce N/m. It follows that if there are more females than males, then the males will on average contribute more offspring. Symmetrically, if there are more males than females, it will be the females who do better on average. The only point at which males and females may expect to do equally well is at a 1:1 ratio.

In this three-generation process, the best strategy for a parent to use in producing grandoffspring is for her to produce offspring who are all of the minority sex. If the next generation is male-biased, produce all daughters; if it is female-biased, produce only sons. The only point at which a parent can't gain an advantage by departing from the population's sex ratio will be when the population is at 1:1. Individual selection yields a 1:1 sex ratio as its equilibrium result.

Note that the above argument predicts that if a population has an uneven sex ratio, a selection process will be set in motion that brings the population to 1:1. That is, the 1:1 sex ratio is a *stable* equilibrium. Not only is that ratio of males and females one at which no selection takes place; in addition, if the population departs from that equilibrium, selection will move it back.[41]

We have imagined that each female produces her set of progeny all at once. The sex-ratio strategy recommended by this line of reasoning concerns the ratio of sons to daughters at conception. Even if one sex runs a higher risk of mortality than the other, the argument holds. If males have a higher death rate than females, then a population that is 1:1 at conception will contain an excess of females at reproductive age. At conception, a son will be less likely to reach reproductive age than a daughter, but if the son survives, he can be expected to have a higher reproductive output than the average female. Taking both these facts into account, sons have the same expected value as daughters, and so the equilibrium sex ratio at conception is 1:1.

Among human beings, males have a higher mortality rate both prenatally and postnatally. In addition, sons are born more frequently than daughters, as if in "compensation." The above model, as just

---

41. In Section 6.1, we will examine a case of *un*stable equilibrium. An equilibrium is unstable if there is no selection at that frequency value, but small departures from that point set a selection process in motion that increases the departure from equilibrium rather than returning the population to its equilibrium value. The difference between stable and unstable equilibria is defined in terms of what is true in the neighborhood of the equilibrium.

noted, makes no such prediction. However, matters change if a female need not produce her progeny all at once but may replace offspring who die. If replacement guarantees that a female may bring a certain number of offspring to reproductive age regardless of the sex ratio of her progeny, then the optimum strategy is one in which the sex ratio is 1:1 at the time at which parental compensation ceases. With greater mortality of males, this predicts that males will be overproduced at conception, so that by the end of the period of compensation the ratio has shifted to 1:1 (Crow and Kimura 1970).

Even when parents cannot replace offspring that die, Fisher's argument allows us to account for the fact that the sex ratio can be uneven at conception and birth. This was the heart of Fisher's idea of "parental investment." Strictly speaking, Fisher's argument was not that a parent should produce equal numbers of sons and daughters but that she should "invest equally" in the two sexes. This means that the equilibrium ratio will depend on both costs and benefits.

The cost of a son or daughter might be measured by the total amount of energy that parents must expend, both prenatally and postnatally, until the offspring lives independently. Among human beings this spans a substantial length of time, but in many other species there is no postnatal parental expenditure. By considering energy costs, we take both these possibilities into account.

We may think of the average mother in the population as having a total package of energy resources (T) available to her. She will devote p (a percentage) of this total investment to sons, and $1-p$ to daughters. If c_m is the cost to produce a son and c_f is the cost of producing a daughter, then her number of sons will equal pT/c_m and her number of daughters will be $(1-p)T/c_f$. Each son or daughter will, on average, provide its mother with a certain return on her investment. The benefit of a son (b_m), as in the previous model, is $N/\#_m$, that is, the total number of grandoffspring (N) divided by the number of sons ($\#_m$) in the population. Similarly, b_f, the benefit that a daughter provides her mother, is $N/\#_f$.[42]

The total benefit that accrues to a mother will then be the sum of the benefits brought by her sons and the benefits brought by her daughters:

$$b_m pT/c_m + b_f(1-p)T/c_f.$$

We now need to consider when a mutant female who departs from the average pattern of investment found in the population at large will

42. Note that all offspring of a given sex are assumed to have equal fitness.

enjoy an advantage. This mutant female will devote p^* of her resources to sons and $1-p^*$ to daughters. Her total return on investment will then be

$$b_m p^* T / c_m + b_f(1-p^*)T/c_f.$$

Hence the mutant female will be more reproductively successful precisely when

$$b_m p^* T / c_m + b_f(1-p^*)T/c_f > b_m pT/c_m + b_f(1-p)T/c_f.$$

Clearly, when

$$\frac{b_m}{c_m} = \frac{b_f}{c_f}$$

any mix of investments in sons and daughters will be equally profitable; at this point, all reproductive strategies are equally fit. But if this equality does not hold, then a mutant female will do best by investing solely in the sex whose ratio of benefits to costs is higher.

If we unpack this criterion further, we find that the equilibrium point is one at which

$$\#_m c_m = \#_f c_f$$

That is, if the population at large invests equally in sons and daughters, no deviant pattern of investments on the part of a mutant female can be at a selective advantage.

How can sons be more profitable to produce? Our first formulation of Fisher's argument, which considered only the benefits that sons and daughters provide and ignored their costs, is a special case of the present model. If sons and daughters are equally expensive ($c_m = c_f$), then sons will provide a greater return when $\#_m < \#_f$. When sons and daughters are equally costly, the minority sex is the one that is more profitable to produce.

Even when there are equal numbers of males and females in the population as a whole, one sex may still be a better investment than the other. Although equal numbers of sons and daughters ensure that they provide identical benefits, it may nevertheless be true that one sex is more expensive to produce than the other. In this case, a mutant female will do better by producing only the sex that is less expensive.

Returning now to our own species, we can perhaps see more clearly why an excess of males are conceived and born. The fact that sons die more frequently is relevant to the compensation argument sketched above, but the present consideration provides what is perhaps a more fundamental perspective. It isn't crucial that a mother can replace a dead child. What matters is that since sons are more apt to die, they will, on average, require a smaller investment. Even if human beings produced a single brood all at once with no possibility of replacement, this argument would still apply; because parents care for offspring both prenatally and postnatally, the equilibrium sex ratio will have an excess of males both at conception and at birth.

It is interesting to contrast Fisher's evolutionary account of the sex ratio with another that comes from a very different point of view. One might account for why the sex ratio is 1:1 in a population by pointing out that daughters are the result of fertilizations from X-bearing sperm and sons are the result of fertilizations by Y-bearing sperm, and the mechanism of meiosis is "fair"—fathers produce equal numbers of the two types of sperm. This kind of explanation might be correct for populations that have a 1:1 sex ratio at conception, but it would obviously fail for populations with other "primary" sex ratios. In those cases, however, it might be true that they are biased one way or the other because fathers produce unequal numbers of the two kinds of sperm.

This explanation is not evolutionary. It describes the present mechanism that generates a population's sex ratio without explaining why that mechanism evolved. Notice that Fisher's argument, on the other hand, makes no mention of the specific physical mechanism responsible for a population's sex ratio; indeed, it is compatible with different populations determining their sex ratios in very different ways. Fisher's argument steps back from the "promixate" mechanism and looks for a more "ultimate" cause (Mayr 1961).

This suggests that the evolutionary explanation is more fundamental than one that focuses on the physical mechanisms found in a population. Selection of the sort Fisher described may cause a "fair" meiotic mechanism (generating 50 percent X-bearing and 50 percent Y-bearing sperm), and that meiotic mechanism in turn may be the proximate cause of an even sex ratio.[43] However, one does not want to overstate

43. That is, the Fisherian account would explain the meiotic account, *if both were true*. Maynard Smith (1978, pp. 148–150) reports that there is little or no direct evidence for the existence of heritable variation in the mix of sons and daughters that a parent produces; he goes on to remark, however, that sex ratio is known to be influenced by environmental factors and "it is a good commonsense principle that if environmental factors can affect some characteristic, it is likely that genes will do so also." Note that if a 1:1 sex ratio were somehow an inevitable physical consequence of the meiotic process, one could not explain a population's sex ratio by positing a selection process that works on variations that all occur within the context of meiosis. Not that meiosis is always "fair," as we will

this difference between evolutionary explanations and explanations that appeal to physical mechanisms. Note that in applying Fisher's argument to the human case, I asserted that sons have a higher mortality rate, both prenatally and postnatally, than females. The Fisherian argument does not explain this fact but assumes it. So it would be a mistake to think of an evolutionary explanation as being prior to *all* facts about the phenotypic and genotypic properties of a population. A physical property of a population may have an evolutionary explanation, but that explanation itself will presuppose that other physical properties were in place ancestrally. The relation between the two approaches is one of mutual illumination.

Fisher's sex-ratio argument establishes a source law for the force of natural selection. It describes a kind of condition[44] that generates selection pressures of a certain sort. The argument shows how a population sex ratio can create selection pressures for and against various possible reproductive strategies. Notice that the theory of the sex ratio approaches the question of reproductive strategy at a very abstract level of generality. The fact that fitness is not identical with a single physical property does not imply that biologists must approach zebras, cockroaches, and anteaters as if they were totally unique. One still may generalize over physically different species so as to make their biological similarities salient.

In Fisher's argument, the fitness of a parent shows up in its expected number of *grand*offspring, not in its expected number of offspring. What counts is not the *number* of offspring a parent produces but the mix of sons and daughters. This means that we must generalize the concept of reproductive fitness described earlier, so that the probabilities take account of the fact that reproductive success is not determined only by the number of sons and daughters. This modification of the concept of Darwinian fitness is only the beginning of the road. As we will see in Part II, the concept needs to be extended in other directions as well.

Specifying source laws for natural selection is one of the major tasks of evolutionary theory. The aim is to show how properties of organism and evironment can render certain kinds of traits selectively advan-

---

see when we discuss meiotic drive (Section 9.1). And even if sex chromosomes are now so constructed that a fair meiotic division is ensured, it is an evolutionary problem to explain how this came to be. Another question, which we will consider in Section 9.3, is whether the assumptions about random mating implicit in Fisher's argument in fact hold of the populations whose sex ratios we wish to explain.

44. The condition cited in the sex-ratio argument concerns a property of the population itself, not of the external, physical environment. The "environment" of a characteristic must not be understood too narrowly.

tageous. Other evolutionary forces—for example, mutation and migration—have their source laws as well. In addition to providing the machinery for calculating the evolutionary *consequences* of mutation, for example, evolutionary theory also provides a considerable body of knowledge about the conditions that can produce mutations. Mutation, like selection, is understood in terms of what it can produce and in terms of what can produce it.

The source laws for selection often are part of the subject matter of theoretical ecology. Fisher's argument was not developed in the context of a genetic model of how the sex ratio is determined.[45] Whether the ratio is due to one gene or many and whether it is under maternal, paternal, or biparental control, Fisher's argument does not say. The assumption in developing such source laws is that the genetics "will take care of itself"—the assumption is that the phenotypic trait is heritable. This is frequently a safe assumption, but sometimes it is not.[46]

Whereas it is mainly ecology that tries to provide source laws for natural selection, the consequence laws concerning natural selection are preeminently part of the province of population genetics. It doesn't matter to the equations in population genetics *why* a given population is characterized by a set of selection coefficients, mutation and migration rates, and so on. These values may just as well have dropped out of the sky. The equations of population genetics merely take these numbers and compute changes in gene frequencies. Richard Lewontin has called population genetics the "auto mechanics" of evolutionary theory. One plugs in parameter values and cranks out predictions.

It is tempting in evolutionary theory, just as it has been in Newtonian physics, to think that consequence laws exhaust our understanding of the idea of force. "$F = ma$" has led many philosophers and physicists to think that force has no conceptual life of its own apart from a definitional connection with mass and acceleration. The analogous point of view in evolutionary theory has been to think that fitness has no more significance than is captured by the consequence law that says that fitter organisms have a higher probability of survival and reproductive success. The importance of source laws in a theory of forces should show us one way in which this understanding of physical and evolutionary forces is an impoverished one.

45. See Crow and Kimura (1970) for a genetic formulation.
46. In this connection, recall the discussion of heterozygote superiority in Section 1.4. The idea of heritability will be clarified in Section 5.2

Chapter 2
The Tautology Problem

It has frequently been alleged that evolutionary theory is defective because "the survival of the fittest is a tautology."[1] Philosophers (Popper 1963; later retracted), biologists (Peters 1976), creationists (Gish 1979), and journalists (Bethell 1976) have held this opinion. The discussion of Darwinian fitness in Section 1.4 provides some indication of how this line of criticism can arise. In the constant-viability model of selection, fitness values are numbers that play a certain algebraic role. Roughly, they allow one to compute the frequency of a genotype after selection from its frequency before. It is not surprising that people look at the way the concept of fitness is deployed in evolutionary theory and conclude that a trait that increases in frequency must, by definition, be fitter. The claim that the fit survive, they reason, must therefore merely be a consequence of how the concept of fitness is defined. Although natural selection may at first appear to provide a genuine empirical explanation of why evolution occurs, it does not; "the survival of the fittest" is not a real scientific claim, but an empty truism—a "tautology." The problem with evolutionary theory, so it is said, is that it is untestable.

In reply to this attack, philosophers of biology and biologists concerned with the philosophical foundations of evolutionary theory have produced a large literature on the question of whether "the survival of the fittest is a tautology." Although this question does have a certain interest, its importance has been inflated. Frequently, when philosophers say they are writing a paper on "the structure of evolutionary theory," they mean that they are writing a paper on the tautology problem. It is essential to see at the outset that one proposition does not a theory make. There is considerably more to evolutionary theory than the definition of fitness. It is a simple mistake, but one encountered frequently

---

1. For the next several paragraphs I will talk about the accusation that "the survival of the fittest is a tautology," postponing for a while the clarifications and reformulations that this exceedingly obscure form of words so insistently demands.

in writers from a variety of disciplines, to conclude that all of evolutionary theory has some characteristic simply because one proposition in it does (a point stressed by Kitcher 1982a). One result is that defenders of evolutionary theory have spent an inordinate amount of time dealing with this criticism at the expense of more fertile areas of inquiry.

Although this mistake can be found in the writings of biologists and philosophers alike (not to mention journalists and creationists eager to explode the myth that evolutionary theory is respectable science), there is a second sort of failing that must be credited mainly to the philosophers. Philosophers should be able to clarify the tautology problem and to bring to bear on it a variety of ideas that have been developed in other areas of philosophy. Yet, philosophers of biology have mainly taken the problem to be clear at the outset and have proceeded to write about it in a vacuum.

For the past thirty years, the idea that there are propositions that are "true by definition" has been vigorously challenged in philosophy, notably by Quine (1953, 1960). Not that it is no longer philosophically respectable to assert that some propositions are *a priori* or *analytic*. But such claims can no longer be blithely advanced, as if the ideas of apriority and analyticity were well understood and as if there were some unproblematic test for finding out whether a proposition has the property in question. Philosophy of biology has chugged along as if this were not a problem.

Another issue that has been insufficiently clarified is *why* the charge of tautology matters. As mentioned before, the problem has frequently been inflated, so that the entire empirical status of evolutionary theory is seen to stand or fall on the status of a single proposition. But why is it important whether this or that proposition is empirical? Sometimes it is alleged that fitness would be an *unexplanatory* concept if the survival of the fittest were a tautology. At other times, it has been assumed that fitness will not be a *predictive* concept if the tautology charge is sustained. And at still other times, it has been held that fitness and natural selection cannot be *causes* of survival and reproduction if fitness is defined in terms of survival and reproduction. To read the literature on this problem is to risk having the concepts of explanation, prediction, and causation dance before the eyes in apparently interchangeable and equivalent formulations.

In the next section, I will describe some of the philosophical difficulties that surround the idea of a priori truth. No full theory of the a priori will be attempted, but I will try to explain a bit of what this concept has meant in philosophy and why it is philosophically problematic. One result, I hope, will be an appreciation of why, even if there are such things as a priori truths, it will often be difficult to show that this or that proposition is a case in point. We need to get beyond the idea that a proposition wears its epistemological status on its sleeve—as if cursory inspection reveals whether it is a priori or empirical.

In Section 2.2, I pursue the question of what defect would accrue to evolutionary theory if it did turn out that certain propositions about fitness were nonempirical. Again, the goal is to get beyond the idea that apriority represents, as Philip Kitcher has put it, some sort of *generalized badness*. As the title of this section suggests, I very much doubt that any bad consequences would follow about evolutionary theory, even if certain characterizations of fitness were nonempirical.

2.1 A Good Tautology Is Hard to Find

Although the problem we are considering here is usually described as the question of whether "the survival of the fittest is a tautology," the concept of *tautology* has a narrower and more precise meaning in philosophy and logic than it has in evolutionary discussions. A sentence[2] is a tautology in virtue of its logical form. For example, "snow is white or it is not the case that snow is white" is a tautology. This sentence has the same structure as the sentence "grass is green or it is not the case that grass is green" and we might identify this common structural feature by saying that any sentence of the form "P or not-P" is a tautology. Notice that if we understand the logical words "or" and "not" in the usual way, we can see that any sentence of the form "P or not-P" must be true, regardless of what particular sentence we substitute for P.

Although I will not try to give a logically adequate characterization of what a tautology is, note that the example just discussed differs from another, related case. Consider the sentence "all bachelors are unmarried." The logical form of this sentence is, we might say, the same as the one exhibited by the sentence "all typewriters are useless." Each says that all the objects of a certain kind have a certain property. That is, both sentences have the form "each object x is such that, if x has F, then x has G." But notice that from the meanings of the logical words "each," "if," and "then," we cannot see that all sentences of that form are true. The reason is that not all of them are. "All bachelors are unmarried" is true, but "all typewriters are useless" is not.

"All bachelors are unmarried" is not classified by logicians as a

---

2. Although I will mainly talk about sentences rather than propositions or statements as being tautological, the points I want to make can, I think, survive the reformulations needed to straighten out certain semantic niceties that I will ignore. The same holds true of whatever violence I do to the use/mention distinction.

tautology. The logical terms it contains do not suffice to show that it is true. Yet it might be suggested that this sentence has something in common with sentences of the form "P or not-P." Once one understands the logical terms in the sentence, *as well as the meanings of the nonlogical terms it contains*, one can see just from this that the sentence must be true. It is this broader feature—still insufficiently characterized—that is relevant to the question of whether "the survival of the fittest is a tautology."

The second thing to note about this question is that the property of being a tautology applies only to items that are true or false. Apples are not tautologies, nor are concepts. It is *sentences*—for example, ones of the form "P or not-P"—that are tautologies. So we must sooner or later get around to asking: what *sentence* is being called a tautology when it is said that the survival of the fittest is a tautology? A first stab, which we will improve upon, is this: "If one organism survives and another does not, then the former is fitter than the latter." The tautology problem has suffered from insufficient attention being paid to the task of saying exactly which proposition is supposed to be a tautology.[3]

The questions that have surrounded "the tautology problem" are best viewed as issues about the supposed a priori status of certain propositions, ones that purport to say what fitness is. The idea is that the sentence "fitness is success at surviving and reproducing" is not something that biologists learned experimentally or by natural observation; they simply agreed to define the word "fitness" in a certain way. The thought expressed by that sentence is not empirical; it is a priori. The natural place to begin examining this issue is with the concept of apriority itself. What does it mean for a proposition to have this property?

Since Kant, there have been two ways of describing the notion of a priori truth—one positive, the other negative. The negative way is exemplified in Kant's idea that a priori propositions are "necessary for the possibility of experience." They are *indispensable*—to do without them is to make coherent, systematic description of nature impossible. Kant thought that propositions like "every event has a cause" have this property.

This negative characterization also is reflected in Quine's (1953, 1960) focus on the question of whether some propositions are empirically unrevisable. Are there any beliefs we have that could not be disconfirmed by sense experience? The arithmetic truth "$2 + 3 = 5$" may seem to be an example. Even when we add two gallons of water to

3. Laudable exceptions to this rule include Brandon (1978) and Mills and Beatty (1979).

three gallons of gasoline and fail to have five gallons of liquid, we do not conclude that the arithmetic truth is mistaken, but only that gas and water fail to combine additively (Hempel 1945). This suggests that we hold certain propositions to be "immune from empirical test." When Quine rejected the idea of a priori truth, he was, in effect, claiming that every belief is vulnerable to empirical disconfirmation. Notice that he was using a negative characterization of apriority.[4]

Negative characterizations of the a priori involve the following question: Given that you already believe some proposition ("every event has a cause" or "2 + 3 = 5," for example), what could make it rational for you to abandon that belief? Positive characterizations, on the other hand, look at the issue of rational belief or knowledge from the other direction. They inquire as to the kinds of conditions that allow you to believe or know the proposition in the first place. It is this sort of approach that says that a priori propositions are, roughly, ones that can be known to be true independently of sense experience. Here, it is the route leading to belief, not the route away from it, that is important.[5]

To say, in this vein, that "bachelors are unmarried" is a priori does not mean that people are born knowing that little fact. A priori does not mean *innate*. Presumably, people are born with no such information and need to acquire a mastery of the relevant concepts before they can know this fact. Acquiring concepts apparently requires a good deal of sense experience. So how can "bachelors are unmarried" be a priori?

The answer is that a proposition can be a priori even when it takes *some* sense experience to acquire the concepts that occur in it. An a priori proposition is one that can be known with only this modicum of sense experience. An empirical proposition, on the other hand, is one that can be known only after (1) the relevant concepts have been acquired and (2) an empirical check has been run. So, "bachelors are unmarried" and "bachelors are unhappy" are alike in that you can't know whether either is true until you have mastered the constituent concepts. The difference is that, in the former case, conceptual competence is sufficient to allow one to come to know.[6]

Before saying more about this positive characterization of what makes

4. When Putnam (1978) and Sober (1981d) argued, for different reasons, that there is at least one a priori truth, it was a negative characterization of apriority that they had in mind.
5. Biologists may recognize here an analogue to their distinction between the reason a trait becomes prevalent in a population and the reason it is maintained there (Chapter 6).
6. See Kitcher (1980) for further discussion of the role of experience in a priori knowledge and Kripke (1972) for discussion of the difference between apriority, necessity, and analyticity.

a proposition a priori, it is well to see how it differs from the negative approach. A plausible case can be made for saying that some sentences are a priori in the positive sense, but not in the negative one. Consider, again, "all bachelors are unmarried." Perhaps we know this proposition because we understand the definitions of its constituent concepts. But it is hardly "necessary for the possibility of experience"; this little belief could be erased from consciousness without undermining the possibility of doing science or of forming a systematic and coherent picture of the world.

Similarly, it is easy to imagine an experience that would make it rational for someone to draw back from this belief.[7] Suppose we know someone who is an extremely insightful and trustworthy authority. We also know this person to be very honest. This Wise One says to us one day, "philosophers are always saying that all bachelors are unmarried. But if you look very carefully at what these concepts mean, you'll see that this doesn't have to be true. And, in fact, there are some bachelors who are not unmarried." Now I suggest that it would be *pigheaded* to simply dismiss the remarks of the Wise One out of hand. People have made mistakes in analyzing concepts before, and if the Wise One is so smart and honest, we ought to take him seriously in the present case. So we may decide to back off from our belief. The Wise One's remarks do not allow us to see how "bachelors are unmarried" could be false, but they do give us evidence that it is. Moreover, the evidence has come to us from the testimony of our senses; we *heard* the Wise One's remarks.

This shows, I think, that we do not hold any of our beliefs to be totally invulnerable to empirical refutation. We place different amounts of confidence in various propositions, but it would be irrational to think that we could not possibly be mistaken in the beliefs we have. People do become pigheaded, of course; they do at times develop a sort of unshakable certainty, but this seems to embody a misplaced confidence in one's own insight.

That there are no a priori propositions in this negative sense does not mean that there are none in the positive sense. It may be that by inspecting the definitions of certain concepts, we can know that all bachelors are unmarried. The proposition would then be a priori in the positive sense. To be capable of knowing that bachelors are unmarried, it isn't essential to be pigheaded. You can know a proposition even if it is possible that someone could talk you out of it. For example, people sometimes know what the temperature is by looking at a reliable thermometer; yet, if a friend were to tell them that the thermometer

---

7. This argument I owe to Philip Kitcher.

is inaccurate, they'd probably give up their belief. Knowledge does not imply unrevisability, and a priori knowledge (in the positive sense) does not imply empirical unrevisability.[8]

We now must scrutinize more closely the distinction that this idea of a priori knowledge presupposes. "Bachelors are unmarried" is supposed to be a priori, because grasping the definitions suffices to show that the proposition is true. But what is a definition? Which properties of bachelors are part of the definition of what it is to be a bachelor, and which can only be discovered by empirical check to attach to bachelors? The idea of a priori knowledge requires this bifurcation in the set of truths. How is it to be drawn?

This distinction has been extremely difficult to clarify, let alone to apply in practice. Consider the ordinary concept of a *cat*. What must you believe about cats in order to have this concept? Do you have to believe that cats are animals, for example? I would say not: Someone with lots of other opinions about cats might also hold the bizarre opinion that cats are robots that have been placed here surreptitiously by Martians. This person would be misguided; yet his aberrant opinion does not show that he lacks the concept of *cat*. He might have lots of other correct beliefs, and these might suffice to establish his conceptual competence.[9] On the other hand, one might hold that this strange fellow does not have *our* concept of a cat, but some other concept, since *our* concept has built into it the idea that cats are animals. I have already recorded my hunch on the matter, but I must add that it is only a hunch. Such seat-of-the-pants judgments are not anchored to any principled philosophical theory concerning what conceptual competence amounts to. Not that such theories won't someday be developed. But as of now, it is very unclear how this distinction, presupposed by the concept of a priori knowledge, is to be clarified.[10]

Matters get even worse when we turn from bachelors and cats to a scientific term of art like *fitness*. I used the idea of Darwinian fitness to provide a preliminary sketch of how fitness is understood in evolutionary theory. The sketch was preliminary because there is more to fitness than Darwinian fitness. In any event, the Darwinian fitness of

8. There is a voluminous literature on what knowledge is. For a theory that tries to flesh out the analogy between knowers and reliable thermometers, see Dretske (1982).

9. The example is Putnam's (1970).

10. Note that I do not draw the radical conclusion reached by Quine (1953, 1960)—that there is no difference between the beliefs about bachelors that we acquire simply in virtue of learning to use that concept and the beliefs about bachelors that we may or may not form by acquiring empirical evidence, once we have that concept in hand. I find Quine's arguments here inconclusive and, in some ways, rooted in a behaviorism that I reject.

an organism was described as having two components. An organism's *viability* is its probability of surviving from the egg to the adult stage. Its *fertility* is its expected number of offspring, divided by two if it engages in biparental sexual reproduction. The overall fitness of an organism combines these two components. The Darwinian fitness of a genotype, or of a trait generally, is the average of the relevant organismic fitness values.

To characterize Darwinian fitness in this way is not yet to comment on whether these facts about it are a priori or empirical. Gravitational force is understood in Newtonian mechanics in terms of the inverse-square law. Force is described in the equation "$F = ma$." But it is one thing for these propositions to play a central role in a theory, quite another for them to provide a priori definitions of the constituent concepts. How to decide whether a proposition in a scientific theory is a priori or empirical?

One might be tempted to advance the following line of reasoning:

> I can imagine an experiment that might, in principle, disconfirm the inverse-square law of gravitation. But I cannot imagine an experiment that could possibly count against the characterization of Darwinian fitness given above. The propositions connecting Darwinian fitness to survival and reproduction are like the propositions of arithmetic. No possible experience could count as evidence against "$2 + 3 = 5$." This shows that arithmetic is a priori, and the same holds true for the above facts about Darwinian fitness.

Although I have several objections to this position, there is an idea here that should not be overlooked. A salient property of the statement "$2 + 3 = 5$"—one that gives it the *appearance* of being a priori—is that we can't imagine rejecting that proposition *just on the basis of the observed result of adding two things and three things together*. Similarly, it is hard to see how we could get evidence against the proposed definition of fitness *merely by seeing which organisms survive and reproduce more successfully than others*. The status we accord these propositions seems to be such that we do not think of them as empirically testable in a certain straighforward way. The anecdote about the Wise One, however, should warn us against concluding that there is no way at all of empirically testing a proposition, simply because no "straightforward" procedure appears to be available. Tests can be devious and circuitous; before we can conclude that this or that proposition is a priori, we need to develop a fuller perspective on what it is about a theoretical statement that makes it empirically testable.

One defect in the above quoted line of reasoning is that it places too much confidence in our spontaneous ability to imagine experiments.

Perhaps physicists before the invention of non-Euclidean geometry asserted that they could not imagine an experiment showing that the sum of the angles of a triangle departs from 180. They perhaps thought that any measurement departing from 180° would merely show that the figure being measured was not a triangle. The conclusion to draw from this, however, is not that Euclidean geometry is a priori true but simply that the scientists' capacities of imagination were limited. It took the invention of non-Euclidean geometry, which was no mean intellectual feat, to make certain possibilities conceivable (Putnam 1962).

Each of us can readily imagine what it would be like for telephones to all be pink. Strenuous intellectual activity is not required to form a detailed conception of an alternative to a known truth (that telephones aren't all pink), when that truth is fairly trivial. But when the truth in question plays a fundamental role in a scientific theory, imagining alternatives can be harder. Unlike the telephone case, it is sometimes a major intellectual breakthrough just to identify an alternative possibility, even if one has no idea whether that possibility is true.

A biologist accustomed to thinking about fitness solely in terms of Darwinian fitness might draw a blank when asked to imagine how organisms could differ in fitness and still have the same expected number of offspring. It is not immediately obvious that this is possible. If one were unable to readily conceive how this could be true, it might be tempting to assert that the definition of fitness implies that this simply isn't possible. This jump from "I can't imagine it" to "It isn't possible" is a dangerous one. Fisher's sex-ratio argument does us the favor of showing how misguided it can be. In Sections 9.1 and 9.4, we will consider concepts of fitness that depart from the Darwinian paradigm in other ways.

What do alternative concepts of fitness show us about the correctness of the idea that an organism's fitness is its viability and fertility? One possibility is that this "definition" of fitness is no definition at all, because it is false. After all, the process that Fisher modeled is clearly a selection process; for it to occur, there must be differences in fitness for selection to operate upon. By examining this process, we were able to see that our earlier "definition" is inadequate. Changes in our understanding of fitness do not reflect new a priori definitions—as if the history of evolutionary theory produced a series of unrelated concepts fitness$_1$, fitness$_2$, etc. This picture is no more adequate than the suggestion that Euclidean geometry is true by definition, and so is non-Euclidean geometry, since each theory defines its own special concepts. In fact, both theories aimed at characterizing the geometry of physical space; they discussed the same kind of object—a "straight line"—and offered different theories about its properties. Similarly, different models of selection processes, both early and late, aim at describing the properties of a single magnitude—*fitness*.

The other possibility is to regard each model of a selection process as stipulating its own special definition of fitness. Darwinian fitness has its own special concept, fitness$_D$, whereas Fisher's model talks about a different property, namely, fitness$_F$. And in each case, we can say that the characterization of fitness is a priori true. Fitness$_D$ is, by definition, an organism's expected number of offspring, whereas fitness$_F$, by definition, takes account of an organism's expected number of offspring *and* grandoffspring.

Whereas the first interpretation sees successive models as honing and improving our grasp of a single theoretical magnitude, the second sees each evolutionary model as enshrining its own proprietary property—fitness$_D$, fitness$_F$, etc. One might be tempted to defend this second interpretation by invoking the memory of Lewis Carroll's Humpty Dumpty, who, in *Through the Looking Glass*, advanced the opinion that every use of words is something that we potentially control and so *everything* is a matter of definition.

There is a very minor point on which Humpty Dumpty was correct; this is what Putnam (1975) has called "trivial semantic conventionalism." If our linguistic community decided to use the word "fitness" to name the color red, then that word, suited out with this novel interpretation, would describe ripe tomatoes, but not melanic moths in environments with soot-covered trees. This is true, but irrelevant. We could use the terms "bachelors" and "unhappy" so that the sentence "bachelors are unhappy" would express some sort of definition. But we don't. The problem is to understand language *as it actually is used*. We want to see whether certain beliefs we have are a priori true, where their constituent concepts are understood to have the meanings they presently do. Furthermore, we need to see why Humpty Dumpty's facile advice is something we do not want to take. There is a very good reason that we don't perpetually redefine terms so as to settle issues by stipulative fiat.

The problem with Humpty Dumpty's point is that it fails to take account of the fact that in science we frequently want to deploy concepts that, in Plato's words, "carve nature at its joints." Characterizing fitness as an organism's expected number of offspring is inadequate, because it is insufficiently general. To reply that there is a special concept of fitness, fitness$_D$, of which this characterization is true by definition is to deploy a concept that fails to obey Plato's maxim.

Scientific inquiry into the nature of a property or magnitude can be brought to a screeching halt by invoking a definitional stipulation. Psychologists attempt to identify the cluster of skills and capacities that we call "intelligence." The question arises of whether IQ tests are adequate measures of this magnitude. This question is not settled at all by remarking that intelligence is, by definition, what IQ tests measure.[11]

A more useful paradigm for the deployment and investigation of scientific concepts than the one offered by Humpty Dumpty is possible. The idea is that terms like "fitness" name real theoretical magnitudes and properties that science seeks to understand. The nature of fitness is not settled once and for all by a linguistic stipulation. Although it is up to us how we shall use the term "fitness," it is by no means up to us to settle what fitness is. Words are under our linguistic control, but the items those words name, usually, are not.[12] It is always possible to stipulate a new meaning for a term like "fitness." But the permanent possibility of Humpty Dumptyism shows nothing about the adequacy of our present understanding of what fitness is. We apply this concept in a variety of contexts, and we perhaps believe that we have some grasp of the nature of the property we discuss. However, it needn't be true that any of our proposed "definitions" of fitness characterize this property with perfect accuracy. The propositions codifying our present understanding of fitness needn't be adequate simply because we call them "definitions."

The idea that evolutionary theory provides both source laws and consequence laws for the various forces it describes is relevant here. The question of "the tautology of the survival of the fittest" arose because evolutionary theory describes the force of natural selection in terms of its effects. The exact form of this relationship has frequently been misunderstood, as when it is thought that the fitter of two organisms must always be more reproductively successful. But the tautology problem is not solved by reformulating the consequence law in a way more in keeping with the actual structure of evolutionary theory. Even after fitness is given its appropriate probabilistic cast (Section 1.4), one may still inquire whether the suitable consequence law is a priori. And even when we see that there is more to fitness than Darwinian fitness, the problem is not laid to rest. Consequence laws there must be. Do such laws a priori define the concept of fitness?

A similar controversy has raged (or smouldered) in the philosophy

11. Not that "operationalizing" a concept does not have its uses. The point is that to operationalize a concept, you usually provide it with a provisional, empirical connection with observations; you are not suiting it out with an a priori definition.
12. My colleague Terrence Penner tells a story about Abraham Lincoln that brings this point out nicely. Lincoln once asked: "If we called a dog's tail a 'leg,' how many legs would a dog have?" Lincoln's answer was *four*, since calling a tail a "leg" doesn't make it a leg.

of physics over the question of whether the consequence law "$F = ma$" is empirical or simply a definition of the idea of force.[13] In some ways the history of physics has afforded more perspective on this question than the history of evolutionary theory has provided for the parallel problem about fitness. Perhaps the reason is that physics has had a larger variety of alternative theories to consider; Darwin's conception is in some ways still contemporary and canonical. This comparative perspective is essential, since, as noted above, the issue of apriority crucially involves possible *changes* in theory, and so cannot be settled by attending to the contents of a single theory.

In the case of "$F = ma$," two considerations have suggested, if not decisively shown, that the proposition is empirical. First, there is the fact, noted in Section 1.3, that Aristotelian physics *denied* that an object subject to zero force would remain in uniform motion. Such objects, thought Aristotle, would not move at all. If Newton's theory is empirically superior to Aristotle's, this suggests that the Newtonian consequence law is empirical. Of course, this piece of evidence is not entirely decisive, since it may be suggested that Aristotle and Newton were not discussing the same magnitude ("force") and that each had his own proprietary consequence law that had the status of a definition. This raises the difficulty of deciding when two theories work with the same magnitudes and disagree over their empirical relationships, and when they work with different sets of magnitudes.[14]

Quine's (1953, 1960) views on epistemology suggest a second *prima facie* reason for thinking that "$F = ma$" is empirical. A theory of forces tests its source laws and its consequence laws *together*. When an object is observed to violate the predictions of the theory of forces, either one can postulate the existence of a new source or modify the consequence laws in terms of which the effects of the known forces are computed. If I think that no forces are acting on a particle and yet observe that it accelerates, I can either postulate a force or give up my assumption that $F = ma$. We revise our theory to give the best overall treatment of the phenomena. The consequence laws we adopt, just as much as the source laws we choose, are the result of this empirical procedure. So, in this sense, "$F = ma$" is empirical.

---

13. See, for example, Earman and Friedman (1973) and the references therein for recent discussion. For a more historical (and resolutely positivistic) treatment, see Jammer (1957).
14. The thesis that scientific theories are radically "incommensurable" (associated, with what exegetical accuracy I will not say, with the work of Hanson 1958, Kuhn 1970, and Feyerabend 1970) often opts for the latter view. Those who reject incommensurability often opt for the former. I suggest that neither of these positions is well grounded theoretically; we have very little understanding of what constitutes the meaning of theoretical terms and what ensures that different theories are "about" the same things.

Fisher's sex-ratio argument can be embedded in a line of reasoning of this sort. Someone who begins by thinking of Darwinian fitness as all there is to fitness must find the sex-ratio argument a bit puzzling. Here is a causal process that can change the frequencies of traits; yet females who differ only in the sex ratio of their progeny do not differ in Darwinian fitness. How, then, to represent this argument within the general structure of evolutionary theory? One could refuse to call the process that Fisher describes a selection process on the ground that Darwinian fitness fails to accommodate it. Or one might stretch the concept of fitness to include the process as a kind of natural selection. It is pretty clear that the latter strategy is preferable.[15] Instead of inventing a new kind of evolutionary force, one modifies the consequence laws needed to characterize a force already enshrined in the theory.

The physical analogues of the problem of fitness bring out an important point about the issue of testability. Highly theoretical claims issue in observational predictions only when they are conjoined with still further assumptions (Duhem 1914, Quine 1953, 1960). If one wants to see what observational consequences Euclidean geometry has, one should not look at the geometry alone but at the consequences of joining the geometry with other physical assumptions. This fact shows why it can be difficult to see whether a theoretical assertion is empirically testable, since one cannot find this out by examining the claim in isolation. Rather, one must have a way of surveying the various theoretical contexts in which it might be embedded. If one has little confidence that *all* relevant theoretical contexts have been examined, then one should be suspicious of the assertion that the claim in question is a priori. All it takes is a single novel theoretical setting to reveal that what seemed to be a mere definition has a good deal of latent empirical content to its credit.

The upshot of this discussion is not that evolutionary theory contains no a priori truths about fitness. It may well have some. Rather, my point is to warn against facile claims that this or that proposition is a priori, made on the basis of a cursory examination of a few evolutionary models or on the knee-jerk reaction that the proposition is untestable. Showing that a proposition in a scientific theory is a priori is very *hard*. One must not only examine the structure of the theory as it now stands but must also have some way of foreseeing how alternatives to the theory might be structured. Never forget how non-Euclidean geometry took philosophers by surprise.[16]

15. Some of the arguments to be considered in Part II will lead in the same direction.
16. At the same time, I want to warn against a Quinean argument that might be advanced to show that this or that characterization of fitness *must* be empirical. It is certainly *conceivable* that what we now might think of as an a priori truth about fitness should

2.2 A Little A Priori Truth Never Hurt Anyone

I now want to consider what problems would follow for evolutionary theory *if* the consequence laws associated with the force of natural selection were a priori. I mentioned before how this problem has been ridiculously inflated. With no connecting argument at all, the whole of evolutionary theory has been judged to be unexplanatory, unpredictive, untestable, or un-something because a single proposition is alleged to have this status. A useful corrective to this overgeneralization is the realization that the convenient term "evolutionary theory" collects together a variety of models, hypotheses, and assumptions, and "evolutionary biology" similarly unites a stunning diversity of subdisciplines and approaches.

The problem at hand is still worth pursuing, however. There is more to evolutionary theory than the definition of fitness, but it still is of interest to identify the theoretical role (or roles) that that concept plays. Suppose, for the moment, that Darwinian fitness is an adequate conception of fitness. There is no need to discuss the "tautology" that asserts that the more fit always outsurvive and outreproduce the less. This is not a proposition of evolutionary theory at all, let alone an a priori one. But its probabilistic counterpart—"the probable survival of the fittest"—plays a role in a variety of evolutionary models. What horrors follow from the supposition that it is nonempirical?

By noticing that evolutionary theory does not identify an organism's fitness with its degree of *actual* reproductive success, philosophers like Brandon (1978) and Mills and Beatty (1979) have cut through one of the strands in this knot of problems. If fitness were one and the same property as actual reproductive success, it is hard to see how fitness could *cause* or *explain* reproductive success. If nothing causes itself and if nothing explains itself, then an object's having the property F cannot cause or explain the fact that it has the property G, if $F = G$.

It is an interesting wrinkle in this sound observation that the question of apriority is entirely beside the point. *Being made of water* and *being made of H_2O* are presumably one and the same property. However, this is not a conclusion that follows from the meaning of the English word "water." The molecular structure of water was an empirical discovery, not a definitional stipulation. Still, since being made of water and being made of H_2O are one and the same property, it follows that an object's being made of water cannot explain or cause the fact that it is made of H_2O.

Hempel (1965) anticipated and diffused one objection that might be

---

turn out later on to be empirical. But this doesn't establish that the proposition in question *is* empirical. The fact that a proposition is conceivable doesn't mean that it is true.

made to this thesis that nothing explains itself. Sometimes we might ask "why is this stuff made of H_2O?" as a request for justification, not explanation. What we want to know is why one should believe that the stuff has that composition. In this case, "it's made of water, and water is H_2O" may be a perfectly correct reply. However, the thesis that nothing explains itself remains unshaken.

Another kind of example is more difficult. Consider the relationship between a gas's temperature and its mean molecular kinetic energy. Let us accept the idealization that these have been found to be one and the same property: temperature *is* mean molecular kinetic energy. There is a sense in which a gas's mean molecular kinetic energy *explains* its temperature, but not conversely. The same may be said of water and H_2O: there is a sense in which a liquid's being made of H_2O explains its being water, but not conversely. How is this possible, if the two properties are identical and nothing explains itself?

My suggestion is that the gas's kinetic energy explains what it is for the gas to have the temperature it does. It does not explain *why* the gas has the temperature at the time in question. The relationship here is much like the one between a dispositional property and its physical basis (Section 1.5). If zebras differ in fitness because of their leg structure, then there is a sense in which leg architecture explains fitness relationships. The relationship is not causal; it does not explain why those fitness values came into existence. Science not only explains why certain states of affairs and events come into existence; it also seeks to explain the nature of those events. In explanations of this sort, we have a violation of the principle just stated: F can explain what G is, even when $F = G$. Perhaps we should distinguish two sorts of explanatory problems: there is the task of explaining *why* certain events occur and there is the task of explaining *what* the nature of those events is. Self-explanation is barred in the former case, but not in the latter. I still conclude that fitness cannot explain why organisms survive and reproduce differentially, if fitness is differential survival and reproductive success.

The correct observation to make here is that fitness is no such thing. Fitness is a probabilistic disposition (a *propensity*) to survive and be reproductively successful. It is connected with actual reproductive success the way a coin's bias is connected with its actually landing heads more often than tails. But, having said this, what follows about the fitness concept's explanatory and causal role? Unfortunately, the problems attaching to the preprobabilistic formulation in some ways simply arise anew. Suppose one organism lives and another dies. Would this fact be *explained* by the claim that the one that survived was the fitter?

And would it be correct to say that the greater fitness of the first organism *caused* it to outsurvive the second?

In Molière's play *The Imaginary Invalid*, a doctor announces that the reason a certain potion puts people to sleep is that it contains a *virtus dormitiva*—a dormative virtue. This always occasions a chuckle (at least among philosophers who have not heard the joke too often), since the doctor's "explanation" is quite empty.[17] The problem is one attaching to any dispositional property—to a lump of sugar's being soluble just as much as a vial of medicine's being soporific. When a lump of sugar dissolves in water, is this explained or caused by the fact that it was soluble?

Looked at in one way, a probabilistic disposition—a propensity—is a bit less empty than a deterministic one. If solubility and insolubility are deterministic dispositions, then an object that dissolves *must* have been soluble. If however, the dispositions are probabilistic, then dissolving does not absolutely guarantee solubility. If so, we can see why it *seems* to be entirely vacuous to explain dissolving by invoking solubility, but is not. It is the assumption that the dispositions are deterministic that makes you say "*of course* the object must have been soluble if it dissolved." But if one is in doubt as to whether the disposition is deterministic or stochastic, or if one believes that it is stochastic, then asserting that the object was soluble is genuinely informative. Under the probabilistic reading, soluble objects *probably* dissolve when immersed; insoluble ones *probably* do not. It does not automatically follow that an object that dissolves must have been such that dissolving was probable.[18]

So doubt about whether a disposition is deterministic renders explanatory even those dispositions that are deterministic. But probabilistic dispositions have additional explanatory force. If we know in advance that solubility is a deterministic property, then the dissolved object must have been soluble. But knowledge that a disposition is probabilistic does not permit this automatic inference. A coin that has landed heads more often than tails in a run of tosses *might* have been biased toward heads, but it need not have been. The improbable *can* happen, after all. In the same way, if the organisms that share one genotype have a higher survival rate than the organisms that share another, this *might* mean that the former group had a higher average fitness, but this need not be true. Fitter organisms can "by chance" have a lower survivorship, and so it is not trivial to claim that the genotype that survived better was the fitter.

17. See Sober (1983a) for an attempt to take this joke seriously.
18. Davidson (1963, pp. 14–15) asserts that "immersing an object caused it to dissolve" does not imply that it was soluble, but he does not say why.

Thus the assertion of fitness differences is not entirely empty, but it seems to be pretty close. To explain why one organism (or group of organisms) lived and another died by saying that the former was fitter is basically to say that what happened happened because it was more probable than the alternative. This may not be true, but when it is true, does it explain? And can this probabilistic property really be taken to be causally efficacious? The suspicion is that fitness is the dormative virtue of evolutionary theory.

One thing that ought not to mislead us is that we frequently prefer to think of events rather than standing conditions as causally efficacious. When we drop the sugar into the water and see it dissolve, what immediately strikes us as causally significant is the immersion. This is an event—it involved a datable change of state. The lump's solubility, on the other hand, was a property it possessed for some time. This was a standing condition and so, it might seem, could not be the cause of the sugar's dissolving. We might be tempted to elevate this preference for events into a metaphysical principle, one saying that only events can be causes. If so, fitness differences cannot properly be said to be causes of evolution.

Such a metaphysical principle goes too far. True, we do point to the lump's being immersed, rather than to its being soluble, when we are asked to cite a cause. But this shows more about our narrative preferences than about the nature of causation. If we visualize the process on a more fine-grained time scale, solubility can come to occupy the foreground. Imagine the lump already immersed in water. We observe that it gradually dissolves, and want to know why. The fact that it was immersed will not answer this new question, since we now are asking why the immersed piece of sugar dissolved. In this case, we want to be told about the lump's physical structure. The causal question will be answered by information about a standing condition, not an event. The property of interest is the one that causes immersed sugar lumps to dissolve. And this property is none other than the property of solubility. So solubility *is* a cause, after all. States of affairs, just as much as events, can be causally efficacious.[19]

The fact remains that we feel somewhat disappointed when told that the sugar lump dissolved in water because it was water-soluble. But if this is not a satisfying explanation, how can I say, as I did in the previous paragraph, that the sugar's solubility was a causal factor in its dissolving? The answer is that we must disentangle issues of causation from issues of explanation.

19. Davidson (1963, p. 12) makes this point by appeal to some clear examples: "the bridge collapsed because of a structural defect; the plane crashed on takeoff because the air temperature was abnormally high; the plate broke because it had a crack."

Consider the question "What caused Y?" One correct answer would be "the cause of Y." However, this answer might rightly be classified as unexplanatory, since it does not provide us with any better understanding of why Y occurred than we started with.[20] Explanations afford understanding; therefore, claims asserting that X explains Y are true or false partly because of the relation of those two terms to a third—namely, *us*. Causality is not similarly "consciousness-dependent." Whether or not X caused Y is not in general influenced by whether *we* might find it interesting to be told this.[21]

Even though solubility is what causes immersed lumps of sugar to dissolve, one does not explain why an immersed lump of sugar dissolved merely by remarking that it was soluble. Far from there being a clash between these two facts, our knowledge of the former explains why the latter is true. We *know*, before we even look at the lump of sugar in question, that solubility is the causal factor that connects immersion with dissolution. Because we know this, our understanding of what happened will not be enhanced by the correct remark that the lump was soluble.

How do these observations about the explanatory and causal powers of dispositional properties apply to fitness? Again, we must disentangle questions of explanation from questions of causation. Defining fitness as the *cause* of survival and reproductive success might pose problems for the *explanatoriness* of the concept but presumably would not disqualify fitness from being one of the *causes* of evolution. If anything, the latter would be *guaranteed*, not *thrown into doubt*, by the definition.

I do not propose to argue that fitness is a cause. Indeed, the main point of Section 3.1 will be to deny that this is so. But the fact that it is a dispositional property per se does not disqualify it from this role. The fact that claims about fitness can, in some contexts, be relatively unexplanatory is similarly inconsequential. Causation is a delicate question that we shall shelve for the moment.

Fitness, we have already seen, is not totally vacuous as an explanation of differential survival and reproduction, if only because the improbable sometimes occurs. But to think that this observation fully circumscribes the explanatory power of the fitness concept is to seriously underestimate

20. Again, the point is made succinctly by Davidson (1963, p. 15): "Suppose 'A caused B' is true. Then the cause of B = A; so substituting, we have 'The cause of B caused B,' which is analytic. The truth of a causal statement depends on *what* events are described; its status as analytic or synthetic depends on *how* the events are described."

21. This claim about the difference between the causal and the explanatory relation is independent of the issue of whether "x causes y" and "x explains y" differ with respect to the semantic properties of transparency and opacity. I am inclined to think that both predicates have a transparent and an opaque reading.

its role in evolutionary theory. We must avoid the mistake of assuming that there is just one type of proposition that a given concept might be used to explain. True, fitness can offer some explanation of why one organism is more reproductively successful than another. However, concepts are not connected one-to-one with the propositions that they may help to explain. We must cast a wider net.

Instead of asking why one organism lived and another died, let us focus on a question about a *population*. Suppose we observe that a population has maintained a balanced polymorphism for some time. Censuses in several generations show that both of the possible alleles at a given locus are present. There may be many explanations for this. One possible explanation that we have already discussed is *heterosis*. If heterozygotes are, on average, fitter than either of the homozygotes, the simple-viability model examined in Section 1.4 predicts a balanced polymorphism. Fitness differences clearly can be explanatory.

The fact that an organism has a certain probability of surviving from egg to adult does not provide a great deal of illumination for why it does so. Similarly, the fact that a given radioactive atom has a certain half-life is not a terribly illuminating explanation of why it disintegrated. But to think that these two examples circumscribe the possible explanatory uses of viability on the one hand and half-life on the other is to seriously underestimate the usefulness of those concepts. Every theoretical concept has its trivial applications. Students of evolutionary theory ought not to be fixated on them.

The concept of fitness may figure in intricate lines of reasoning that lead from empirical assumptions about a population to certain deductive or statistical predictions about it. When these lines of reasoning are short and transparent, they offer scant illumination. Make them a bit more sophisticated, and drawing out their logical consequences can be very explanatory indeed.

Fisher's sex-ratio argument and the simple-viability model of heterozygote superiority can each be represented as a line of reasoning in which all the generalizations are a priori and all the empirical assumptions are isolated in claims about the particular population considered. If a Mendelian population experiences constant-viability selection as its only evolutionary force, then the population will remain polymorphic if the heterozygote is the fittest genotype. Some such generalization (perhaps with a few amendments thrown in) may well be a priori, but it is explanatory nonetheless. It is explanatory because it can be used to account for balanced polymorphism. The explanation *as a whole* is empirical because *other* components of it are.

The logical positivists as well as those, like Wittgenstein, who influenced them were fond of saying that a priori propositions are "uninformative." Such remarks should, I think, be viewed as somewhat arbitrary redefinitions of the idea of "information," since it is hard to believe that the whole of mathematics and logic "asserts nothing." Wittgenstein said that "in mathematics there are no surprises." Yet everyone, including Wittgenstein, knows that there are surprising mathematical theorems.

This idea found its way into Hempel's (1965) classic theory of explanation, according to which an explanation must include at least one empirical law of nature. No wonder that some chains of reasoning in evolutionary theory should seem inadequate, since they often appear to consist in the algebraic manipulation of singular starting assumptions. Physical theory, on the other hand, seems a startling contrast. Here, at least, there obviously are generalizations, like Newton's law of gravitation, that are empirical. Physics-worship can then lead to the conclusion that if evolutionary explanations do not contain empirical generalizations, something must be wrong with those explanations.

Certainly, one wants an ideal explanation to contain generalizations, since this is how we understand which properties of the system under study are the ones that matter. Generalizations help us grasp what *kind* of system we are dealing with, and this is important in understanding why it behaves as it does. It is also not hard to see why one wants an explanation to contain empirical elements, if the phenomenon being explained is itself empirical. But no argument was ever offered for why there must be a part of the story that is both general and empirical at once. I suggest that this requirement on explanation should be discarded.[22]

We will return to the explanatory power of the fitness concept in a moment. Right now I want to consider another way that the tautology problem has been formulated. Sometimes the complaint is that we have no way of *finding out* how fit an organism is other than by seeing how successfully it survives and reproduces. Our judgments about fitness are therefore *ex post facto* rather than predictive.

What defect does a theoretical concept exhibit when it lacks this sort of "independent criterion"? If the only way to find out whether C is true is by looking at E, which is its effect, then we won't be able to predict E by first finding out that C is true. But this is a fact about *us*, not, in the first instance, a fact about the relation of C and E. Even when C is accessible only by knowing E, it may still be true that C *causes* E. It may even be true in this circumstance that C *explains* E. Even when you use the reading of the Geiger counter to infer that the

22. I am grateful to Steven Kimbrough for raising this issue with me and for useful discussion.

rock is radioactive, it is perfectly correct to say that the radioactivity causes and explains the Geiger counter's clicks.[23]

The problem of whether there is some "independent criterion" for fitness differences, other than observation of actual survivorship and reproductive success, is somewhat different from the problems surveyed so far. The question of whether fitness can explain or cause actual survival and reproductive success depends on whether "x is fitter than y" is defined in evolutionary theory to mean that x is more successful at surviving and reproducing than y. But this line of argument is not needed to address the issue of whether there is an "independent criterion" for fitness differences. Even if fitness were correctly defined in this way, there still might be other ways of measuring fitness besides attending to actual survivorship.

Here's an analogy. Let's assume that "bachelors are unmarried" is a priori true—a "tautology" if ever there was one. This fact doesn't prevent us from finding out in perfectly empirical ways whether a particular person is a bachelor. It is one thing to worry whether some particular characterization of fitness is a priori, quite another to show that even if it is, there are empirical ways of measuring fitness besides those mentioned in the purported definition.

Gould (1977a) accepts the challenge of finding an "independent criterion" of fitness on its own terms. He argues that there are ways of finding out about fitness besides measuring actual survivorship and reproduction. Gould points out that an engineering-style design analysis of organisms allows us to make plausible, though fallible, predictions about differences in reproductive success. If we discover that one zebra is faster than another, we might infer that the former is fitter. The inference might be based on the hypothesis that speed makes it easier for zebras to escape from predators and on the rather sweeping *ceteris paribus* assumption that the leg structures that make for a faster zebra don't also have other, deleterious effects. It is by now well known, thanks in part to the efforts of Gould himself, that one cannot blithely assume that everything else *is* equal; the "solution" of one design problem may bring ten others on its heels. Still, a careful engineering analysis can in principle permit us to make reasonable judgments about fitness differences in advance of finding out who lives and who dies.

Another strategy for obtaining an independent criterion for fitness differences in a population involves "bootstrapping." One may begin

23. I suggest that this could be true even if Geiger counters were the only *possible* ways of detecting radioactivity. One must not confuse the problems attaching to finding out if an object has a property (fitness, radioactivity) with the question of what that fact, *were it known*, could be said to explain or cause.

by investigating fitness *ex post facto*—that is, by looking at differences in survival and reproductive success in one population and reasoning that these were a consequence of certain sorts of fitness differences. Then, when confronted with a second, similar population, one may be able to use one's previous experience to reason in the opposite direction. The hypothesis would be that the characteristics that determined fitness differences in the first population also do so in the second. Not that this extrapolation from previously investigated populations to new ones is infallible; the point is that it can be reasonable. *Ex post facto* investigations of fitness can themselves contribute to the discovery of independent criteria.

It certainly would be disturbing if the only way to find out about the fitness of organisms were to see how they actually succeeded in surviving and reproducing. As noted before, this would not prevent fitness from causing or explaining survivorship and reproduction. But it would mark off fitness as a rather peculiar property. In general, we expect theoretical magnitudes to be *multiply accessible*; there should be more than one way of finding out what their values are in a given circumstance. This reflects the assumption that theoretical magnitudes have multiple causes and multiple effects. There is no such thing as the only possible effect or cause of a given event; likewise, there is no such thing as the only possible way of finding out whether it occurred. I won't assert that this is somehow a necessary feature of all theoretical magnitudes, but it is remarkably widespread. If fitness were a counterexample to this pattern, that fact would call for special explanation.[24]

The fact of supervenience has implications for the idea that a design analysis of organisms solves this version of the tautology problem. Granted, fitness can be measured by carefully investigating the design features of organisms on a case-by-case basis. One can even generalize over physically distinct organisms, as Fisher did in his sex-ratio argument, and formulate general principles about the fitness advantages conferred by various behavioral, physiological, and morphological characteristics. But, apparently, no perfectly general *physical* characterization of fitness is available. When it comes to saying what fitness

---

24. Operationalism asserts as a matter of principle that a given theoretical magnitude *cannot* be multiply accessible; each new way of measuring "temperature," for example, defines a new theoretical magnitude—temperature$_1$, temperature$_2$, etc. One notorious problem for this position is that it cannot explain why temperature$_1$ and temperature$_2$ are correlated; operationalism cannot assert that there exists a single physical magnitude that may be measured in different ways. Although this philosophy tried to wrap itself in the mantle of scientific precision, its ideas about how theoretical magnitudes are to be investigated are in fact antithetical to the way science works. All this is not to deny the truism that theoretical claims ought to be testable—i.e., rendered "operational."

is *in general*, the characterization given in terms of survival and re-production seems to stand alone. This surely should not be taken as a criticism of the fitness concept. Supervenience is perfectly compatible with the existence of alternate ways of finding out about fitness differences in a population besides observing actual survivorship. All supervenience implies is that no *single* physical characteristic will be the *universal* key to fitness. This hardly means that biologists can only know about fitness differences by an *ex post facto* inference from actual survivorship.

The fact that fitness is a supervenient property also has consequences for the kinds of explanatory roles it can play. Within a single population, fitness is grounded in a set of physical attributes. The fitness differences noted in the model of the sickle-cell trait were based on the properties of resistance to malaria and anemia. In that example, the concept of fitness explains the same fact that the physical basis explains—namely, the persistence of polymorphism in populations living in malarial regions.[25]

In this case, it is hard to see how fitness explains something that the physical facts could not explain. This is not to deny that fitness has explanatory power; the suspicion is that there is no phenomenon that fitness *alone* can explain. Within a single population, there appears to be no explanatory problem whose solution the supervenient property can uniquely claim for itself.

However, at the same time that the supervenience of fitness predicts that fitness will be explanatorily redundant in some contexts, it also predicts that the concept will have a unique explanatory utility in others. Recall that supervenient concepts allow us to generalize over physically distinct systems. They are an antidote to the overly fine-grained analysis that a purely physicalistic vocabulary would impose. The physical basis of heterosis in the sickle-cell system explains the balanced polymorphism we observe there, but in other populations, heterozygote superiority will have a quite different physical basis. Looking at them in isolation, we see that the concept of fitness will help explain the maintenance of the two alleles, but so too will the physical basis of the fitness differences. The vocabulary of fitness and its physical bases will be on an explanatory par, if we take the examples one at a time. But if we want to explain what these various polymorphic populations have in common, the idom of fitness will be indispensable. Only from the

25. For expository purposes, I assume here that *anemia* and *resistance to malaria* are "physical" properties. In fact, they may themselves be supervenient. Perhaps the right way to put the point about fitness is to say, not just that it supervenes on physical properties but that it supervenes on a lot of biological properties as well.

perspective of this supervenient property do certain regularities come into explanatory focus.[26]

The tautology problem is now two-thirds solved. If one assumes, waiving the questions raised in Section 2.1, that this or that consequence law of natural selection is a priori, one then faces three problems: (1) How can fitness be explanatory, if it is defined in terms of its probabilistic connection with an organism's survival and reproductive success? (2) How can one have knowledge of the fitness values of organisms or traits independently of observing their actual survivorship and degree of reproductive success? (3) How can fitness (or fitness differences) be a cause of evolution?

The first of these—the problem of explanatoriness—I answered by describing a context in which fitness does explain. The example considered was one in which heterozygote superiority explains a balanced polymorphism. In addition, the fact that fitness is a supervenient property was used to argue that there are some explanatory problems in which fitness has an irreducible and unique explanatory utility. The physical bases of fitness are just as explanatory as fitness itself when selection processes are looked at one at a time. But if the point is to explain what physically distinct selection processes have in common, fitness provides a perspective that fine-grained physicalistic description does not.

The second challenge—the problem of finding "independent criteria"—I addressed by invoking the idea of design analysis. Evolutionary biology has developed, and continues to elaborate, a system of models that describe the consequences for fitness of various traits. Supervenience excludes evolutionary theory from finding a single physical characterization of the basis of fitness. But the alternative is not an ad hoc collection of anecdotes about fast zebras, striped moths, and so on. As Fisher's sex-ratio argument shows, one can generalize over a wide range of populations and identify characteristics that have a systematic bearing on fitness relations.

The third problem—the issue of causality—remains. The fact that an organism has a certain level of fitness is not an *event*. An organism's fitness is more like a sugar lump's solubility than its sudden immersion in water. However, the fact that fitness is a dispositional property does not show that it lacks causal efficacy. Nor is the definitional connection of the fitness concept with (probable) survival and reproductive success a reason for holding that it is casually inert. Davidson (1963) has remarked that sunburn is, by definition, caused by exposure to the sun.

---

26. This point about the explanatory power of supervenient properties has been explored in the philosophy of mind, for example, by Block and Fodor (1972) and Dennett (1980).

But this "definitional connection" doesn't somehow make it false that the sun causes sunburn. Even if fitness were definitionally connected with survival and reproductive success, this wouldn't show that fitness is causally inert. Yet, for reasons to be developed in the next section, this is precisely what needs to be said.

The position I will develop may seem paradoxical. Natural selection is one of the causes of evolution, but an organism's fitness is not a cause of its survival and reproductive success. Not the least of the puzzles to be sorted out is the idea that selection causes, but fitness is causally inert. This is bound to be confusing, since "natural selection" and "fitness" are usually thought of as interchangeable descriptions. What could be more standard in evolutionary theory than the idea that natural selection just *is* the existence of fitness differences? I hope that biological readers will continue on for a few more sections before slamming this book shut with the conclusion that I have plunged into irredeemable sophistry. I will argue that fitness and natural selection in fact play different conceptual roles in evolutionary theory. This will begin an analysis of the causal structure of selection processes that will move us into the units of selection controversy, to be investigated in Part II.

Chapter 3
Survival, Reproduction, Causation

At first glance, the concepts of fitness and natural selection seem to play interchangeable and equivalent roles in evolutionary theory. True, we usually talk of the fitness of the *organism*, rarely of the fitness of the *environment*. And we describe the *environment* as selecting organisms, less frequently of an organism as selecting itself. But beneath this superficial difference, biologists recognize an underlying unity. An organism enjoys whatever degree of fitness it has in virtue of its relation with its environment. The environment exerts whatever sort of selection pressure it does on a population in virtue of the relations that organisms stand in to their environment (and to each other). The concepts of fitness and selection seem to provide equivalent ways of describing the web of relationships connecting organism and environment.

At a deeper level, the impression of equivalence continues to be sustained. Apparently, if a population is subject to a selection process, there must be variation in the fitness values of its member organisms. What is more, the implication runs in the other direction; variation in fitness implies selection. This idea is directly reflected in the organization of mathematical models of selection, both simple and complex. In the constant-viability model of selection described in Section 1.4, three fitness values (w_1, w_2, and w_3) were assigned to the three genotypes (*AA*, *Aa*, and *aa*) that can occur at a single locus with two alleles. It is merely a notational variant on this mode of representation to talk about the *selection coefficient* (or the *selective value*) of a genotype. This idea is usually defined by first "normalizing" the three fitness values, converting each to a ratio that it bears to the fittest of the three genotypes. For example, if we were examining a case in which *AA* is superior, we would replace the three fitness values in our models with the quantities 1, $1-s_2$, and $1-s_3$, respectively. This reformulation is, of course, purely terminological, not substantive, since the selection coefficient of a genotype is simply 1 minus its (normalized) fitness value.

With fitness and selection linked together so simply, it may appear perverse to try to pry them apart. There is some plausibility in the declaration that "what the algebra has joined together let no one rend asunder." However, I will argue in this chapter that a distinction needs to be drawn between the conceptual roles of fitness and natural selection.

In Section 3.1, I argue that an organism's overall fitness should not be regarded as the cause of its survival and reproduction. Since a trait's fitness (e.g., the fitness of a gene, genotype, or phenotype) is just an average of certain organismic fitnesses, it follows that the fitness of a trait is not the cause of either the mortality rate or the degree of reproductive success it enjoys. The line of thinking I pursue in this section involves making some subtle distinctions, ones that I wish I could supplement with more knockdown arguments.

In contrast, the argument of Section 3.2 is, I think, much more clearcut. We often talk about there being *selection for* and *selection against* various properties. We also speak of the *selection of* some objects and traits as opposed to others. Finally, we talk about the *fitness of* this or that characteristic. These locutions differ in a modest way—it is a mere preposition that separates "selection *for* a trait" from "selection *of* a trait." However, those prepositions reflect a world of difference. To understand how issues of causality arise in the theory of natural selection, it is essential that we bring this distinction into focus.

3.1 Fitness is Causally Inert

What makes it true that a given organism has a certain probability of surviving and a certain expected number of offspring? An organism is bombarded by a multiplicity of causal influences. Predators menace. Disease and food shortages may similarly threaten. Its prospects for reproductive success are likewise influenced by a variety of causal facts, reflecting the organism's ability to attract a mate, its capacity to collaborate with that mate in the production of viable offspring, and then to further ensure that those offspring themselves survive to reproductive age.

Let's consider only the factors affecting mortality and how they determine that part of an organism's fitness that concerns its chance of surviving.[1] Each of the factors may be thought of as corresponding to a certain probability. The fact that an epidemic is spreading through a population may give different organisms different *chances of dying from the epidemic*. An organism's vulnerability will be influenced by its own robustness, its proximity to the disease, and the degree of immunity it possesses. Similarly, the presence of predators provides

---

1. Points similar to the ones that will be developed here concerning an organism's viability also apply to the component of fitness that pertains to reproduction.

different organisms with different *chances of dying from predation*. An organism's vulnerability to this source of mortality may be influenced by its speed, its protective coloration, its sensory acuteness, and so on.

If an organism is exposed to a predator that subsequently kills it, we can view its fitness with respect to predation as a link in the causal chain leading to its demise. This component of fitness reflects several probabilities. First, there is the chance that a predator will come in range of the organism. Second, there is the probability that the predator will detect the presence of the organism, given that they are close to each other. Finally, there is the probability of the organism's being killed by the predator, given that the predator has spotted it and is nearby.

Similarly, when an organism dies of a disease, its antecedent vulnerability to the disease (i.e., its fitness with respect to disease) plays a causal role. This state of vulnerability is, of course, dispositional. It's like the physical state of sugar that makes sugar cubes water-soluble. This state plays a causal role in connecting the immersion of the sugar cube with its actually dissolving. But unlike the trait of water-solubility, which, I take it, has a single physical basis, an organism's susceptibility to disease is composed of a complex constellation of physical properties.

If an individual contracts a disease to which it is highly vulnerable and dies from it, its *susceptibility to the disease* is reasonably cited as a cause of its death. Conversely, if an individual has a high level of immunity from a disease, its relative *invulnerability to the disease* is appropriately viewed as having causally contributed to its living through an epidemic. The same holds true of vulnerability to predation. When a slow, conspicuously colored, and tasty prey is killed by a predator, we see how its *vulnerability to predation* contributed to its downfall. Likewise, when an organism with a high degree of invulerability to predation escapes a predator's attack, it is reasonable to point to its invulnerability as causally contributing to its having survived. Notice that no mention has yet been made of the causal role played by *overall fitness*. In this section, I want to argue that high overall fitness does not cause survival and low overall fitness does not cause death. Thus, I will deny that overall fitness is causally efficacious.

When an individual survives from one time to another, it survives *all* the risks of mortality that impinge. The organism has a chance of encountering predators and a chance of being exposed to disease. It may face both, one, or neither, of these threats during the period of its life we are thinking about. For each of these possible circumstances, there is a probability that the organism will survive. Assuming that predation (P) and disease (D) are the only possible causes of mortality, we then can define the organism's overall chance of surviving (S) as the sum of its chances of getting through each of these different possibilities, weighted by their probabilities of occurrence:

$$Pr(S) = Pr(S/P \text{ and } D)Pr(P \text{ and } D) +$$

$$Pr(S/P \text{ and not-}D)Pr(P \text{ and not-}D) +$$

$$Pr(S/\text{not-}P \text{ and } D)Pr(\text{not-}P \text{ and } D) +$$

$$Pr(S/\text{not-}P \text{ and not-}D)Pr(\text{not-}P \text{ and not-}D)$$

We may further simplify the example by assuming that predation and disease are independent, both with respect to their chances of impinging on the organism and with respect to their chances of killing the organism, if they impinge. Suppose, just for the sake of an example, that an organism has a 0.5 chance of running into a predator and 0.8 chance of surviving if it does. Similarly, imagine that it has a 0.4 probability of being exposed to the disease and a probability of 0.7 of surviving if it does. Then there are four possible situations the organism may face:

	D and P	D and not-P	not-D and P	not-D and not-P
Probability of exposure to	0.2	0.2	0.3	0.3
Probability of surviving, given exposure to	0.56	0.7	0.8	1.0
Probability of surviving	0.112	0.14	0.24	0.3

The *overall* probability of survival is the sum of the entries in the last row, namely, 0.792.

Rather than thinking of these probabilities as applying to a single individual, we might view them as attaching to a huge population of clones—a population in which the frequencies reflect the probabilities mentioned above. Supposing that there are 1000 individuals at the beginning of the process, 200 encounter both predation and disease, and 112 survive. Three hundred encounter neither of these threats, and they all survive. Combining the four scenarios, we see that of the 1000 original individuals, 792 are still alive at the end of the process.

It is essential to see that an organism's overall chance of surviving takes into account the probability of its surviving all the *possible* causes of mortality. Since predation and disease are facts of life for the population, each organism has a chance of encountering these risks and then of dying from them. This doesn't imply that each individual actually meets up with a predator or is exposed to the disease. An individual with a high probability of surviving has this in virtue of its capacity to cope with whatever contingenices (within a given range of possibilities) *may* occur.

When an individual survives because it manages to avoid being caught by a predator, I can see how its *invulnerability to predation* was a causal factor contributing to its survival. Likewise, when an individual is infected by a disease but survives because of its immunity, I can see how its *invulnerability to disease* helped keep it alive. But an organism's overall fitness—its high probability of surviving, no matter what cause of mortality may present itself—strikes me as being causally inert.

Suppose an organism has a certain chance of surviving predation and another of surviving a disease. Suppose, further, that, as it happens, the individual is attacked by a predator and escapes but is never exposed to a disease. That is, *exposure to disease is a possibility that is never actualized*. In this case, to cite overall fitness as a cause is to blend together a true account of what the actual cause was with an irrelevant account of what the cause might have been, but was not.

Moriarty realizes that Holmes and Watson are on his trail. To confound their efforts to kill him, Moriarty puts on a bulletproof vest and swallows a potion that makes him immune from a certain sort of poison. The probability of surviving being shot at, if he wears the vest, is 0.8. The probability of surviving the poison, if he takes the potion, is 0.7. The probability that Holmes and Watson will shoot is 0.5; the chance that they will administer poison is 0.4. The vest and the potion together confer a probability of surviving of 0.56, if Moriarty is both shot at and poisoned. We can calculate his chance of surviving if Holmes and Watson choose only one method or, mercifully, decide to spare Moriarty's life.

Holmes and Watson then show up and fire away. They leave the poison at home. Moriarty survives. What was the cause? Wearing the vest causally contributed, but the immunity-conferring potion played no role. In addition, I suggest that the overall probability of surviving— the probability that sums up what would happen if two, one, or zero risks of mortality had impinged—is likewise causally inert.

If Holmes and Watson had poisoned Moriarty as well as shot at him, then the relevant probability to consider would be Moriarty's chance of surviving, given that both causal factors are present. But again, Moriarty's overall invulnerability to dying from various possible causes (most of which were not actualized) is not what keeps him alive.

The same holds true if Moriarty dies. If Holmes and Watson shoot but do not poison, Moriarty's level of vulnerability to *bullets* would certainly be a cause of his death. The fact that he was or was not immune to poison would be irrelevant. And the overall chance of mortality—the probability of dying from various possible causal factors—likewise did not cause Moriarty to expire.

Another analogy will help bring into focus the reason for doubting that an organism's overall chance of mortality causes its death. When a cube of ice melts after having been placed in a warm room, one can correctly say that its being made of H_2O was a causal factor.[2] But suppose that there were a different physical substance—call it X—that happens to have the same melting point as H_2O. Now, it is obvious that the cube didn't melt in virtue of its being made of X, since it wasn't even made of that stuff. What I want to deny is that the disjunctive property—that of being made of H_2O or of X—was causally efficacious. True, if the cube had been made of X, *that* would have caused it to melt. But in spite of this "invariance of the effect across the disjunction of possibilities," I am asserting that the disjunction is not itself causally efficacious.

The claim I am advancing needs to be distinguished from another that is perhaps less controversial. It is a familiar and uncontroversial fact that the cause of an event may be one thing *or* another. If, for example, an animal is killed by a disease but is quite unaffected by predation, then disease *or* predation caused its death. The disjunction is true because one of its disjuncts is. Perhaps causation obeys a distributive law: A or B causes something if and only if A causes it or B does. But the position to be developed here is independent of the question of whether this distributive principle is universally correct. For one thing, the principle is consistent with thinking, in the above example, that since being made of H_2O caused the cube to melt, it must also be true that the cube's being made of H_2O or of substance X was causally efficacious. The distributive principle does not speak directly to the issue of which properties are causally efficacious. Although the examples concerning the organism's death and the ice cube's melting do not make waves for the distributive principle, they do help illustrate my thesis that "causation abhors a disjunction."

Here's another way to see the difference between the question of whether the distributive principle is correct and whether disjunctive properties are causally efficacious. Consider *kicking*. If Sam or Aaron

---

2. I argued in the previous section that causal factors need not be events; states of affairs—standing conditions—are also causally efficacious. However, the points to be developed here about disjunctive causes are independent of this view about events, so that someone who finds it misguided to describe the cube's being made of H_2O as a causal factor may wish to substitute a more event-like example.

kicked the ball, then I suppose that either Sam kicked it or Aaron did. However, this platitude doesn't force us to say that "disjunctions" of kickers are themselves kickers. "Sam or Aaron" doesn't name an entity that does any kicking.

Now for an example, similar in structure to the ice cube example, that is apparently much less favorable to my thesis. Suppose a toy gets crushed because it was on the ground next to a large tree, which came crashing down on top of it. What appears to be the causally efficacious property—being under the tree—apparently is disjunctive. To be under a tree is to be *somewhere* under it—in one location *or* another within a certain territory. Suppose we divide the ground that the fallen tree covered into two halves, which we'll call location A and location B. It seems reasonable to describe the toy's demise as due to its presence in location A. It is not permissible to describe the cause as having been that the toy was in location B, since the toy was not in fact there (though if it had been, it would have been crushed by the tree fall). The difference with the ice cube example is that in this case the disjunctive property— being under the tree (i.e., in either location A or location B)—seems to have been causally efficacious.

We now have three examples—(1) the ice cube that melted because it was made of H_2O, not because it was made of X; (2) the organism that dies because of its vulnerability to disease, not because of its vulnerability to predators; (3) the toy that was crushed because it was in location A. The causal structure of these three cases might be diagramed as follows. The solid arrow represents actual causal influence; the broken one represents possible, but nonactual, causal influence. I have placed the selection example in the middle to facilitate comparison with the other two.

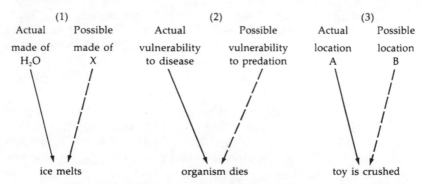

(1)		(2)		(3)	
Actual	Possible	Actual	Possible	Actual	Possible
made of	made of	vulnerability	vulnerability	location	location
H_2O	X	to disease	to predation	A	B
ice melts		organism dies		toy is crushed	

In case 1, I suggest that the disjunctive property of being made of H_2O or of X was causally inert. In case 3, however, it is much less objectionable to attribute causal efficacy to a disjunctive property.

Surely it is correct to say that the toy's being under the tree caused it to get crushed, and as noted above, being under the tree is a disjunctive property. How are we to think of case 2? To attribute causal efficacy to the organism's overall risk of mortality is in fact to say that a certain disjunctive property—its susceptibility to dying from predation *or* from disease *or* from food shortages, etc.—was causally efficacious.

Should case 2 be assimilated to case 1 or to case 3? I suggest that the different intuitions we have about cases 1 and 3 can be diagnosed as follows. If the disjunctive property strikes us as causally inert in case 1, this is because we imagine that the melting of objects made of H_2O is a *different kind of causal process* from the melting of items made of X. And if we have no qualms about attributing causal efficacy to the property of "being under the tree" (i.e., of being in location A or in location B) in case 3, this is because we imagine that the difference between location A and location B would make no difference in the kind of causal process that would lead to the toy's getting crushed. My thesis, then, is that disjunctive properties will appear to be causally efficacious only to the degree that their disjuncts strike us as subsuming similar sorts of possible causal processes.

This claim is supported if we flesh out these two examples with a different set of details. Suppose, contrary to what was stipulated above, that H_2O and X are identical with respect to a structural property (S) that determines how melting proceeds, and that this structural property is found only in objects that are made of H_2O or X. Then, it would not be particularly objectionable to say that being made of H_2O or of X caused the melting, since this would be a way of picking out the common structure S. The mirror-image sort of modification can transform example 3, in which a disjunctive property appears to be causally efficacious, into an example in which the same property seems to be causally inert. If we imagine that a toy in location B would have been crushed by a process very different from the one that was at work in location A, we will be much less happy with thinking of the property of being under the tree as causally efficacious. Suppose location A is under the tree and the toy was crushed by the tree falling on it, whereas if the toy had been in location B, this would have meant that the tree fall would have missed the toy but that the toy would have been crushed by a crushing machine. If we knew this, and wanted to know what caused the toy to get crushed, we probably would not be satisfied with the answer that it was under the tree. This would not tell us what the cause is, because the disjunctive predicate unites two quite distinct kinds of causal processes.

If this proposal correctly marks the difference between cases 1 and 3, it supports the claim that overall fitness is causally inert. Predation, disease, and starvation are biologically distinct kinds of causal processes, each of which can result in the same effect—death. A disjunction linking them together no more names a causally efficacious property than does the disjunction of being made of H_2O or of X, where these substances are quite dissimilar, save for the fact that both melt at 32°F. A further reason for this conclusion will be offered in Section 6.2.

Overall fitnesses are very much like the actuary's idea of *life expectancy*.[3] In assigning you a life expectancy, an actuary assembles an overall picture of your chances of surviving another n years. Your life expectancy is the sum of the number of years of additional life you might have, weighted by the probability that you have them. These probabilities are supposed to take into account your chances of mortality due to all *possible* causes, even ones that, as it turns out, will not do you in. You should be so lucky as to live another n years. But your longevity will not be caused by the fact that you have a particular life expectancy.

Your life expectancy is, of course, a reflection of the causal situation in which you find yourself. You run certain risks of encountering various life-threatening circumstances. However, it is one thing for life expectancies to *reflect* potential causal factors; it is quite another for them to actually do the causing.

In the previous section, fitness was assigned an explanatory role, but here it is being denied a causal one. Is this consistent? Although I argued that an organism's fitness offers some explanation (though perhaps a rather meager one) of its survival and reproductive success, I claimed that the concept of fitness reveals its more robust explanatory character when it is asked to account for two other sorts of facts: changes in the proportion of traits in a population and commonalities between physically different populations. My intent here, however, is to consider only the relation of an organism's fitness to its survival and reproductive success and to claim that that relation is not a causal one. Let us consider the contrast between explanation and causation directly: how is it possible for an organism's overall fitness to explain (or help explain) a fact that it does not cause?[4]

Consider an organism that is extremely vulnerable to numerous possible sources of mortality. If predators come near, it is very likely to be killed. If an epidemic sweeps through a population, it will very probably die. In addition to grasping these conditional facts about the

3. The actuarial analogy is explicit in Fisher's (1930) definition of Malthusian parameters as an extension of the idea of a life table. It is also suggested briefly by Hodge (1983, pp. 58–59) as a reason for denying that fitness causes or explains anything.
4. The discussion in this section is not intended to show that fitness *never* plays a causal role, but just that there is one particular causal role that it does not play.

organism ("if such-and-such happens, its chances of dying are thus-and-so"), I may also know what the probability is that predators will approach, that an epidemic will occur, and so on. I might summarize these facts in the judgment that the individual has a very low overall chance of surviving the next year. Now suppose the individual dies shortly after this overall assessment is made. Presumably, these facts together offer *some* explanation of why the individual failed to live through the year. Granted, they do not specifically single out the *actual* cause of death, and therefore may seem to be less than ideally complete.[5] Still, life expectancy describes the overall causal framework of factors that may impinge, and thus may afford *some* understanding of why the individual dies.[6]

So it isn't that explanation and causation are absolutely irrelevant to each other. For now, I will describe their connection a bit vaguely: The explanation of an event describes the "causal structure" in which it is embedded.[7] Explanation, to be illuminating, need not single out the events that actually did the causing. Even when a description does this—as when the cause of Y is described simply as "the cause of Y"—it may fail to be explanatory. The concepts of causation and explanation have yet to be fully clarified, but I hope that the contrast I have in mind is beginning to come into focus.

In this section, I have argued that an organism's overall fitness and the overall fitness of a trait (which, recall, is the average fitness of the organisms possessing it) do not cause survival and reproduction. Not that this constitutes any victory for those who disparage evolutionary theory or the theory of natural selection. As we will see in the next section, natural selection is causally efficacious, even if overall fitness is not.

---

5. Although in Section 6.1 we will examine a style of explanation—equilibrium explanation—in which the specification of the actual cause seems to contribute little or nothing to our level of understanding.

6. Also relevant here is the point made in Section 1.5 that fitter organisms need not outsurvive less fit ones; so when they do, the fact that they are fitter offers some illumination of the kind of causal process at work.

7. This claim concerns the explanation of single events, not laws of nature or mathematical truths. It would be desirable to have a theory of explanation that could subsume all these cases in a univocal format. However, I will not propose a view at this level of generality here.

3.2 Selection Of and Selection For

The overall fitness of an organism (with respect to those factors influencing viability) is one minus the sum of its chances of dying from numerous possible causes. Two organisms A and B may therefore have the same overall chances of dying, even though they have different vulnerabilities to specific possible causal factors. For example, this may be true because A is more vulnerable to disease than B, while B is more vulnerable to predators than A.

Let's imagine a large population of organisms each of which is a clone either of A or of B. The organisms in this population would be identical in overall fitness. Does it follow that no selection process can occur here, since the requisite variation in fitness is absent? In a sense that I will try to clarify, the answer to this question is *no*. When we look at the population, it wouldn't be surprising to learn that the percentage of A's that die of disease exceeds the percentage of B's that die from this cause. Nor would it be a shock to learn that the percentage of B's that fall to predators exceeds the percentage of A's that get eaten. Disease selects against type A and favors type B, while predation selects against B and favors A. It's true that there is no *overall* selective difference between the two types, because there is no overall difference in fitness. Nevertheless, there's a good deal of selecting going on.

We can think of the two sources of mortality at work in this population as component forces. One favors A at the expense of B; the other favors B at the expense of A. The net force is zero; so there is no net change in the frequencies of the two traits. The situation here is like the billiard ball case in wich you push the ball north and I push it south with the same vigor. There are component forces at work, but no net force, and so no change.

The concept of overall fitness offers a rather coarse-grained description of the various selection processes at work in a population. It summarizes the various selection processes in the way that the vector addition used in billiard ball physics combines the component forces into a single net force. The summary constructed involves a loss of information, in this sense: From the component forces, you can calculate what the net force is,[8] but you can't recover the components from the net. Given that the billiard ball has a zero net force acting on it, you can't tell whether there were lots of component forces that canceled out each other (and what these various influences actually were) or whether no forces impinged on the object at all. Given that A and B are equal in

8. Assuming, of course, that the theory of forces provides a rule for calculating the effects of forces when they act together. See the discussion in Section 1.2 concerning the nontrivial character of this compositional problem.

overall fitness, you can't tell whether there were various component selective forces at work in the population that canceled out each other, or if the population was simply not touched by selective forces of any kind.

These different possibilities will make no difference if one is interested only in calculating the *consequences* that the present state of a system has for its future. Whether the billiard ball is acted on by no forces at all or by a suite of forces that are mutually nullifying, the same thing happens—nothing. The same holds for the population that has no variance in overall fitness. It was pointed out in Section 1.4 that population genetical models are usually interested only in the consequences for gene frequencies of various configurations of forces. It is no wonder that such models make use of the concept of overall fitness and do not need a more fine-grained description. However, there is more to selection than its effects. One also wants to consider its sources. One frequently wants to know *why* the net selective force has the value it does, and this requires decomposing the net into its components. In the sickle-cell example (Section 1.4), we were interested not only in the overall fitness of each of the three genotypes but also in *why* the genotypic fitness values have the ordering they do. The mathematical model was written in the idiom of overall fitnesses, but the explanatory story that accompanied it provided further information.

There is another way in which the concept of overall fitness provides an impoverished representation of the causal structure of a selection process. The fitness of a trait, recall, is the average fitness of the organisms having that trait. This implies that two traits attaching to precisely the same organisms must have the same fitness value. But to treat such traits as equivalent in fitness is to ignore the fact that one of them may have great selective importance, whereas the other may simply be "neutral." In Section 1.2, I mentioned the example of the human jaw and chin. Imagine that individuals with a certain jaw structure always had chins (and conversely). Hence the overall fitness of the two traits must be the same. Yet there was selection for jaws, but not for chins. To express this idea, we clearly need a concept that allows us to say more about a trait than simply describe its overall fitness.

Thus the concept of overall fitness is limited in two ways. It describes the net force of selection, not its components. And it cannot identify differences in the selective significance of traits that apply to the same individuals. These inadequacies are consequences of the deeper fact that fitness is causally inert. But in its division of explanatory labor, evolutionary theory provides a causal concept to do the work that fitness cannot perform. This is the idea of *selecting for and against properties.* In the example considered above of the population composed of two types of individuals *A* and *B*, the traits were identical in overall fitness, even though there was *selection for* the ability to evade predators and *selection for* immunity from disease. In the second example, although jaws and chins are indistinguishable in terms of their overall fitness, the present concept identifies a difference: there was *selection for* having a jaw of a certain kind, but none for possessing a chin.

We must explore the logic of this new concept—the idea of selection for and against characteristics. It is pleasant that the commonsense notion of "selecting" has the structure and marks the distinctions we need to identify. Figure 2 shows a toy that my niece once enjoyed playing with before it was confiscated to serve the higher purposes of philosophy. Each horizontal level contains holes of the same size. The holes on each level are larger than those on the level below. The balls also vary in size. If the balls are at the top, shaking the toy distributes them to their respective levels. This is a selection machine. Balls are

Figure 2. A selection toy in which the name of the game is getting to the bottom. The green balls are the smallest and therefore have the best chance of descending. After a thorough shaking, the small balls—i.e., the green ones—are selected. There is selection for being small, not for being green. This illustrates the difference between the concepts of *selection of objects* and *selection for properties*.

selected for their smallness. The smaller a ball is, the more successful it is at descending. Balls of the same size happen to have the same color. The smallest balls, and only they, are green. So the selection process selects the green balls, because they are the smallest.

There are *two* concepts of selection that we must pry apart. There is *selection of objects* and there is *selection for properties*. The smallest balls are the objects that are selected; it is equally true that the green balls are the objects that are selected. However, the concept of selecting for properties is less liberal. There is selection for smallness, but there is no selection for being green.

"Selection of" pertains to the *effects* of a selection process, whereas "selection for" describes its *causes*. To say that there is selection for a given property means that having that property *causes* success in survival and reproduction. But to say that a given sort of object was selected is merely to say that the result of the selection process was to increase the representation of that kind of object.

When the green balls reach the bottom more frequently than the blue ones, we think that there must have been a reason why. Green balls were selected; so they must have had some property that was selected for. But the property in question was not their color. There was *selection of* green objects, but no *selection for* greenness. I offer the following slogan to summarize this logical point: *"selection of" does not imply "selection for."*[9] This idea will be important when the question of the "selfish gene" is considered in Part II.

"Selection for" is the causal concept *par excellence*. Selection for properties causes differences in survival and reproductive success, even though (as argued in the previous section) overall fitness is causally inert. An organism's overall fitness does not cause it to live or die, but the fact that there is selection against vulnerability to predators may do so. Overall fitness gives a summary picture of an organism's vulnerability to *possible* selection forces. There being selection for a particular property, on the other hand, means that a certain causal process is *actually* in motion.

So far, I have discussed the fitness *of* objects, the fitness *of* properties, selection *of* objects, and selection *for* properties. The *of* concepts all aim at describing the effects of selection for properties but do so in slightly different ways. There can't be selection of objects without there being some sort of selection for properties.[10] But one can have selection for

9. Larry Wright (1973, pp. 163–164) draws the same distinction. I am indebted to Steve Kimbrough for discussion on this point.
10. This doesn't mean, of course, that there can't be selection of green objects unless there is selection for greenness. It needn't be *that* particular property that drives the causal engine.

properties without the total selection process favoring one kind of object over another. The component forces may cancel out each other and generate no net selection effect at all. In the previous example, there was selection for predator avoidance and for disease resistance, but no overall selection of A individuals as opposed to B.[11]

The distinctions drawn here are not limited in their application to children's toys. The idea that there can be "free riders" on selection processes depends on the same bifurcation. Selection can result in a characteristic increasing in frequency without there being selection for it. As noted in Section 1.2, *pleiotropy* is one way that this can happen. If two phenotypic characteristics are both caused by a common underlying gene (or gene complex), then selection for one may increase the representation of both. Recall the example of the human chin; it emerged via a selection process but, apparently, without there being selection for it. A second way in which there can be free riders on a selection process is by way of *gene linkage*. A neutral gene and an advantageous gene may be close together on the same chromosome. Selection for one may increase the frequency of both. Whether the free rider is phenotypic (as in the pleiotropy case) or genetic (as in the case of gene linkage), we say that there was no selection *for* the character in question, meaning that its increase in frequency was not caused by its conferring an advantage.[12]

When a free rider characteristic increases in fitness owing to natural selection, its overall fitness (i.e., the average fitness of the organisms possessing it) must have been greater than the average fitness of the alternative characteristic(s). And, looking back on the way the process proceeded, we may correctly note that the free rider characteristic was selected (*cf.* the green balls in the toy). The concept of overall fitness and the idea that there was *selection of objects* fails to capture the causal structure that the concept of *selection for properties* is designed to characterize.

So natural selection—that is, selection for characteristics—is one of the causes of evolution. When selection for and against various properties of organisms produces evolution, it must be true that the organisms differ in overall fitness. However, there can be selection for and against

11. And even when there is a net selection effect, other evolutionary forces, like mutation and migration, may cancel *that* out, with the result that no evolution at all takes place. The *ceteris paribus* clauses are nested inside each other like so many Russian dolls.
12. Note that it may nevertheless be true that there is selection for linkage to advantageous characteristics. A gene's possessing this property may cause it to increase in frequency. This is quite consistent with saying that the gene is itself selectively neutral (i.e., fails to confer reproductive advantages on the organisms in which it occurs). Selection of the former sort will be discussed in Section 9.1.

properties without this being reflected in differences in overall fitness. Conversely, the mere fact that there is variation in overall fitness does not yet establish why it exists. Fitness differences among organisms or traits do not by themselves reveal which properties are selected for and which are selected against.

The contrast drawn in this chapter between fitness and natural selection does not depend on the claim, developed in Section 3.1, that fitness is causally inert. Even if an organism's fitness were taken to cause its survival and reproductive success, a conceptual difference between fitness and natural selection would still remain. To say that one organism is fitter than another is not yet to say *why* this is so. To say that there is selection for one characteristic and against another, however, is to give a more detailed description of the causal facts. We could abandon the claim that fitness is causally inert and still maintain the thesis that the fitness concept is causally *oblique*. Without wading into hard questions about whether (or in what circumstances) disjunctive properties are causally efficacious, we can still acknowledge that the proposition "C_1 caused E" gives us a more concrete picture of a causal process than does the proposition "C_1 or C_2 or . . . or C_n caused E." Not that this difference shows any *defect* in the fitness concept. As we saw in Section 1.5, that concept's lack of specificity allows it to carve out general patterns of explanation that would be invisible to more concrete concepts like "selection for predator avoidance" and "selection for disease tolerance." Evolutionary theory's conceptual economy deploys different concepts with different functions. Concepts complement each other; they do not put each other out of business.

Chapter 4

Chance

What is the role of the concept of chance in evolutionary theory? This question often elicits one of the two following responses:

1. Evolutionary theory says that the variation on which selection acts is due to mutations occurring in the genetic material. And it is a central proposition of the theory that such variation occurs "at random." Hence, mutation is a principal way in which the concept of chance enters evolutionary theory.

2. Natural selection is itself a process with an important chance component, and this constitutes a significant difference between the theory of natural selection and, say, the theory of gravitation. Whereas classical physics provided a deterministic theory of the phenomena it subsumes,[1] the theory of natural selection was an early example of the power of statistical thinking in the natural sciences.

Both 1 and 2 contain elements of truth, but both require clarification and supplementation.

In Section 4.1, I analyze the standard remark that mutation is "random" and discuss the difference between Lamarckian and Weismannian theories about the origin of variation. In Section 4.2, I explain how the terms "deterministic" and "stochastic" are used in evolutionary theory. In one very central sense, mutation and natural selection are conceived of as *deterministic* processes: They do not themselves incorporate an element of chance but stand in contradistinction to the stochastic element in evolutionary processes, namely, *random genetic drift*.

In Section 4.3, I address a question that has more traditional philosophical credentials. In evolutionary theory, probabilities are assigned to a variety of events. What is it that makes these probability assignments true? Until the twentieth century, one widely accepted answer was that

---

1. For a detailed examination of what is defensible and indefensible in this standard remark, see Earman (1984).

only our ignorance of detail justifies our talk of probabilities; if we only knew enough, we could say what *would* happen, not just what would *probably* happen. Quantum physics is widely believed to have undermined this point of view. Quantum theory—an immensely well-confirmed theory—suggests that chance is part of the fabric of nature. Probabilities are in the world; they are not merely reflections of our ignorance.

The question I explore in Section 4.3 concerns how the interpretation of chance in evolutionary theory is related to judgments about determinism established in physics. If nature were deterministic, would evolutionary theory's only excuse for using probability be our relative ignorance of how biological processes proceed? Is the biologist's assignment of a probability to an event mistaken if it fails to coincide with the one that would derive from taking *all* the physical facts into account? Does physics establish ground rules that determine the correctness of probability assignments in the rest of science? These questions concerning the relation of chance in evolutionary theory to chance in physics allow us to consider yet another facet of that many-sided issue known as *reductionism*.

4.1 The Randomness of Mutation

A discussion of the "randomness" of mutation should begin by specifying what it does *not* mean. It doesn't mean that in a given environment, a gene has equal chances of changing or of remaining in the same state. Typically, the physical copies of a gene found in a particular cell in a given organism will not mutate; a common mutation rate for a gene might be 1 in 10^6 to 1 in 10^8 per cell division.[2] Genes are remarkably stable entities.

Second, the randomness of mutation does not mean that mutations are inherently unpredictable. Facts about the environment—for example, ones concerning the presence of certain chemicals and kinds of radiation—may make some sorts of mutation far more probable than others. Given a large enough population (e.g., the human race immediately after a nuclear war, perhaps) biology can predict that certain kinds of mutation events will occur.

A third misreading of the idea of randomness is that a mutation is just as likely to be deleterious as it is to be advantageous. Fisher (1930) observed that mutation's randomness does not imply this sort of even-

---

2. A different ballpark figure would be relevant if one wanted to consider mutation rate in a population of organisms, or the mutation rate among gametic, rather than somatic, cells.

handedness. Random tinkering with a watch is more apt to make it run worse than better; by the same token, the odds are that a mutation will diminish rather than enhance an organism's fitness.

Fisher (1930) fleshed out this idea with an ingenious graphical argument for thinking that the probability that a mutation will be beneficial increases to 1/2 as a limit, as the magnitude of the mutation approaches 0. Imagine that an organism's present state is represented as a point on a plane and that it would be better off if it could get closer to some other point, which represents an optimal situation. The direction of change that the organism will experience is random; each direction is equiprobable. For an organism to improve, it must end up inside the circle whose center is the optimal point and whose radius is the distance from that point to the one representing its present configuration.[3] A very large change (one greater than twice its present distance from the optimal point) will certainly make the organism worse off. As the magnitude of the change decreases, the odds improve that the random change will reduce the organism's distance from the optimal point. In the limit, as the magnitude of change becomes tiny, ending up inside the circle (improvement) and ending up outside of it (a decline in fitness) become equiprobable.

The defensible idea in the claim that mutation is random is simply that mutations do not occur *because* they would be beneficial. Organisms might be glad to have a detecting device that would channel their mutations in an advantageous direction. Although a Lamarckian mechanism of this sort probably would not increase the organism's own chance of surviving, it might make the organism's offspring better able to survive and reproduce, and the genes' coding for this hypothetical mechanism would thereby spread.[4] There is nothing "unscientific" about contemplating such a mechanism. The point is just that Weismann's (1889) doctrine of "the continuity of the germ plasm" has turned out to be empirically correct, whereas Lamarckian ideas did not. It is a little misleading to summarize this result by saying that "mutations occur at random." One might just as well say that the *weather* occurs at random, since rain doesn't fall because it would be beneficial.

3. Since we are talking about change of location in a plane, we are imagining that the adaptedness of the organism consists in its values for *two* magnitudes—one for each of the two spatial axes. Fisher's argument, of course, is not limited by this assumption, since the same argument could be advanced for a space of arbitrarily many dimensions.
4. They could spread by causing advantageous mutations at other loci and then hitchhiking on them. The advantageous mutants would enhance the fitness of the organisms in which they occur, and the reproductive success of those organisms would increase the frequency of both the advantageous mutation and the correlated Lamarckian device that caused it. See Maynard Smith (1978, pp. 112–114) for a similar suggestion for how genes that regulate the recombination rate could evolve by selection.

Lamarckism, like most *isms*, is a family of doctrines, often imprecisely formulated. The portion of it pertaining to the mechanism of inheritance provides a two-part thesis, one describing a pattern, the other prescribing a causal process that connects the elements of that pattern. First, there is the idea that characteristics that are "acquired" within the lifetime of a parent may become "genetically encoded" for its progeny. Second, there is the thesis that the trait is genetically encoded in the offspring *because* it was advantageous in the lifetime of the parent. Weismannism and the "central dogma of molecular biology" that is its successor (Watson 1976) contradict the claim about process, but not the claim about pattern. Waiving momentarily the question of what "acquired" and "genetically encoded" mean, it is not terribly controversial that acquired characteristics benefiting a parent may occasionally be the same as genetically encoded characteristics found in the offspring. It violates no matter of principle that a mother giraffe should once have made her neck longer by stretching it and that her offspring should fortuitously contain a "gene for a longer neck." Weismannism doesn't prohibit this sort of coincidence from happening. Indeed, if mutations really are random in the sense intended, one would expect to find a certain number of parents and offspring who exhibit this very pattern.

What is controversial in Lamarckism is the second, causal, claim connecting these two events—that the mutation event producing the "gene for a longer neck" occurred because of the mother's stretching, or because her longer neck was good for her. Weismann's denial of this causal thesis might be stated probabilistically: The chance of a mutation of this sort occurring is the same whether or not the acquired characteristic occurred in the parents or was beneficial to them.[5]

The idea that a trait may be "acquired" or "genetically encoded" requires clarification. An organism's phenotype arises out of the inter-action of its genotype with the environment in which it develops. This is true of *all* the characteristics it has. The mother giraffe we have been considering has a long neck in virtue of the genotype with which she was born and the career of stretching she pursued. Likewise, her new-born offspring has a long neck in virtue of its genotype and the uterine environment (including factors like nutrition) in which it developed. What would it mean for a characteristic (e.g., a "long neck") to switch from being "acquired" to being "genetically encoded"?

A genotype's *norm of reaction* describes the different phenotypes it will produce as a function of the sort of environment in which it de-velops. The y axis represents possible phenotypic values (in this example,

---

5. The probabilistic account of causality to be developed in Chapter 8 will allow this thesis to be stated more precisely.

neck length). The x axis represents variation in the environmental parameter of interest. Where the environmental variable considered is
amount of nutrition, the norm of reaction might display a significant
slope. On the other hand, if we compare environments that differ only
with respect to amounts of neck stretching, we might expect the norm
of reaction line to be rather more flat.

The norm of reaction provides two ways of clarifying the idea that
a phenotypic characteristic is "genetically encoded"—one intrinsic, the
other comparative. If a mother giraffe "acquires" her long neck by
stretching, this suggests that her stretching made a difference—if she
had not stretched, or had not stretched quite so much, her neck would
not have been so long. In graph (a) in Figure 3, the mother's neck
length is shown to depend on the environmental parameter of interest
in just the way one would expect for the mother's long neck to be
"acquired". Her offspring, according to the Lamarckian scenario we
are considering, does not need to stretch to have a long neck. The trait
has become, as we say, "genetically encoded." This suggests that the
offspring's norm of reaction is flatter relative to the environmental
variable considered. This also is shown in graph (a).[6] Here we interpret

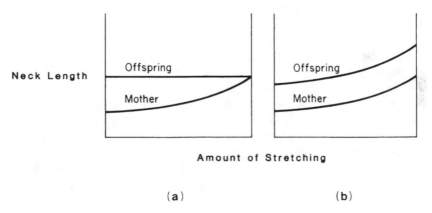

Figure 3. Two pairs of parent/offspring norms of reaction that display a "Lamarckian"
pattern. In both (a) and (b), the mother has a long neck only if she stretches, whereas
the offspring has a long neck with or without stretching. In (a), but not in (b), the offspring
genotype is less labile.

6. In using the idea of a genotype's norm of reaction to make sense of the distinction
between "acquired" and "genetic" in this example, I am ignoring another standard way
of explicating that idea—namely, the idea of *heritability*. This concept, which I'll examine
in Section 5.2, is of use when the question concerns the explanation of variation in a
population. The more individual-level question considered here makes the norm of reaction
the more appropriate tool of analysis. Confusion of these two ways of conceptualizing

the idea that a phenotype is genetically encoded to mean that the genotype will produce that phenotype across a wide range of environmental variation. This provides an "intrinsic" definition of what "genetically encoded" means—one need only determine whether the genotype's norm of reaction is flat.

Another sense of "genetic encoding" is worth mentioning, one that involves comparing the norm of reaction of one genotype with the norm of reaction of another. Suppose that regardless of what the environment is in which mother and offspring develop, the offspring comes out taller. That is, suppose the norms of reaction are as shown in graph (b) of Figure 3. Although both genotypes are labile, there is nevertheless an intuitive sense in which the offspring has a genotype that makes it taller.

The graphs in Figure 3 conform in different ways to the idea of "inheritance of acquired characteristics." Graph (b), however, is closer to what Lamarck required. Lamarck thought that giraffes took many generations to arrive at the long neck they currently have. Stretching must be a way of manipulating the phenotype for both parent *and* offspring; when an offspring itself becomes a parent, the process must be repeated. A flat norm of reaction, as in graph (a), brings this evolutionary process to a grinding halt. Similarly, if a trait can change through use and disuse in parents and offspring, as Lamarck thought, then *both* norms of reaction must be nonflat.[7]

Although these are important differences between graphs (a) and (b), there is something they have in common. In both cases, it may be true that the mother exceeds, say, eight feet in height only if she stretches. The offspring exceeds this height whether or not it stretches. I will gloss over the distinction between (a) and (b) in what follows by saying that the offspring's genotype is less labile than the parent's. Where lability is understood qualitatively (an individual either has a certain property or not, depending on the environment) rather than quantitatively (how tall an individual is depends on the environment), this should occasion no confusion.

Lamarckian "inheritance of acquired characteristics" suggests this relationship between parental and offspring norms of reaction but imposes an additional requirement. It is certainly not controversial that different genotypes may have different norms of reaction. Since sexual reproduction makes it unsurprising that offspring differ genetically from their parents, one expects that there will be cases in which the genotype

---

the difference between "nature" and "nurture" has been rampant. See Lewontin (1974a) and Block and Dworkin (1974) for discussion.

7. I am grateful to Kristin Guyot for pointing out to me the importance of relationship (b).

of a parent will be more labile than the genotype of the offspring. But as mentioned above, Lamarckism encompasses a claim about causal process in addition to a thesis about pattern. It is required not only that the parental and offspring genotypes have different norms of reaction but also that the offspring have the genotype it does *because* of the fitness advantage the parent enjoyed.

Weismannism denies that this causal setup is a real one. However, there probably are *some*, perhaps highly unusual, confluences of causal factors that exhibit the Lamarckian structure. Someplace, somewhere, there probably is an acquired characteristic whose superior fitness triggers an environmental condition that functions as a specific mutagen. A perverse laboratory arrangement could probably create a Rube Goldberg arrangement of this kind. The experimental protocol would dictate that if an organism has a certain acquired characteristic, it is to be administered a certain mutagen known to act specifically in organisms of that kind to increase the chance of a gene coding for the characteristic. There is no threat that "Weismannism" will topple if such an experiment is actually performed. That doctrine is better construed as maintaining that this kind of causal process occurs, if at all, very rarely. Evolutionary biology provides no absolute prohibition against it. Here, as in the controversy over the units of selection to be discussed in Part II, questions about causal mechanisms in evolutionary theory usually concern the importance and frequency of a kind of process, not the problem of whether the process is in principle possible.

Weismannism and Lamarckism, then, disagree over whether the deployment of variation in a population is "directed." But this disagreement does not bear on whether mutation is represented as a stochastic or a deterministic evolutionary force. The equations of population genetics, which describe the evolutionary *consequences* of mutation, are unaffected by this dispute over the *causes* of mutation. This reflects the division of labor in a theory of forces between consequence laws on the one hand and source laws on the other (Section 1.5). It is not just natural selection that evolutionary theory characterizes in two ways at once.

Although I have contrasted Lamarckian ideas with Weismannian ones, it is rather standard to think of Darwin as the preeminent historical figure who opposed Lamarckism. This interpretation is quite anachronistic, if the point is to consider divergent views about the origin of variation. In *The Origin of Species*, Darwin repeatedly avails himself of the mechanism of "use and disuse" as a cause of evolution additional to the action of natural selection. It is only this century's version of Darwinism that has so resolutely assimilated the Weismann doctrine. Nevertheless, there is an important truth in the idea that Lamarck and

Darwin had fundamentally opposing points of view. But the proper locus for their disagreement was not the origin of variation but the mechanism of evolution itself, a subject I turn to in Section 6.2.

The slogan that "mutation is random," I have argued, is misleading. The issue of whether variation is directed is an important one, but it has little or nothing to do with how ideas about determinism and chance figure in evolutionary theory. Weismannism and Lamarckism are perfectly consistent with mutations' having a deterministic origin, and they are each compatible with the idea that mutations are the result of irreducibly probabilistic processes. The time has come to see the role chance really plays in evolutionary theory.

4.2 Deterministic and Stochastic Processes

In evolutionary theory, mutation and selection are treated as *deterministic* forces of evolution. *Chance*, as the term is used, is not a property of the selection and mutation processes but stands in contradistinction to them. *Random genetic drift*, on the other hand, is the source of the stochastic element in evolution. To understand the concept of chance in evolutionary theory, we must grasp what random genetic drift amounts to.

In evolutionary theory, the word "chance" is used to cover the possibility of "sampling error." However, this sampling error is not something that biologists commit as part of the activity of investigating nature. Rather, sampling error is itself a natural process that goes on in populations whether they are subject to biological scrutiny or not.

Imagine a population composed of two types of individuals—A and B. Suppose we begin at the egg stage with 50 individuals of each type, and that A individuals have a 0.9 chance of surviving, whereas the B individuals are much more vulnerable and have only a 0.8 chance of making it to the adult stage. If we census the population after these individuals mature, what mix of the two types will we find? There is no guarantee that there will be $50 \times 0.9 = 45$ A individuals and $50 \times 0.8 = 40$ B individuals. As mentioned in Section 1.3, the frequencies of traits cannot be deduced from their fitnesses when populations are finite in size. All one can extract from this information is a probability distribution of the different possible outcomes.

Natural selection is typically claimed to be a deterministic process in this sense: *When it acts alone, the future frequencies of traits in a population are logically implied by their starting frequencies and the fitness values of the various genotypes.* This pattern of inference was the one we seemed to pursue in applying the constant-viability model of selection in Section 1.4. A stochastic process, on the other hand, is one in which starting frequencies plus the relevant coefficients imply only a probability distribution of possible future states.

An analogy will help clarify, and also correct, this standard formulation. The frequency of heads in a run of independent tosses of a coin cannot be deduced from the probability of heads. But, as noted in Section 1.4, Bernoulli's theorem assures us that the larger the number of tosses (assumed to be independent), the higher the probability is that the frequency of heads will fall within some fixed percentage of the probability of heads. In the limit, the probability is arbitrarily close to unity that the frequency will be arbitrarily close to the probability. But even in this case, the connection of frequency and probability is not deductive. As Van Fraassen (1977) and Skyrms (1980) have pointed out, each infinite sequence of heads and tails is possible, though each has only an infinitesimal probability of occurring. One of these infinite sequences would be actualized were the coin tossed an infinite number of times. Each is infinitely unlikely, although the probability is unity that one of them will occur. So it is possible that a fair coin should land heads up each time it is tossed, and this is still possible even if it were tossed an infinite number of times. Frequencies don't *have* to converge on probabilities, even at the limit.

We can now correct the characterization offered above of why natural selection is a deterministic evolutionary force. Fitness values plus starting frequencies do *not* permit the deduction of changes in gene frequencies, even on the supposition of infinite population size. They do, however, permit a probability inference of almost unbeatable strength: the probability is arbitrarily close to unity that the frequency will be arbitrarily close to the expected value. This is, I imagine, a fine point that may interest philosophers of probability; it is not something that will rattle practicing population geneticists.

So infinite population size makes it "virtually" certain that frequencies will reflect probabilities. As population size decreases, there is increasing scope for frequencies to depart from probabilities. It is here that we talk of "sampling error." Ten thousand independent tosses of a fair coin are reasonably certain to be between 45 and 55 percent heads, but a run of only a hundred tosses has a better chance of falling outside that margin. If the coin is tossed only ten times, the opportunity for sampling error to play a significant role is greater still.

One standard biological context in which this sampling process is important is the sampling of gametes for the formation of zygotes. Organisms almost invariably produce many more sperms and eggs than are needed to create the organisms of the next generation. If there are to be N zygotes, one sample of size N must be drawn from the sperm pool and another from the egg pool. Selection may influence how many gametes a given organism produces. It also may be reflected in differences in the viabilities of the gametes themselves. But once a set of viable gametes is available, zygote formation may proceed by a process of random sampling. Each sperm has the same chance of making a zygote as any other; similarly for each egg. Yet the frequency of a gene among the zygotes may differ from its frequency in the gamete pool. Frequencies can fail to mirror probabilities because the zygotes are finite in number.

Sampling error appears in speciation theory in the idea of the "founder effect" (Mayr 1963). A single breeding population can be split in two when some geographical barrier interposes itself (like a river that changes its course). The result might be that a few organisms are isolated from the rest. Even when organisms have the same chance of being in the isolated group, "sampling error" may make the two populations that result from this fission differ in gene frequencies. If differences in selection pressures make the populations diverge further, the isolating event may turn out to have been the beginning of a speciation process.[8]

Since biological populations must be finite, there is no doubt that chance plays a role in evolution. The question that interests biologists is not whether this is true but how important the role of chance has been. The long-standing difference between R. A. Fisher's (1930) and Sewall Wright's (1931, 1945, 1978) approaches to population genetics concerns this question. One large population of size N will permit a smaller role for random processes than a set of small semi-isolated but interacting populations that add up to N. Wright's ideas on population structure will be discussed in Section 9.2 in connection with the group selection controversy.

Random genetic drift can exert a predictable and systematic influence on the composition of a population. If I draw a small random sample from a large urn containing 90 percent red balls and 10 percent green, there is a better chance that the sample will be all red than that it will be all green. Minority traits have a higher probability of "extinction" under a process of random sampling. But this difference in probability is quite compatible with the sense in which the sampling proceeds at random: each individual *ball* has an equal chance of being drawn.

8. Whether the isolates were in the same species as the organisms from which they were isolated depends on what happens later. What species an organism is in is not settled by its "internal," nonrelational properties; nor is it settled merely by the totality of facts that obtain during its own lifetime. There was a skit on Sid Caesar's "Your Show of Shows" in which someone in 1914 is shown reading a newspaper whose headline is "World War I declared!" The joke was not that the headline was false but that the conditions that made that world war the first (but not the last) came into existence only twenty-five years later. See Kitcher (1984b) for a different view of the matter.

Although we may talk informally about "selecting at random," this idea is something of a contradiction in terms in evolutionary theory. In Section 1.2, I observed that the change in gene frequency that is generally regarded as necessary and sufficient for evolution must be computed per capita; gene frequencies don't increase by organisms getting fatter, for example. In just the same sense, zygote formation is random when each *gamete* (not each *trait*) has an equal chance of being represented in the next generation of zygotes.

In order to visualize the systematic effect of random genetic drift on a population, it is useful to think of a large number of identical populations all experiencing drift as its sole evolutionary force. Each population passes through a number of generations. We will follow this process's impact on the frequencies of the two alleles (A and a) that can occur at a given locus. Each population begins with the same mixture of A and a. Gamete formation and random sampling of gametes then produces the next generation of zygotes. Those zygotes then reach the adult stage and the process is repeated. If p is the starting frequency of A in each population, then in the short run (i.e., after relatively few generations of this sampling procedure), most of the populations will still maintain a frequency of A somewhere around p. But as the process continues, the number of populations that drift to fixation increases, and the distribution of gene frequencies among the remaining populations flattens out. Given enough time, all populations will go to fixation, having reached the absorbing state of 100 percent A or 100 percent a. If p is the starting frequency of A in each population, then one expects A to reach fixation in p of the populations and a to reach fixation in the remaining $1-p$.

Two characteristics influence the pattern displayed in this ensemble of populations. First, there is the starting frequency of the gene in question. Second, there is the size of each of the populations.[9] Intuitively, the larger the population, the smaller the role that sampling error will play. That is, if we compare two ensembles of populations, each undergoing this process for the same number of generations, we expect more dispersion in the ensemble whose populations have the smaller size. Figure 4 illustrates this pattern.

Another pattern we can expect random genetic drift to produce concerns the level of heterozygosity found in a population. Again, let us imagine an ensemble of populations, each of which contains the same

9. A relatively fine point: It isn't census size that influences the power of random drift but the effective population number, which is a number that takes into account not only the census size but also the proportion of males and females. When there are as many males as females, the effective population size is the same as the census size.

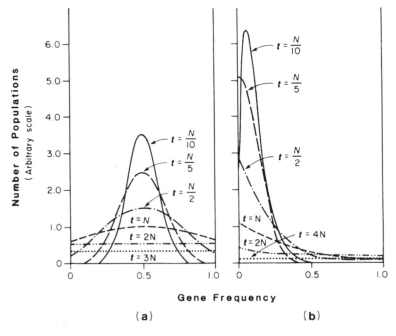

Figure 4. Predicted results of random genetic drift. In (a) the initial allele frequency is 0.5; in (b) it is 0.9. Areas under the curves represent the proportion of populations yet to reach fixation. At an early stage of the drift process, a population's probability distribution of possible states shows relatively little dispersion around its initial allele frequency. As the process continues, the dispersion increases, with the probability of fixation increasing and the intermediate frequencies approaching a flat distribution (adapted from Crow and Kimura 1970).

frequencies of the genes *A* and *a*. A certain percentage of individuals in each of these populations will be heterozygotes. As drift proceeds, this percentage will decline to zero, as more and more populations reach fixation for one or the other gene (Crow and Kimura 1970). Randomness in evolutionary theory has not evaded the scientist's search for laws. Quite the contrary: Population genetics allows us to describe the systematic consequences that chance may have for a population's evolutionary fate.

Notice that if the frequency of a gene is *p* in a large ensemble of populations, and each evolves by a process of random genetic drift, the frequency of the gene in the ensemble may still be *p* after each subpopulation has gone to fixation. Although gene frequencies in the ensemble have not changed, the frequency of heterozygosity has; it has gone to zero. This is one reason that the definition of evolution as change in gene frequency, discussed in Section 1.2, should not be viewed as engraved in stone. Evolution can occur when there is a shift in the frequencies of gene *complexes*: a change in the frequencies of single genes is not essential.

Philosophers may find it a little exasperating to learn that selection, mutation, and migration are "deterministic" in only the sense described here. To begin with, the biological distinction is entirely independent of whether micro-theories in physics are deterministic or stochastic. Drift would be stochastic and selection deterministic even if quantum mechanics were someday discarded and replaced by a deterministic theory. Further, the decomposition of an evolutionary process into its deterministic and stochastic components may seem rather contrived. Could one not just as easily construe coin tossing as having two components, one "deterministic," the other "stochastic"? It is not controversial that two "factors" influence whether the percentage of heads one gets on a run of independent tosses will fall in a given interval. These are (1) the coin's probability of landing heads and (2) the number of times the coin is tossed. But it would be bizarre to treat these two factors as constituting separate processes or forces. Why, then, is drift separated from selection, mutation, migration, and other deterministic factors in evolutionary theory?

The reason is to facilitate comparison among different populations. Two populations may be characterized by the same set of selection coefficients and yet differ in size. Two other populations may have the same effective size and yet differ as to the selective forces that impinge on them. Separating selection and drift yields concepts that are needed to mark important biological distinctions.[10]

Coin tossing affords the same distinction. Two coins may have the same chance of landing heads but may be tossed a different number of times. Two other coins may be tossed the same number of times but may differ in their chances of landing heads. The first two coins have something in common, and so do the second. Probability theory provides us with the conceptual machinery needed to describe what these points of similarity amount to.

Stochastic and deterministic evolutionary forces may also be distinguished in terms of how each contributes to the *magnitude* and *direction* of change. Again, let us begin with coin tossing. Suppose two coins are both biased in favor of heads; coin A has a probability of landing

---

10. In addition there is the sheer convenience of ignoring stochastic effects; with large enough population size, the fit of model to reality may seem good enough. However, this pragmatic reason is distinct from the one cited above, according to which there are biological facts that are captured by separating drift from selection that would be rendered invisible if this distinction were not drawn.

heads of 0.8, while coin *B* has a probability of landing heads of only 0.7. If we toss each coin a large number of times, it is virtually certain that *A* will land heads more frequently than *B* will. If we toss the coins a smaller number of times, our certainty that this will be the outcome is reduced. But no matter what the sample size, the probability that *A* will come up heads more often that *B* exceeds the probability that *B* will produce more heads than *A*, or that they will have an equal frequency of this outcome. Sample size affects our confidence that *A* will win; it does not affect our opinion that it is more likely that *A* will win than that *B* will.

Now imagine a population subject only to selection and drift, where *A* and *B* are the two types of organisms; they reproduce asexually, with like always producing like. If *A* is fitter than *B*, then, regardless of population size, the odds are that *A* will increase in frequency. How certain this outcome will be depends on population size. If the population is huge, then we can be quite confident that *A* will increase in frequency. But even if the population were very small, this would not *reverse* our expectation of which trait will increase; it would only lessen our confidence that this will occur. If we were forced to wager, we still would bet on *A*, but with much less assurance than we would have with larger sample size.

Thus, when selection and drift are the only forces at work, the selective advantage of *A* over *B* favors the prediction that *A* will increase in frequency rather than decline or remain the same. Regardless of the role that chance plays here, the *direction* of change is something that the model tells us something about.[11] On the other hand, the *magnitude* of change is not something that the mere fact of selection allows us to judge; we must know what role drift plays in order to take this further step.

Drift, on the other hand, says something about the *magnitude* of change, although in a sense it remains silent about its *direction*. A population subject to drift alone beginning with a gene frequency of *p* has *p* as the expected value of its frequency throughout the process. Initially, the probabilities are less dispersed; later, they spread and cluster around the absorbing states of 100% and 0%. If the process lasts a given number of generations, information about population size predicts how dispersed the probability distribution of possible frequencies will be. Magnitude of probability dispersion, if not direction of frequency change, is entailed.

I have chosen to describe random genetic drift as a "force" of evolution

---

11. Not that the model implies that *A must* increase in frequency. The point is that this prediction is always better supported than its negation.

to emphasize the causal role it plays. Nothing much hangs on this terminology, however. In any case, it is clear that drift is a very different kind of force from its deterministic counterparts. When a population evolves under the impact of a suite of deterministic forces, it makes perfect sense to ask what contributions the component forces made. If group selection favors an altruistic characteristic and individual selection favors the selfish alternative, we can see which force is more powerful by seeing whether altruism increases over selfishness. Similarly, when two forces are in the same direction—for example, when mutation and selection both favor a given gene—we can estimate the relative contributions of each by seeing how the trait would evolve in a population in which only one of them is at work. Not that this is always a feasible problem to investigate empirically. The point is that we know what it means for deterministic forces to make varying contributions to an evolutionary outcome.

Matters are very different when we ask about the relative importance of selection and drift. When traits differ in fitness in a finite population, there is no way to make sense of the idea that selection has been more influential than drift in producing the evolutionary result. If a fair coin lands heads up six out of ten times it is tossed, there is no saying how much its probability of landing heads contributed to this result as compared with the fact that it was tossed only ten times. We can say, of course, that the probability and the number of tosses made some outcomes in the run of tosses more probable than others. And we can observe how those probabilities would have been different, if the coin had had a different probability or had been tossed a different number of times. But the vector addition and subtraction that underwrites our understanding of how deterministic forces combine cannot be used here. If drift is an evolutionary force, it is a force of a different color.

The difference between deterministic and stochastic processes in evolutionary theory has nothing to do with predicting the traits of individual *organisms*. A deterministic model of selection does not provide a deductive inference (or generate a probability inference arbitrarily close to unity) as to how reproductively successful any particular organism will be. In a sense, individual organisms are not really what the theory is *about*.[12] The theory of natural selection provides a deterministic treatment of the frequencies of characteristics in populations. This highlights the fact that when we ask whether a theory is deterministic, we must always specify the magnitude or property in question. The theory of natural selection is deterministic with respect to trait frequencies but

12. This point will be important in Section 5.3 when we consider how "population thinking" (to use Mayr's 1976b phrase) has figured in evolutionary biology.

not with respect to the survival and reproduction of individual organisms. The theory of random drift is not deterministic with respect to trait frequencies, much less with respect to what happens to individual organisms.

4.3 What Laplace's Demon Would Be Missing

With this background, let's reexamine the example (mentioned in Section 1.4) that Scriven (1959) used to illustrate the probabilistic character of the idea of natural selection. Imagine identical twins walking on a mountaintop. One of them is struck by lightning and dies; the other is unscathed and goes on to reproduce. It is intuitive to ascribe this difference in reproductive success to "chance." The twins, we may want to say, were identical in fitness, and so natural selection played no role in shaping their different fates. In what follows, I want to address the problem of explaining the basis of the strategy underlying this analysis. Suppose the twins differed in some way that would explain the fact that one lived and the other died. Perhaps the one struck by lightning was walking on a slightly higher path and this more exposed position slightly increased the chance of a lightning strike. Would this fact force us to redescribe the event as involving natural selection? If not, why not?

As mentioned in Section 1.4, this problem does not disappear when we shift from single organisms to traits (e.g., genotypes) as the bearers of fitness values. True, evolutionary models of the sort described in Section 1.4 do not assign fitness values to organisms one at a time. But the fitness of a trait is the average fitness of the organisms possessing it; the idea of genotypic fitness inherits whatever conceptual problems attach to the idea of individual fitness. In addition, the twin problem can easily occur for genotypes. The fact that individuals with one trait happen to outsurvive or outreproduce the individuals with a second trait does not force the conclusion that the first trait is fitter. After all, individuals in the second group may have been struck by lightning more often than individuals in the first. Accidents can happen to groups of similar organisms just as easily as to single organisms.

Returning now to the twins, let's consider what consequences follow from the assumption that the twins are *genetically identical*. Is it a principle of evolutionary theory that genetically identical individuals must be identical in fitness? The answer is *no*. Even genetically identical individuals may be phenotypically different. Two genetically identical corn plants may differ in their resistance to drought because one grew strong and the other remained frail; they have different phenotypes because they developed in different environments. Evolutionary theory does not demand that the twins be assigned the same fitness value simply because they are genetically identical.

Suppose that the twins are not only genetically identical but phenotypically identical as well. It may perhaps seem plausible to think that the fact that the first twin was at one spot on the mountain while the second twin was somewhere else when lightning struck does not count as a difference in phenotype. If phenotypically identical organisms cannot differ in fitness, it would follow that the twins must be identical in fitness.

I do not doubt that the concepts of *phenotype* and *fitness* are connected in this way: Fitness differences among organisms must be reflected in their phenotypic differences. But why say that the difference between the twins concerning where they walked that afternoon is not a difference in their phenotypes? Phenotypes don't stop at the skin. The sort of environment an organism inhabits is part of its phenotype. If one organism habitually lives out in the open and another habitually lives under cover, we might think of the first as more vulnerable to predators. In this case, an ecological difference makes for a difference in phenotype. Why is it any different with the twins?

A third tack suggests itself. Part of what makes us say "chance" rather than "selection" in this example may be the idea that where the twins walked on the mountain was due to momentary whim rather than long-standing habit. If I had embellished the story by saying that the first twin *usually* walked on the highest, most exposed, part of the mountain, but the second twin tended to walk along less lofty paths, the idea that there is a difference in fitness might have seemed more plausible. We can even assume that there is a causal explanation for why the twins followed the paths they did on the day in question. But if their difference in behavior does not reflect a *regular* difference in behavior, we may be inclined to say that there was no difference in fitness. So we now need to ask whether evolutionary theory implies that differences in mortality that have *unusual* causes are due to chance while differences in mortality due to *usual* circumstances are due to selection.

To be sure, there are models of "stochastic variation in selection pressures." Sewall Wright (1978), in his preferred taxonomy of evolutionary forces, lists chance variation in selection coefficients, mutation rates, and so on as a kind of stochastic process. I don't doubt that it may be theoretically illuminating to think of the values of deterministic forces as themselves subject to variation around some long-term mean value. But if one knows the actual values of the selection coefficients at each instant, why not treat the process as a deterministic one in which the selection coefficients change value? In the present example, the death of the twin had an obvious cause. Why not say that if habitual lightning strikes can exert a selection pressure on a population, unprecedented ones can too?

At bottom, we may feel certain that the twin that lived must have somehow differed from the twin that died. Why not stipulate that fitness differences must always precisely mirror differences in actual survival and reproductive success? Perhaps it is only our ignorance of the underlying causal facts that makes us rely on probabilities. If so, talk of "chance" does not single out anything distinctive in nature; rather, it merely excuses our lack of information.

The idea here—that probability statements must be confessions of ignorance—received its canonical formulation in Laplace (1814, p. 4):

> Given for one instant an intelligence which could comprehend all the forces by which nature is animated and the respective situation of the beings who compose it—an intelligence sufficiently vast to submit these data to analysis—it would embrace in the same formula the movements of the greatest bodies of the universe and those of the lightest atom; for it, nothing would be uncertain and the future, and the past, would be present to its eyes.

A few paragraphs later, Laplace concludes that "probability is relative, in part to [our] ignorance, in part to our knowledge."[13]

Laplace's conclusion is, of course, based on the assumption that nature is deterministic. A full specification of the state of a closed system at one time would, in conjunction with the laws of nature, *logically imply* its state at the next instant.[14] The heartfelt conviction that the twins *must* have differed in some way before lightning struck one of them rests on the assumption that determinism is true. Indeed, *deterministic systems* that differ in their future state must have differed somehow in their previous state. However, the same does not hold true for systems that are irreducibly stochastic. If the lightning strike were a fundamentally chance event, the twins and the respective "microenvironments" they occupied could have been precisely the same, down

13. In *The Origin of Species*, Darwin (1859, p. 131) endorses this Laplacean interpretation: "I have hitherto sometimes spoken as if the variations . . . had been due to chance. This, of course, is a wholly incorrect expression, but it serves to acknowledge plainly our ignorance of the cause of each particular variation." See Schweber (1983) for a discussion of the conceptual affinities of Laplace's demon, Maxwell's demon, and Darwin's sorting being.

14. Notice that here I use the idea of determinism in a way that is standard in philosophical discussion, not in the sense I developed in the previous section for saying that natural selection and mutation are "deterministic" forces in evolutionary theory. The philosophical idea, of course, requires various sorts of clarification, but it will be adequate for the purposes at hand.

to the last molecule. This would be perfectly compatible with their occupying different states at the next instant.

In physics, quantum mechanics has upset the assumption of determinism. If quantum mechanics is true and complete, as it well may be, then nature is irreducibly probabilistic. We are forced to rely on probabilities not out of ignorance but because of the way the world is.

Philosophers and scientists sometimes seem to think that the indeterminism discovered by quantum mechanics is limited to the micro-level. The idea seems to be that organisms and populations are composed of too many fundamental physical particles for chance at the micro-level to "percolate up" to the macro-level that evolutionary biology describes. If true, this thesis would justify the belief that the twins *must* have differed before the strike, since one was hit by lightning while the other was not.

I fail to see how the law of large numbers (or anything else) can make the macro-world safe for determinism, if chance is an irreducible ingredient of the micro-level. Presumably, if *enough* elementary particles had behaved differently, the behavior of the macro-object (the organism, the population) that they compose would have also been different. And there is no deterministic guarantee that the ensemble of particles *must* have behaved the way it did. The most that the ensembles of particles we call organisms can do is exhibit an impressive degree of predictability. But, so long as they are made of particles that have an irreducible chance component in their behavior, they too must be indeterministic systems. If chance is real at the micro-level, it must be real at the macro-level as well.

Although the death of determinism in physics answers one part of the question raised by Laplace, it does not resolve the problem I now want to consider about evolutionary theory. What does the idea of chance in evolutionary theory have to do with the issue of determinism in physics? The Laplacean view asserts that the twins in our example are equally fit only if it is an irreducibly chance matter as to who lives and who dies. If we were to assume that there is *some* (deterministic) physical cause of what happened on that afternoon, the Laplacean position implies that although assigning identical viabilities to the twins may be *convenient*, it is really just a *useful fiction*.

What I am calling "the Laplacean position" is of interest here not because of its false assumption that determinism is true but because it has reductionistic implications that do not depend on the truth of determinism. Before stating the Laplacean thesis explicitly, I want to illustrate how it works in an example. I toss a coin. It lands either heads or tails. What probability should I assign to the two possible outcomes? I might begin by noting that the coin was evenly balanced, and use this as a justification for assigning each outcome a probability of 0.5. But then I might note that the impulse my thumb imparted to this evenly balanced coin made heads more probable than tails. I would then use this augmented data base to revise my previous probability assignment. Then, if I were told that the coin was magnetized in a certain way and that the coin toss took place near a magnet, I would take this fact into account along with the other data to revise my probability assignment once more. The Laplacean position asserts that this pattern of revision is mandatory:

> If you are to assign a probability to an event H where circumstances F_1 and F_2 both obtain, and if $Pr(H/F_1) \neq Pr(H/F_1\&F_2)$, then $Pr(H/F_1\&F_2)$ is the preferable probability assignment.

Notice that if determinism were true, this principle would imply that the only reason you could have for failing to assign H a probability of 0 or 1 is that you lacked the requisite information. Even if determinism is false, the only excuse for failing to assign the probability that all the facts dictate is ignorance of them. The idea is that a probability assignment is justified only if it reflects every bit of known information. This is the Laplacean idea that is independent of the truth of determinism.[15]

I want to argue that this methodological thesis is mistaken because it fails to take into account the different purposes we have in assigning probabilities. The reasons I will initially state to support this criticism are somewhat obvious but do not bear directly on the relation of probability assignments in evolutionary theory to probability assignments in physics. Subsequent reasons, which merely elaborate on the pattern of the initial objections, do home in on this question about reductionism.

Suppose I happen to know that the coin I tossed landed heads. You ask me what the outcome was. Presumably, I should consult my entire body of information to answer your question. This total corpus logically implies that the coin landed heads; relative to this body of information, the probability of heads is 1.0. Clearly, I should not set aside my information about what happened when I assign the event a probability. When I am asked about what happened, I consult all the information I have available.

However, suppose you asked a different question. Suppose you

---

15. As stated, the thesis is epistemological or methodological, not ontological. It concerns what probability assignments we ought to *use*, not which ones are in fact *true*. Consistent with this principle are two views about the ontological matter. The first asserts that an event has different probabilities, depending on the reference class in which the event is placed. The second asserts that there is a unique *true* probability of an event, perhaps the one that in some sense reflects all the facts, whether known or unknown.

wanted to know *why* the coin landed on the side it did. I would then point out, as I did before, that the coin landed heads. However, assigning that outcome a probability would now proceed differently. I would not compute the probability relative to everything I know. I would take into account only those facts that pertain to the conditions of the coin and the tossing situation that obtained *before* the coin landed heads. Only those facts would be *explanatorily* relevant.

Suppose the coin is tossed at time t_1, lands heads or tails at time t_2, and remains in that state through t_3 until t_4. We want to compute the probability that the coin is heads up at t_4. The new wrinkle is that at t_3, the result of the coin toss is inscribed in a book. The causal structure is as shown below:

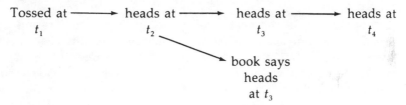

How should we assign a probability to the event at t_4? If we wanted to explain why the coin was heads up at t_4, the inscription in the book would be irrelevant. The coin isn't heads up *because* of what the book says. On the other hand, if we wanted only to predict the state of the coin at t_4, it would be perfectly legitimate to take the testimony of the book into account.

I claim that it makes perfect sense to compute the probability of heads at t_4 relative not only to earlier events on the causal chain leading up to it but also relative to events that are not on the causal chain.[16] There may be nontrivial physical constraints on what this system must be like for both these probabilities to make sense—a point that also holds for computing the probability of a past event relative to facts about the present. My point is that there is no a priori reason why these requirements should not be satisfied. When they are, there will be two ways of assigning probabilities. If we wish to explain, only one of them may be used; if we wish to predict, either may do.[17]

16. It may be thought that the propensity interpretation of probability precludes this sort of pluralism. As I understand that idea, however, this does not follow. One might claim that probabilities are always grounded in propensities and still maintain that there are circumstances in which Pr(A/B) makes sense, even though B is not a cause of A. See Fetzer (1981) for a detailed discussion of the propensity interpretation.

17. It might be thought that whatever information the inscription in the book provides about the coin's state at t_4 is provided by the coin's state at t_2. That is, it might be maintained that a correlated effect cannot be a better predictor of an event than its cause

A similar bifurcation between prediction and explanation shows up even when we restrict our attention to the probabilities an event has relative to various other events that lie on the causal chain leading up to it. The probability of an event can *evolve* as a causal process proceeds. Consider random genetic drift. A neutral allele begins, let us say, with a frequency of 0.6 in a population. At this point, the probability of that allele's going to fixation is 0.6. The random sampling process then proceeds, and the frequency of the allele may go up or down. Supposing that the random walk takes the allele to a frequency of 0.45, we would then say that the fixation of the allele has become less probable than it was before. If the sampling process then took the frequency to 0.95, this would mean that the chances of fixation for the neutral allele had increased.

Suppose that at the next instant, the allele goes to fixation. Now imagine someone asking what the ultimate fate of this gene was that some time ago had a frequency of 0.6. If we wanted to predict this result—whether it was driven out of the population or went to fixation—we presumably would want the *latest* information on its frequency. In the above story, we would start by conferring a probability of fixation of 0.6, and then we would replace that estimate by 0.45 and then, finally, by 0.95. Later probabilities render earlier ones *predictively ir-relevant.*[18] This way of updating probability assignments is in perfect conformity with the Laplacean principle stated above.

However, when we change the task from predicting the eventual fate of the gene to explaining why it went to fixation, our view of these evolving probabilities changes. Later probabilities do not render earlier ones *explanatorily irrelevant.* When the gene went to fixation, it is ex-planatory to note that the gene began at a frequency of 0.6 and random genetic drift was the only evolutionary force at work. It also is ex-planatory to note that immediately before fixation, the frequency under the influence of drift had become 0.95. And the full story of the evolving probabilities is a third explanatory account in addition to these two piecemeal remarks. Does accepting the explanation given in terms of the latest probability assignments mean that we should simply toss out the others? I would say no; this negative answer is enough to establish a difference between how we update predictions and how we update explanations to take account of evolving probabilities. In explaining an event we are interested in where it came from, and this interest may lead us to look farther back along the causal chain; even though

---

is. However, this need not be so: Common causes need not "screen-off" one joint effect from the other. This will be discussed in Chapter 8.
18. This is because random genetic drift has the so-called Markov property.

more ultimate causes may confer a lower probability than more prox-imate ones, they offer an indispensable perspective on why the event occurred. The task of prediction, however, is much more forward-looking: we want to secure the best possible basis for ascertaining whether the event will occur. In this case, looking farther back into the causal background may be quite pointless.

These three examples have something in common. When it comes to explaining an event, one may discard data that are probabilistically relevant. Typically, one does not compute the probability of the event relative to everything one knows or even relative to everything one knows about the state of the world before the event occurred. Fur-thermore, there is no requirement that one compute the probability relative to the *latest* event on the causal (Markov) chain leading up to the event we want to explain. It now is time to show another respect in which the task of scientific explanation can justify ignoring information.

Although I have argued against the Laplacean position so far by distinguishing the prediction of an event from its explanation, there is another dimension to the Laplacean position that must be examined. It may at first appear that whether one explains or predicts an event, one assigns the event a probability that should not be influenced by how other, similar events are to be explained or predicted. The shared idea here is that the scientific treatment of an event is governed only by considerations about how it is to be treated in isolation. I now want to suggest that this "atomistic" point of view (to choose a handy label for it) fails to consider how the scientific interpretation of an event may properly be influenced by the desire to bring out what it has in common with other events.

In Section 1.5, I noted that there is a great methodological advantage in the fact that fitness is not a physical property. The supervenience of fitness allows that concept to characterize systems that are physically quite different from each other. The general concept of probability possesses the same explanatory virtue. Suppose two coins are tossed and one comes up heads and the other tails. In spite of the difference in outcome, the coins may yet have something in common; one pos-sibility is that they had identical chances of landing heads. By luck, it may turn out that the physical basis of the first coin's probability is the same as the physical basis of the second coin's. They may both be bent in precisely the same way. On the other hand, this need not be true; the coins may have their probabilities grounded in different phys-ical bases. The concept of probability allows us to bring out what systems have in common even when the details of their physical mech-anisms and behaviors are distinct.

Fitness is useful when it comes to describing the organisms in a single population, but it is even more useful when it comes to saying what a zebra population and a cockroach population have in common. The usefulness of the concept extends beyond its allowing us to characterize organisms one by one. Its irreducible utility arises from its contribution to the construction of theories that allow us to subsume a huge variety of single events within a common framework. This explanatory virtue of the fitness concept is a consequence of that property's *supervenience* (Section 1.5).

I have already mentioned in this context how the constant-viability model of heterozygote superiority can provide a *uniform* explanation for why each of a set of populations remains polymorphic. This model can be modified slightly to bring the present point about probability into clearer focus. Change the model so that it takes account not only of selection but also of two ways that sampling error can play a role. There is first the possibility that actual survivorship may depart from the probabilities represented in the genotypic fitness values. Second, sampling error can cause zygotic gene frequencies to depart from gametic gene frequencies. That is, chance can play a role in both survival and reproduction.

This model can now be applied to a wide range of populations. They will differ in the phenotypic basis of the fitness relations that obtain among the three genotypes. They may differ with respect to the genetic locus the selection process affects. They may even differ with respect to whether they remain polymorphic. The concept of probability allows us to treat this wide range of populations within a single explanatory framework. This is an explanatory advantage of the probability concept that it possesses regardless of whether determinism is true. If determinism were true and we chose to forsake the use of probabilities, we might be able to explain each population's trajectory of change. But these separate explanations would likely fail to bring out what all the populations have in common.

A similar structure is at work when random genetic drift is invoked on its own. Consider any population with a neutral allele that is at a frequency of 0.6. The theory of random genetic drift tells us that the chance that the gene will go to fixation is 0.6 and the chance that its alternative allele will go to fixation is 0.4. This applies to *any neutral allele*, regardless of what species it is in and regardless of what its physicochemical identity might be. Such populations may pursue quite different random walks and will, of course, probably differ in their evolutionary fates. The probability concept allows us to model their dynamics in a uniform way. If nature were deterministic, we could, in principle, give up that concept and still explain the different trajectories one by one. But I very much doubt that we would be able to recover what the various populations have in common.

So it isn't just ignorance that motivates the use of probability. Granted, if you do not know the full details of the deterministic backdrop, a stochastic model may be a way to make do. But even with full knowledge of details, stochastic modeling may retain its point. Besides excusing our ignorance, the probability concept is a tool for carving out generalizations.

My remarks to this point have been methodological. I have tried to pinpoint a use that probabilities have over and above the one the Laplacean position has identified. However, there is a slightly different question I want to consider—one that is ontological in character.

Laplace led us through a thought experiment to argue that probabilities do not reflect objective features of the world. Imagine a being who is unconstrained by ignorance. Would such a being have any need to postulate chance processes? Laplace—imagining that determinism is true—answered this question in the negative. I have identified another use that probability statements have beyond glossing over our ignorance. Does this use of probability mark an objective feature of the world, or does it merely reflect another aspect of our subjective, human perspective? Again, we can formulate this question in terms of Laplace's demon. What would a being who can provide a deterministic explanation for each event taken alone think of probability models that unite such apparently different systems within a common structure?

Think of the populations mentioned earlier in which a neutral allele has a starting frequency of 0.6. These populations pursue numerous different random walks; some take the allele that was initially in the majority to fixation while others eliminate that gene from the population. Laplace's demon would presumably give each of these different processes its own detailed deterministic explanation. The demon would lack the probabilistic concepts needed to bring out what I have alleged they have in common. The demon, were he informed of my argument about why the usual population genetics treatment of random genetic drift is illuminating, might reply that what I claim to be a common thread running through these populations is in fact an illusion. He, the demon, has no need of this "common thread" and so it must merely be our limited human perspective that prevents us from immediately grasping the detailed deterministic explanations.

It is difficult to bring this science fiction thought experiment to a decisive conclusion. Perhaps the demon might thank us for pointing out generalizations that he had not noticed. Perhaps he would deny this, and we would then insist that he was missing something objective—that he was failing to see the woods for the trees (Lucas 1970).

Although Laplace was able to *stipulate* how his demon was able to perform, I do not think any comparable stipulation about the problem I now have in mind cuts much ice.

The kind of question we are considering here about probability has been explored in another connection. Supervenience, as I have mentioned, has interested philosophers of psychology. Organisms that are physically different can have mental characteristics in common. It is perhaps only a matter of time before computers exhibit mentality. If so, it would appear that psychological properties—ones involving perception, memory, and problem solving, for example—are supervenient.

A question then arises: Why should we take psychological properties seriously, if a Laplacean demon, given knowledge of the elementary particles we are made of, could predict what we do? It may simply be our ignorance of these detailed physical goings-on that drives us to talk about minds. However, there is another hypothesis, namely, that supervenient mental properties allow us to identify what these various physically distinct systems have in common.[19]

We can imagine a Laplacean demon denying that psychological properties have any reality. He might point out that he has no need to postulate mental states when it comes to predicting and explaining behavior. The proper reply, I think, is that the demon's atomistic pursuits do not exhaust what science aims to do. Science *does* need to attribute abstract mental states to organisms if it wishes to carve out generalizations at that level of abstraction.

The same point holds with respect to the use of probabilities in a deterministic universe, but there is a novel wrinkle. In the case of the supervenience of psychological properties, the issue can be described even on the assumption that both physicalistic and psychological theories are deterministic. We can, in principle, predict and explain your behavior from the vantage point of your current mental state and also from the vantage point of your current physical state. Moving to the supervenient psychological perspective need not represent a decline in predictive power.

However, our present puzzle about the point of probabilities in a deterministic universe is a bit different. The choice between deterministic and stochastic modeling involves a certain trade-off. The former brings with it enhanced predictive power; the latter provides greater scope for generality. By allowing a role for chance, we concede that physical systems may be in the same state at one time, even though they may

19. Although this explanatory virtue of supervenient properties is stressed in Block and Fodor (1972) and in Fodor (1975), the formulation in Dennett (1980) is the one that makes the analogy with Laplace's demon most explicit.

differ in subsequent state. By specifying the causal factors so completely that chance is eliminated from the model, we end up describing the system in such a way that it would be remarkable if many other systems proceeded in the same way. If two deterministic systems have identical inputs, then they count as instances of the same sort of deterministic system only if they have identical outputs. But stochastic systems can be in the same state even if they produce different outputs for identical inputs. Stochastic modeling, in this sense, casts a wider net.

Laplace's demon should not convince us that there really are no such things as psychological states. Granted, he might not need them to predict or even to explain behaviors one by one. But there is more to science than this. It would be a precipitous application of Occam's razor to deny the existence of the mental, simply because Laplace's demon does not need to recognize the existence of psychological properties in his pursuit of atomistic explanation and prediction. I suggest that it is equally implausible to think that probabilities are unreal in a deterministic universe.

This antireductionist conclusion holds when we drop Laplace's mistaken assumption of determinism. If we were able to analyze the trajectory of a population that begins with a gene at a frequency of 0.6 from the point of view of elementary physical particles, we would obtain a rather different explanation and prediction of that gene's ultimate fate. The micro-level explanation would still be probabilistic, but it would probably isolate that population from other apparently similar ones. The micro-explanations of what happens to different populations would almost certainly differ substantially from each other; the macro-explanation—the one provided by population biology—would be able to treat these different populations within a single framework. The probability concepts of evolutionary theory have an autonomous explanatory power, whether determinism is true or not.

Let us return, finally, to the parable of the traveling twins. This story has frequently been misinterpreted; it is often supposed that evolutionary theory is somehow *committed* to treating examples like this as nonselective. However, nothing about the genetic identity of the twins, the rarity of lightning strikes, or the infrequency with which the one twin walks on more exposed paths forces us to view them as identically fit. This example does not illustrate any matter of principle about how fitness values are to be assigned. Let's call the characteristic of the first twin that inclined him toward the high road on that fateful afternoon "quality K."[20] I see nothing wrong with saying that there was selection

20. Here, again, I assume for the sake of argument that there was *some* difference between the two twins that helps account for why one but not the other was hit by lightning.

against quality K that afternoon. Of course, this momentary selection event may have had little impact on gene frequencies in the population. But there is nothing in the notion of a force that requires a force to have a substantial effect each time it impinges.

The same situation appears to obtain in the case of coin tossing. Suppose two coins are tossed, with one coming up heads, the other tails. Under what conditions is it true that they had identical probabilities of coming up heads? One answer might be that they should be regarded as equally biased if their internal physical properties are the same. They were of identical weight, balance, and so on. Here we ignore the fact that they may have been tossed in slightly different ways or that a breeze affected one but not the other. This difference between the coins would be glossed over if we assigned them identical probabilities of landing heads. On the other hand, we might in some circumstances want to say that they had different probabilities because they were tossed in slightly different circumstances. Here we allow our assignment of probabilities to be influenced by differences in context as well as by "internal" properties.

Which of these assignments is correct? It strikes me that this question is ill-formed. Both are correct, and which characterization we use depends on our purposes. Neither science nor philosophy provides any general principle for saying whether the two coins "really" had equal or unequal chances of landing heads. The feeling that the real probability is the one that takes account of all information about the initial conditions is, I think, a misreading of the fact that in predicting an event from its causes, one wants to come as close to this complete data base as possible. But prediction from causes is only one use of probabilities among many; it has no special claim on identifying an event's "true" probability.

If we assign the two twins identical fitness values, this implies no commitment that their difference in survivorship must be an irreducible matter of chance. Micro-physics may perhaps have something to say about this, but evolutionary theory is in no position to take a stand. What is crucial to the outlook of population biology is something different—a certain methodological stance. One must be willing to treat organisms that differ in their reproductive success as identically fit in order to capture biologically significant generalizations. We already have the concept of actual reproductive success in evolutionary theory. In a deterministic universe, the Laplacean proposal would have the effect of making fitness *redundant*, since it would entail that fitness and actual reproductive success must generate precisely the same groupings of organisms. The Laplacean point of view collapses the distinction between random genetic drift and deterministic forces like selection, mutation, and migration in a way that is theoretically unfruitful.

Consider an analogy. I show you a box and ask you how many things there are in it. You see *one* chess set, but you also see *thirty-two* chess pieces. The question of "how many" is relative to a kind; "how many" is short for "how many *what*."[21]

The fact that questions of number have answers only relative to a choice of kind doesn't imply that facts about number are "subjective" or "conventional." It is a perfectly objective matter that there are thirty-two chess pieces, but only one chess set, in the box. Probability is no more "conventional" than this. The probability of an event is relative to a set of propositions: The probability of P is always the probability of P, given Q, for some set of propositions Q. But once the value of Q is fixed, a perfectly determinate probability of P can be expected to follow.[22]

Philosophers have labored long and hard to find rules that dictate what the reference class of an event must be. The Laplacean principle discussed above is a (rough) formulation of one of the guiding ideas in this project. But, if I am right, this endeavor cannot hope to discover what the unique probability of an event is, since there is no such thing. Rather, the principle should be understood as saying that although an event may have many probabilities, we ought to use the probability that takes account of all background information. Here we are talking about pragmatics—about what probabilities we ought to *use*—not about what *the* probability of an event actually is.

Even given this pragmatic gloss, it remains to say why a principle like this is needed. There is no comparable "problem of the reference class" for the assignment of number. It isn't for philosophers to dictate which concepts should be used in the activity of counting. We recognize that science may see a single organism, a collection of cells, and an even larger multiplicity of atoms all occupying the same volume of space. Each way of counting has its use in science, and there is no

---

21. Frege (1884) misconstrued this relativity by taking the position that number is a property of concepts, not of objects. All his examples show is that number is not a property *only* of the contents of the box (Kessler 1980). One also could quarrel with the assumption in his argument that the one chess set is *identical* with the thirty-two pieces. A particular chess set may perhaps continue to exist even when one of the pieces is destroyed and replaced. This at least suggests that the set may, at a given time, be *made of* those thirty-two pieces, without being *identical with* them (Wiggins 1980).

22. Not that we are always, or even ever, fully explicit about the set of propositions Q relative to which the probability of P is to be evaluated. When we talk about "the probability that the coin will land heads, given that it was tossed," we need to supply a great deal more information about what properties of the tossing situation are to be held constant and what the distribution of the variable conditions is supposed to be, as Sklar (1970) has stressed.

overarching principle that dictates which we ought to use, much less which is really correct.

It may, of course, be true that there are theoretical and practical advantages that a cellular description of an organism has that an organismic description cannot offer. Conceivably certain sorts of predictions can be made more accurately by this more fine-grained level of analysis. But this would not show that there is a uniquely correct way of counting—one performed by subsuming living things under the concept *cell*, rather than *organism*. Regardless of certain predictive advantages that the fine-grained description possesses, one can describe "one and the same physical phenomenon" (Frege 1884) as either one organism or an ensemble of cells. In the same way, the fact that a probabilistic *prediction* should take account of all background information does not show that the "true" probability of an event is the one it has relative to that complete corpus of background information. Coarse-grained description has its point, and there is no reason to deny that it captures something objective in nature.

I have advanced two theses in this section, one methodological, the other ontological. The former concerns the uses that the probability concept has in theoretical science. The Laplacean position asserts that the only use for probabilities in a deterministic universe is to compensate for ignorance. The thought behind this claim does not depend on the assumption of determinism; the idea is that the only justification for assigning a probability that fails to take account of all information is ignorance. I have tried to suggest a second methodological motive for using probabilities, one that has a claim on scientific activity even in the face of full information.

The second, ontological thesis I have suggested concerns the reality of probability. It can perhaps be summarized by saying that probability is an irreducibly relational notion and that the relation in question is one that an event bears to different sets of propositions. I argued for this in several ways. First, looking only at the causal chain leading up to an event, we can see that the probability of the event *evolves*. Relative to what is true at different earlier times, the event may have different probabilities as an objective matter of fact. This same relativity is found when we broaden our perspective and take account of events that do not lie on that causal chain. Other earlier events may confer probabilities on the event; and so may events that are simultaneous with or later than the one in question.

The Laplacean position is initially attractive because we think of the probability of an event as something it has solely in virtue of the earlier events on the causal chain leading up to it. If determinism were true, the probability of an event would not evolve as we move from one link on the chain to the next. This further encourages the idea that, in reality, there is only one probability that an event has—the one it has had at all earlier moments. Dropping the assumption of determinism forces us to abandon the idea that probabilities do not change. But this, I think, is only the beginning of what is wrong with the Laplacean perspective.

If we are interested in having a probability assignment reflect only the earlier causal factors at work at some previous time, we use one probability. If we want a probability that facilitates accurate prediction, we may make a different choice. However, there is a kind of explanatory project that leads to a different perspective on probability. We may wish to subsume a variety of different physical systems within a single explanatory framework. The probabilities we then assign may involve a refusal to take account of all the detailed differences that obtain among the systems in question. We obtain a probability that tells us something about all the systems by abstracting from the vagaries of each.

Any such probability must, as a matter of mathematics, be equivalent to a weighted average of more specific probability assignments. Suppose we think that all the systems have in common the property C, although they differ with respect to $B_1, B_2, \ldots, B_n$ (which are mutually exclusive and collectively exhaustive). We wish to assign each system a probability of displaying the property E. Now it may well be that we would end up assigning the different systems different probabilities if we took into account not only their common feature C but also the fact that they differ with respect to $B_1, B_2, \ldots, B_n$. If we nevertheless talk about all the systems as having a single probability, relative to C alone, this probability must be expandable into the following summation:

$$\Pr(E/C) = \sum_{i=1}^{n} \Pr(E/C\&B_i)\Pr(B_i/C).$$

In claiming that the univocal probability expressed by "$\Pr(E/C)$" can be explanatory, I am asserting that the summation probability with which it is equivalent can be too.

This leads us back to issues concerning the explanatory utility and causal inertness of fitness (Section 3.1). When we see that Moriarty is still alive, it may be explanatory to be told that he had been wearing a protective vest and had swallowed an antidote to poison. This is explanatory, even though we do not know whether Holmes and Watson chose two, one, or no method of assassination. The actual event (Moriarty's still being alive) is explained by situating it within a range of possibilities, each with its separate probability.

Moriarty's overall probability of surviving is a sum of his more specific chances of surviving the various arrays of causal factors that may impinge, weighted by their chances of occurring. It is the same overall probability we would associate with Moriarty regardless of which of the possibilities in fact occurred. To put the same point another way, the model would apply equally to a Moriarty who is shot and poisoned, to one who is shot but not poisoned, to one who is poisoned but not shot, and to one who is neither shot nor poisoned.

The fitness values assigned in evolutionary theory obey the same logic. The point is not to isolate the unique constellation of factors that most precisely circumscribes the fate of a single organism. Rather, the idea is to bring out patterns that apply both within and among populations. Perhaps Laplace's demon, equipped with the equations of a deterministic particle physics, would not miss a trick when it comes to predicting and explaining the characteristics of single events. But the demon's perspective on *structure*—on the similarities and differences that obtain among single events—would be seriously impoverished. It is here that we see what Laplace's demon would be missing.

Chapter 5
Explanation

This chapter combines general issues from the philosophy of science concerning the concept of explanation with more specific questions about the structure of explanations that deploy the concept of natural selection. Philosophers have standardly divided explanations into two parts: a proposition that is explained (the *explanandum*) and a set of propositions that do the explaining (the *explanans*). The philosophical analysis of scientific explanation, since Hempel and Oppenheim's (1948) classic paper, has focused on the problem of describing what sorts of information must be provided by the *explanans* propositions and what the connection between *explanans* and *explanandum* must be for the resulting whole to count as an explanation. In Section 5.1, I pursue some of the philosophical issues that have arisen in this connection. Although I consider the extent to which evolutionary theory's explanatoriness depends on its being predictive, my main concern at first is with questions that are not specific to evolution or natural selection.

In Sections 5.2 and 5.3, however, I examine properties of evolutionary explanations that are specifically biological. Elaborating on some ideas developed by Richard Lewontin, I try to circumscribe one of the fundamental conceptual shifts that Darwinian evolutionary theory introduced into biology. Natural selection is not just a novel explanation of evolution, it is a novel *kind* of explanation. Darwin did not simply expropriate the phenomena that earlier biologists wished to account for and supply them with a new explanation; he reformulated the very propositions that required explanation. This transformation, I will argue, constituted a change in the logical structure of biological explanation.

In Section 5.3, I examine Ernst Mayr's distinction between typological or essentialist thinking on the one hand and population thinking on the other. As in Section 5.2, the goal is to identify one of the salient structural properties of Darwinian and post-Darwinian evolutionary explanations. I argue that at the very heart of population thinking is the idea that variation in nature is not only something that can be explained but is also an explanatory principle in its own right. The difference between typological and populational modes of thought is developed by examining differences between Aristotle's views on reproduction and variation and ones that are now standard in evolutionary biology. The point, of course, is not to become fixated on how Aristotle was wrong on points of factual detail but to show how the *kind* of theory he tried to develop is no longer part of the evolutionary world view.

Sections 5.2 and 5.3, then, are complementary. The former concentrates on the structure of the propositions that evolutionary theory attempts to explain; the latter focuses on the way population thinking has led evolutionary biologists to deploy certain sorts of theoretical claims in the propositions that do the explaining. Both in its *explanandum* and in its *explanans*, evolutionary explanation has a great deal of philosophical interest.

5.1 Explanation and Prediction

The connection between explanation and prediction has been important to evolutionary theory in various ways. In Chapter 2, we saw that one formulation of the tautology problem faults the concept of fitness for its *ex post facto* character. This criticism alleges that we have no way of knowing which organisms are fitter aside from observing their actual survivorship and reproductive success. The problem, so it is said, is that fitness is not *pre*dictive but can only be *retro*dicted.

My reply to this objection was in two parts. First, I argued that even if fitness had this property, this would not imply that fitness is *unexplanatory*. A physician may decide that a certain disease is present because some characteristic symptom is observed. The fact that the disease is diagnosed only after one of its effects has been observed does not mean that the disease cannot explain the presence of the symptom. Evidential inference often goes from effect to cause without preventing explanation from relating cause to effect.[1]

The second part of the reply suggested that there in fact *are* numerous ways of assigning fitness values aside from observing who survives and reproduces. Of course, the supervenience of fitness implies that there will be no single physical property that can be used in all cases as a measure of fitness. But an engineering-style "design analysis" can nevertheless provide a variety of independent evidential sources for assigning fitness values. Nor does supervenience imply that this design analysis has to be pursued piecemeal on an organism-by-organism or population-by-population basis. As the example of Fisher's sex-ratio

---

1. The example of the Geiger counter in Section 2.2 made the same point.

argument showed (Section 1.5), supervenience in no way prevents biologists from developing source laws for natural selection that are extremely general.

This criticism of the fitness concept and the reply to it focus on a matter of *principle*. Does evolutionary theory have certain kinds of conceptual resources? But there is another question about predictiveness and explanatoriness that has considerable biological importance. Can biologists *in practice* find out the values of the parameters they need to know in order to construct explanations of evolutionary change? That biologists have this practical problem may, in itself, reflect no defect in the theory they use. The world we live in may simply be such that certain questions are very hard to answer. If so, the difficulties that biologists find in applying a theory may actually be a sign not of the theory's imperfection but of its *adequacy*.

This practical problem is perennial in population genetics. The observation of changes in gene frequencies in a population can be very hard to explain. When the phenotypic consequences of the genes studied are known, an engineering analysis may be possible. But in the absence of this kind of independent access, it may be very difficult to identify the causal factors responsible for a population's evolutionary trajectory.

The problem of disentangling possible causes of an observed effect is not unique to evolutionary theory; in principle, it confronts any theory of forces. But there is a special feature of this problem that plagues evolutionary theory. The usual pattern for natural selection is for it to act on very small fitness differences. "Superstar" mutations that double an organism's fecundity or halve its risk of mortality are hardly the stock-in-trade of evolutionary change. More typical would be a fitness difference between organisms possessing a trait and those in the population with the alternative on the order of 0.01 percent. But if this is the sort of modest margin that one expects to find in nature, how is one to tell whether differences in actual survival and reproductive success reflect fitness differences or are due to chance?

A coin-tossing analogy may help here. I have two coins and suspect that one may be more biased toward heads than the other. I know, however, that if this bias exists, there is unlikely to be a difference of more than, say, 0.01 percent. How many times will I have to toss the two coins to be reasonably confident that the coin with the higher frequency of heads is more biased toward that outcome? The answer is that a very large number of tosses is needed to discriminate between this hypothesis and the null hypothesis that asserts that the difference in frequencies of heads is due purely to chance.

The problem of discriminating among alternative hypotheses also arose in the debate between the Dobzhansky and Muller schools over the origin and maintenance of genetic diversity, described in Lewontin (1974b). Muller saw mutation as the principal source of variability, with selection acting to destroy variation by driving the most advantageous genes to fixation. Dobzhansky thought that mechanisms like heterozygote superiority (Section 1.4) play a significant role in maintaining genetic variability. When a population is observed to have several alleles present at a given locus, does this mean that selection is working against all but one, but has not yet driven the population to fixation, or that selection is acting to preserve the variability?

We can see in this epistemological problem a *practical* limitation on evolutionary theory's explanatory power. In a finite population, a deleterious gene may go to fixation and a modestly advantageous one may disappear. If we knew the fitness values and other evolutionary forces impinging on a population, we could explain why it pursued the evolutionary trajectory it did. But nature has conspired to impede our access to the required parameter values. Population geneticists would welcome a clever bit of theorizing that would enable them to do an end run around this and similar problems. Perhaps there are interesting ways of looking at natural populations that permit a more powerful discrimination to be made among alternative hypotheses. But, in the first instance, this limitation on evolutionary theory should not be taken as a sign that the theory is immature or that it fails to measure up to some lofty standard of IDEAL SCIENCE. It is not the scientist's fault that nature has made some of its secrets relatively opaque to human scrutiny.

I should mention that this problem would not exist if an organism's or trait's fitness were simply defined as its actual survivorship and reproductive success. An exhaustive census of a population would then show whether one genotype on average produced more offspring than another. But the epistemological problem of interest here is more profound: even an *exhaustive* census may fail to discriminate among the hypotheses considered. That this is a real problem in evolutionary theory shows that fitness is not equated with actual survivorship and reproductive success (Mills and Beatty 1979).

There is another way of looking at the connection between explanation and prediction, one that is more a part of mainstream philosophical discussion of the idea of scientific explanation. Let's not worry about the fact that knowledge of the initial conditions that obtain in a population (in particular, the genotypic fitness values) is often very difficult to obtain and, when obtained, is sometimes *ex post facto* in character. Suppose that this information is known. To explain a later state of the population, we want to relate that later state to the population's initial state and to the relevant laws of evolution. Likewise, to predict a population's later state, we endeavor to effect a relation between it and the population's present condition and the laws governing its evolution. So both explaining an event and predicting it involve relating the event to earlier conditions and to laws of nature. The question that has occupied center stage in philosophy of science involves the character of these two relations. What is it to explain? What is it to predict? And are these two relationships at bottom the same?

Prediction has not appeared to present much difficulty. Prediction is inference. We predict an event E from information about a system's initial conditions I and the relevant laws of nature L by showing that E follows, either as a deductive consequence or with high probability, from I and L. Explanation, on the other hand, has raised philosophical questions. Is every explanation of an event a prediction of that event? Is every prediction an explanation? Hempel's (1965) model of explanation answered both questions in the affirmative. But more recently there have been several objections to both halves of this "symmetry thesis."

Not every prediction of an event explains it (Scriven 1959, Bromberger 1966). I can predict tomorrow's weather by consulting today's barometer reading, but the barometer reading does not explain why it will rain. A second example makes the same point: You may be able to deduce how tall a building is from information about the length of its shadow and the position of the sun (and laws governing the movement of light). But surely this information doesn't explain why the building has the height it does.

Why are these predictions unexplanatory? One attractive answer (defended by Salmon 1975 and Lewis 1984) is that the explanation of an event must identify its cause. Today's barometer reading is not the cause of tomorrow's rain; rather, they have a common cause—today's weather conditions. The sun's position and the shadow's length don't cause the building's height; rather, the sun's position and the building's height cause the shadow's length.

But we have already seen reasons for balking at this simple solution. To begin with, it was argued in Section 3.1 that an organism's overall fitness may explain why it lives or dies, but the organism's having the overall fitness it does does not cause it to live or die. The reason overall fitness is explanatory is that it provides a summary picture of the constellation of *possible* sources of mortality that are arrayed against an organism. However, citing an organism's overall fitness fails to pinpoint its *actual* cause of mortality. Explanation must describe causal structure; but it needn't do so by "saying what the cause is."

The example of Fisher's sex-ratio argument (Section 1.5) leads to the same conclusion. If a population has the kind of structure described in Fisher's argument, this helps explain why its sex ratio is 1:1. But Fisher's argument does not say what the sex ratio actually was at an earlier time. It may have been male-biased, in which case selection for parents who produce all daughters would have pushed the population toward equilibrium. It may have been female-biased, in which case selection for parents who produce all sons would have achieved the same result. Finally, the population may have remained at a 1:1 ratio throughout its whole history, in which case no selection at all would have been at work.

The sort of explanation we are now considering might be termed "equilibrium explanation," to constrast it with "causal explanation" (Sober 1983b). Equilibrium explanation shows why the actual cause of an event is, in a sense, explanatorily *irrelevant*. It shows that the identity of the actual cause doesn't matter, as long as it is one of a set of possibilities of a certain kind. Disjunctive properties do not cause, but they can explain.

Two philosophical remarks are in order here. First, it should be clear that I reject the idea that causes are necessary for their effects, a thesis defended by Lewis (1973). The idea of equilibrium explanation requires that causes not be necessary for their effects, since such explanations provide a survey over possible causes, none of which is necessary for its effect. But this thesis about causality is plausible, I think, independently of the picture of explanation I am defending. When Holmes and Watson simultaneously shoot Moriarty through the heart, it will be true that each caused him to die, though neither did anything that was in fact necessary for that effect.[2] And if Watson stands poised to poison Moriarty if Holmes fails to shoot him, it will be true to say that Holmes' pulling the trigger caused Moriarty to die, although, again, his action was not necessary to produce the effect. What may have been necessary in the circumstance was that either Holmes pull the trigger or Watson give the poison; Moriarty would not have died without one or the other of these events. But that disjunctive state—Holmes' shooting *or* Watson's poisoning—is not causally efficacious (Section 3.1). Not only are causes usually not necessary for their effects; frequently the things that are necessary are not causes.

A second clarification is required for the thesis that equilibrium explanation and causal explanation are different. What does it mean for an explanation "to say what the cause is"? I claim that some explanations do this, while others do not. Those who think that all explanation is causal explanation assert that genuine explanation always "says what the cause is." What does it take to do this?

2. The example is due to Skyrms (1980).

To understand this locution as meaning only that an explanation of an event must provide *some* information about the cause is to make the thesis I reject trivially true (on the assumption, which I am not questioning here, that every event has a cause). Indeed, it makes the thesis too true to be good. Wednesday's storm was caused by whatever caused Tuesday's barometer reading; the building's height was caused by whatever allowed it to cast a shadow of a certain length when the sun had a particular elevation. These truths fail to explain the events in question, although they provide information about the cause. The thesis that all explanation is causal explanation, just as much as the contrary thesis that I wish to defend, requires some more contentful notion of what it means "to say what the cause is."

Saying what the cause is is like *knowing where the treasure is buried.* Success or failure in such tasks depends on the context of inquiry. In one context, knowing that the treasure is on the north side of the island may count as "knowing where it is." In another, this information may be insufficient. Knowing-*wh*, that is, knowing who, or what, or why, or how or where, is a discrimination task. To know what a hawk is, you've got to know a hawk from a handsaw (or from some other contrasting alternative); to know what shinola is, you have to be able to distinguish it from some alternative substance. To know what the cause of an event is, you have to be able to say which of a set of alternative answers is correct. If the alternatives are miles apart from each other, discrimination may be easy, and knowing what the cause is will be a piece of cake. But if the alternatives considered are difficult to pry apart, knowing what the cause is may be less easy to achieve.

In evolutionary theory, the forces of mutation, migration, selection, and drift constitute causes that propel populations through a sequence of gene frequencies. To identify the cause of a population's present state, in that theory's proprietary sense of "saying what the cause is," requires describing which evolutionary forces impinged. If all explanation were causal explanation, the epistemological problems facing evolutionary theory would be especially acute. A population's past history of gene frequencies is rarely known, and the specific forces that have impinged are also hard to discover. The possibility of constructing equilibrium explanations is often what saves evolutionary theory from being incapable of providing explanations. Fortunately, the details of a population's past often do not matter to its present configuration. As in Fisher's sex-ratio argument, it often transpires that a model will imply that which causal process was at work is explanatorily irrelevant. Explanation sometimes proceeds not by identifying the cause but by obviating the need to identify it.

Causal explanation and equilibrium explanation have in common something that I can only describe vaguely. Both explain an event by showing how it is "embedded in a causal structure." They differ insofar as the former emphasizes the actual trajectory of the population whereas the latter shows how the event to be explained could have resulted in a number of possible ways.

Whether or not equilibrium explanation is acknowledged as a separate style of explanation, it seems clear from the barometer and the shadow examples that one-half of the symmetry thesis is false: The propositions that allow an event to be predicted may fail to explain why it occurs. What of the other half of the symmetry thesis? Must every explanation of an event be such that the event could have been predicted from the propositions cited? The main challenge to this thesis (Greeno 1971, Jeffrey 1971, and Salmon 1971) asserts that an event can be explained by a set of propositions even when those propositions say that the event in question was improbable.

Let's consider random genetic drift as an example. Suppose a mutant *a* gene is introduced into a population that was previously monomorphic for the allele *A*. Assume that no more mutations occur and that *a* and *A* are equal in fitness; drift will be the only force that influences gene frequencies. In this circumstance, it is overwhelmingly probable that the novel mutant will disappear because of sampling error and the population will return to its initial configuration in which *A* is universal. Still, there is a finite chance that the novel mutant will sweep through the population and go to fixation (100 percent). If this happens, is it explained by the facts just described? Explanation and prediction part ways if this is true, since the fixation of *a* could never have been predicted on the basis of those propositions that allegedly explain it.

Let's be clear on exactly what is being asked here. There is no doubt that the story about drift explains why the novel mutant *may* go to fixation. And it seems pretty clear that the story explains why the fixation of the novel mutant is improbable. Also, it is not at issue that the propositions show why novel mutants will go to fixation under random drift *occasionally*. But, in this particular population, is the fixation of the novel mutant explained by a set of propositions that imply that that event was very improbable?

Neither am I asking whether the sweep of the novel mutant to fixation *confirms* the hypothesis that it was neutral—that random genetic drift was the only evolutionary force at work. If we observed the mutant proceed smoothly to fixation, we might well prefer the hypothesis that asserts that selection for the mutant trait was the driving force. The rationale for this preference would be that the selection hypothesis makes the observation more probable; that is, this hypothesis has a higher *likelihood* relative to the observation. It is relatively uncontroversial that likelihood is one factor determining a hypothesis's overall plausibility (Hacking 1965, Edwards 1971). I draw twenty balls at random from an urn of unknown composition. You see that all twenty are green. You infer that this observation supports the hypothesis that all the balls in the urn are green better than it supports the hypothesis that only 50 percent are. The reason the former hypothesis is better supported is that it makes the observation more probable.[3]

So an observation confers greater support on the hypothesis that provides it with the higher probability. Although we will return to the idea of likelihood in a moment, it is important to see that the problem we are now considering is not one of confirmation. The problem at hand involves the status of propositions that are already thought to be adequately confirmed. Granted that a hypothesis is true; can it explain an event it says is relatively improbable? The question is not how we know that the neutrality hypothesis is true, but whether its truth would explain the mutant's going to fixation, given that the hypothesis says that this event was very improbable.

Those who have argued that an explanation of an event may say that the event was very improbable have done so mainly by appeal to plausible examples. In the famous single-slit experiment in quantum mechanics, the probabilities of a particle's hitting different locations on a detector screen, given that it passes through the slit, are predicted by the laws plus the initial conditions of the system. A particle's hitting an improbable location is explained by the same facts that explain hits at probable locations. There is nothing inherently mysterious about improbable events; they are rendered intelligible in the same way their probable counterparts are. An event is rendered comprehensible in this case by showing how it is the effect—whether probable or not—of a given causal process. Explanation need not show that the event was to be expected.

If we assume that nature is deterministic, then an explanation that confers a low probability on the event being explained will, in one sense, be *incomplete*.[4] But to say that it is incomplete is not to say that it is no explanation at all. On the other hand, once we give up the

3. The likelihood of the hypothesis should not be confused with its probability. In the circumstance described, you may have no idea how probable it is for an urn to contain only green balls. To have an opinion on this, you presumably would want to know about the manufacturing procedures of the Acme Urn Company, which has a monopoly on furnishing philosophers with visual aids.
4. But recall the argument of Section 4.3: Probabilistic explanations in a deterministic universe, although they may in one sense be incomplete, may nevertheless have the virtue of embedding particular systems in general patterns, an advantage that more fine-grained deterministic explanations may be unable to provide.

idea that nature must be deterministic, we may have to grant that even an ideally complete scientific explanation may imply that some events are improbable. In either case, an event is explained by embedding it in the causal structure that produced it; and this causal structure need not be deterministic.[5]

Philosophical discussions of the concept of explanation, whether they endorse the symmetry thesis or dissent from it, have focused on the question of whether an account is an explanation *as opposed to being completely unexplanatory*. It is plausible to grant that explanations need not say that the event being explained was probable; it is undeniable that stories like the one given above about the single-slit experiment furnish *some* understanding of why the event occurred. Yet, at the same time that philosophers have concentrated on the explanatory/unexplanatory distinction, they also have been interested in characterizing the idea of an "ideally complete" scientific explanation. It is certainly true, as Scriven (1959), Hull (1976a), and others have said, that some verbal reports that do not explicitly mention any law count as explanations. But Hempel (1965), Salmon (1971), and others have not been much troubled by this point, since they view such accounts as incomplete "explanation sketches." Here the ground has shifted from the question of whether the story is explanatory to the question of whether it is an ideally complete explanation.

These two distinctions are by no means equivalent. In the examples drawn from genetic drift and quantum mechanics, we counted stories that say that an event is improbable as explanations because they obviously offer *some* illumination of the events considered. We said this because the stories are not totally devoid of explanatory power. But this doesn't show that an ideally complete scientific explanation can say that the events it treats are improbable. Here, the question is not whether the story provides some nonzero amount of illumination but whether it measures up to some scientific ideal.

Is an explanation that says that an event is improbable "ideally complete"? Certainly, if nature is indeterministic, scientists cannot be blamed for saying that some events are inherently improbable. Such accounts may be "ideally complete" in the sense that they are the best that any scientist may ever hope to provide. But this leaves it open that our concept of explanation may label as less than ideal those accounts that say that the target event was highly improbable. It's just that in this

5. On the idea that causality does not require determinism, see Dretske and Snyder (1972), Anscombe (1971), and the probabilistic theory of causality developed in Good (1961–1962), Suppes (1970), Cartwright (1979), Skyrms (1980), and Eells and Sober (1983). This probabilistic theory will be discussed in Chapter 8.

case, we may want to blame *nature* rather than *scientists* for the imperfection.

The use of the concept of likelihood as a measure of evidential support offers some indication that our concept of explanation has this property. We regard as less than ideal those explanations of an event that say that it is improbable. The likelihood of a hypothesis, relative to an observation, provides some indication of how well the hypothesis explains the observation.[6] A deterministic explanation of an event (one that confers a probability of unity on the event) thus serves as a kind of explanatory ideal. Not that improbable events are not explained *at all*; the point is that a theory explains less fully or less well those events that it implies are improbable.

There is another sign that the power of a story to explain an event increases with the probability it confers on the event. An event is explained by showing why it, *rather than some contrasting alternative*, occurred. A priest once asked the bank robber Willie Sutton why he (Willie) robbed banks. Willie replied that that was where the money was. Evidently, the priest had a different contrasting alternative in mind; he wanted to know why Willie *robbed* rather than not. Willie took the question to be why he robbed *banks* rather than some other institution. A third question that neither had in mind was why he—Willie Sutton—was the person who robbed the banks; why didn't someone else do the stealing? Vary the contrasting alternative and you vary the explanatory problem at hand.[7]

If we allow that an event may be explained by propositions that say that it was improbable, what becomes of the idea that events are explained by contrasting them with alternatives that failed to obtain? In the example considered before, suppose we allow that the fixation of the novel mutant *a* is explained by the mechanism of random genetic drift, even though this story says that the event was very improbable. When we explain the fixation of *a*, with what does this event contrast? Presumably, we are explaining why *a*, *rather than A*, went to fixation. But how does the account explain why *a* became universal rather than *A*, since it says that the latter event was by far the more probable?

6. Some indication, but not a perfect measure. The likelihood of a hypothesis, relative to itself, is unity. But surely a hypothesis does not explain itself, let alone provide the best explanation of itself. Nor am I claiming that likelihood is the *only* factor influencing the plausibility of an explanation. As Bayesians and non-Bayesians alike have pointed out, the magical deterministic hypothesis that an event happened because God decreed that it should has a likelihood of unity. The point is that likelihood is one factor influencing the overall plausibility of an explanation.
7. The example is Garfinkel's (1981), although the point about explanation was first made, as far as I know, by Dretske (1973).

How can an explanation say why one event rather than another took place, if it would equally well explain whichever alternative came true?

The idea that likelihood is a measure of explanatory power helps provide an answer. An account explains why one event rather than another occurred only if it says that the former was more probable.[8] This leaves it open that even events said to be improbable can have their contrasting alternatives. The fact that the population began with $2N-1$ copies of the A allele and only a single copy of a and then experienced random drift explains the novel mutant's subsequent universality. It does *not* explain why a went to fixation rather than A. But it does explain why it was a rather than some third (hypothetical) allele that spread. Thank heavens that improbable events have still more improbable events with which to contrast!

Our concept of explanation does not presuppose that determinism is true. Events may be explained by fitting them into causal structures that are probabilistic in character; even improbable events may be explained. But consistent with all this is the fact that our notion of explanation is oriented toward a deterministic ideal. Although propositions may explain an event even when they say that it was highly improbable, there is nevertheless an element of truth in one-half of the symmetry thesis. An explanation need not be a prediction, but one property that makes an explanation good is that it facilitates prediction. That is, likelihood is one dimension along which explanatory power increases.

A common thread unites the present point about explanatory power and the idea developed at the beginning of this section concerning why fitness differences can be difficult to establish. Philosophers of science have often inferred that a science's inability to explain or predict a particular event must reflect some logical defect in the theories deployed. For example, if evolutionary biologists find it difficult to assign fitness values to characteristics, this must imply some imperfection in the theories they use. This same idea, turned on its head, asserts that if a theory that is beyond reproach issues in explanations of a certain sort, then such explanations cannot fall short of any explanatory ideal we might reasonably maintain. So we find it asserted that there is nothing wrong with explaining an event by showing that it was very improbable, since quantum mechanics produces explanations of just this sort. There is only one thing wrong with this point of view: it ignores the way *the world* can guarantee that explanations fail to live up to reasonable standards. Besides blaming the victim (the scientists stuck with the fact

---

8. Van Fraassen (1980, p. 128) notes that this idea was suggested in an unpublished manuscript by Bengt Hannson.

that the world has forced them to accept certain theories) and readjusting the standards of evaluation to mask the fact that there is a shortcoming, another response is possible. We may acknowledge that the world plays a crucial role in determining the extent to which the theories we accept will permit us to construct accounts that measure up to our standards of good explanation.

5.2 Variational and Developmental Explanation[9]

The idea, developed in Section 5.1, that an event is explained by showing why it rather than some contrasting alternative occurred throws an interesting light on the concept of natural selection. Darwin not only offered a new explanation of old phenomena—namely, the observed adaptedness of organisms to their environments. Additionally, he redescribed the world so that the very propositions that called for explanation were different from the ones that earlier theories had focused upon.

It is standard for a scientific innovation or revolution to change our picture of what phenomena require explanation. Typically, this transformation occurs because new empirical concepts are introduced. For example, evolutionary theory now seeks explanations for the amount of heterozygosity in a population, whereas Darwin did not. The reason is simply that we now possess a descriptive vocabulary that Darwin lacked. The explanatory innovation I want to explore in this section is different. In a sense I will try to make precise, it involves a change in *logic*, not a change in *concepts*.

As noted in Section 1.1, Darwin's was certainly not the first evolutionary theory. The revolution was not the idea of change but the mechanism that produces it.[10] Seventeenth- and eighteenth-century evolutionary theories have been described by Lovejoy (1936) as "temporalizing" the Great Chain of Being. The hierarchy from "lower" organisms to "higher" ones, with human beings at the pinnacle, was thought to correspond to a chronological pattern. The evolutionary theory associated with the name of Lamarck (1809) was in this mold (Mayr 1976a). For Lamarck, there is a predetermined path that evolution tends to pursue. Each lineage begins with very simple life forms and

9. The contrast between variational and developmental theories elaborated in this section is due to Lewontin (1983).

10. This is not to deny that part of Darwin's achievement was showing how the hypothesis of evolution could explain certain phenomena better than the doctrine of special creation. It is important to separate Darwin's defense of the hypothesis of evolution from his defense of the idea that natural selection has been the principal cause of evolution. This is done with admirable lucidity in Kitcher (1984a).

gradually gains in complexity. The evolutionary progression continually begins anew, with spontaneous generation constantly replenishing the store of primitive organisms. Within a lineage, lower rungs on the evolutionary ladder are vacated and higher rungs are eventually occupied.

This postulated progressive tendency predicts that if the whole history of life consisted in the development of a single lineage, we would not observe any diversity at all; there would be just one species. But since different lineages begin at different times, diversity is possible; different lineages, having begun at different times, have ascended to different heights. *Our* lineage must be the oldest, since human beings are the "highest" living forms.[11]

Lamarck's theory postulates a second, and secondary, evolutionary force. Besides being guided by a progressive tendency, a lineage may be influenced by "forces of circumstance." Local conditions may produce differences between populations that are at the same evolutionary level of development. This means that Lamarck did not have to arrange the entire diversity of life in a single linear ordering, as if each pair of species had to be related as more or less advanced. But since the central upward tendency was given the greater emphasis in his thinking, Lamarck was committed to thinking of the main features of morphology, physiology, and behavior as unfolding in a one-dimensional sequence.

It is significant that in his notebooks Darwin reminded himself "to never say higher and lower" (Gruber 1974, p. 74). I will not presume to say what this piece of advice meant to Darwin at the time, but its relevance to his mature theory is worth pondering. The point is not that there can be no differences in adaptedness between organisms or between species. As Ospovat (1981) has argued, Darwin had a genuine interest in and commitment to finding a measure of perfection. Rather, the idea that this remark makes salient is that the notion of a hierarchy of stages through which evolution proceeds plays no role whatever in the theory of natural selection.

Lamarck's theory was *developmental*. It explained the evolution of species by laying down a sequence of stages through which life forms are constrained to pass. Species evolve because the organisms in them are gradually modified. In contrast, Darwin's theory of the evolution

11. Pietro Corsi pointed out to me that Lamarck drew back from this explanation and eventually reached a position that was purely descriptive, in that no mechanism was suggested for generating the patterns he posited. As is usual, what comes to be known as "X's position" is, *at best*, X's position at a certain time. In the same vein, Mayr (1976a, 1982) argues that the rather standard interpretation of Lamarck as thinking that evolution is propelled by a kind of psychological striving on the part of organisms is the result of a mistranslation of "besoin."

of species is not developmental. Darwin explained change in a species by a mechanism that permits (and, in a sense to be made clear, even *requires*) *stasis* in organisms. In addition, the Darwinian paradigm views evolution as opportunistic, not preprogrammed. Selectional theories and developmental theories have fundamentally different explanatory structures.

A simple example helps bring out the contrast. You observe that all the children in a room read at the third grade level. What could be the explanation? Two strategies of analysis are possible. A developmental account will take the children one at a time and describe the earlier experiences and psychological conditions that caused each to attain that particular level of reading proficiency. These individual stories may then be *aggregated*. You may explain why all the children in the room read at the third grade level by showing why Sam, Aaron, Marisa, and Alexander each do.

A selectional explanation would proceed very differently. Suppose it were true that individuals would not be admitted to the room unless they could read at the third grade level. This would explain why all the individuals in the room read at that level. But, unlike the developmental story, the selectional account would not explain the population-level fact by aggregating individual explanations. The selectional theory explains why all the people in the room read at the third grade level, but not by showing why Sam, Aaron, Marisa, and Alexander do.

Notice that in the selectional explanation the individual children *do not change*. We may imagine that candidates for admission to the room are evaluated in an antechamber. Some read at the third grade level while others do not. A selection is then made for the ability to read at the third grade level. This was the criterion for admission, *and it is assumed that the individuals do not lose their ability to read at the third grade level as they enter*. In this account, all individuals are static, yet a selection among static individuals can produce change in the composition of a population. The change in the population is not due to the fact that the individuals in it *develop*; rather, what is crucial is that they *vary*. This is the essence of variational explanation.

Both the developmental and selectional accounts explain why all the children in the room read at the third grade level. But they do so by placing that proposition into different contrasting contexts. The developmental story says why each individual has one reading level rather than another. The selectional story, on the other hand, shows why the room is filled with individuals reading at the third grade level rather than with different people with different reading abilities. It would be misleading to describe the two explanations as both discriminating between the following alternatives: ⟨All people in the room read at the third grade level, Not all the people in the room read at the third grade level). This representation glosses over the fact that the developmental explanation construes "all the people in the room" as encompassing the same set of individuals in both contrasting propositions. The selection account, however, interprets that expression as picking out different individuals in the two contrasting alternatives.

Part of Darwin's revolution was to embed the problem of "explaining organic diversity" in a new contrastive context. Lamarck and Darwin both could have interested themselves in explaining why giraffes have long necks. Lamarck's developmental theory would have interpreted that problem as requiring that one show why existing giraffes have long necks rather than short ones. A progressive tendency would be invoked to show how individual ancestral giraffes were modified, thereby producing a change in the giraffe population. Darwin's selectional account proceeded differently. Rather than aggregating individual developmental stories into an explanation of the population-level fact, Darwin took his question to have an irreducibly population-level character. Population change isn't a consequence of individual change but of individual stasis plus individual selection. The explanatory question became one of saying why the giraffe population is composed of long-necked individuals rather than of other individuals who are not long-necked. The theory of natural selection created a new object of explanation by placing the population fact in a new contrastive context.

In the selectional explanation of why the children read at the third grade level, the individuals do not change their reading abilities as they cross the threshold into the room. Selectional explanations in a more explicitly biological context do not require this sort of absolute stasis, but they tolerate it quite comfortably. To see why, we need to examine how mortality and fertility selection require certain "stability assumptions," if selection is to produce change.

What does it take for mortality selection to produce a change in trait frequencies in a population? Suppose the transition from zygote to adult in a population has individuals with characteristic F dying twice as frequently as individuals with the alternative characteristic $not\text{-}F$. Does this mean that the frequency of F individuals will decline? For this to follow, we must make the usual *ceteris paribus* assumption: Migration exerts no decisive influence, nor does mutation, and so on. However, in addition to these, there is another assumption we need to make. Roughly, we need to assume that an F individual at one time is an F individual later on (and similarly for those who are $not\text{-}F$). Individuals need not be absolutely stable in their characteristics for selection to produce change, of course. But stability there must be;

otherwise, selection may produce no net change, even in the absence of mutation, migration, drift, and so on.

The same sort of stability assumption is more obviously required in the case of fertility selection. If F individuals have twice as many offspring as *non-F* individuals (and no other evolutionary forces impinge), this will produce an increase in the frequency of F's in the next generation only if the trait in question is *heritable*. This doesn't mean that the trait has to be "genetically encoded," in the sense described in Section 4.1 having to do with the norm of reaction. Heritability simply ensures the right sort of *resemblance* between parents and offspring.[12] Will a greater birthrate among cowboys in one generation increase the frequency of cowboys in the next? Only if there is a resemblance between the occupations of parent and offspring. This "matching" may be due to the cultural fact that parents teach their children. Heritability does not require a "gene for cowboyhood."

I have already noted (Section 1.1) that selection doesn't imply evolution. The present point goes beyond this: Selection does not imply evolution even when selection is the only evolutionary force at work. Selection implies evolution only when no evolutionary force counteracts it *and the trait being selected is heritable*.

Issues of selection and heritability are often confused. To check on the selective advantage of a trait in an evolving population is one thing; to see whether it is heritable is another. For example, if you wanted to know whether heterozygotes have a higher fertility than homozygotes, you might count up the number of offspring that heterozygotes produce per capita and compare this with the number of offspring that each of the two homozygote forms produces. In doing this calculation, it would be irrelevant what the genotypes of the offspring were. A parent who is heterozygotic may produce offspring of all three genotypes, but *all* its offspring, regardless of their genotypes, are counted as part of the parent's reproductive output.

Another way of estimating genotypic fitness values—a fallacious one—is to compare the frequency of heteroyzgotes in the parental generation with the frequency of heterozygotes among their offspring and to reason that if the trait declines it must be at a selective disad-

12. A quantitative trait like height is heritable when parents who are above or below the mean tend to produce offspring who deviate in the same direction. Imagine a "scatter diagram" of midparent heights (i.e., the average of the parents' heights) on the x axis and offspring heights on the y axis. If parents and offspring are related in the way just described, a regression line drawn through these data points will have a positive slope; indeed, the heritability (h^2) just is the slope of this regression line. For an explanation of what heritability is and how it influences the response to selection for a quantitative character, see Roughgarden (1978).

vantage. A simple example of what is wrong with this procedure is furnished by the balanced lethal system described in Section 1.4. Suppose that all homozygotes die in the passage from egg to adult stage and that all heterozygotes survive and then reproduce. The frequency of heterozygotes at the adult stage is 1.0. The frequency of heterozygotes among the next generation of offspring at the egg stage is 0.5. Does this show selection against heterozygotes? Absolutely not.

To see whether a characteristic is being selected against, we discover whether organisms with the trait on average produce fewer offspring. But to see whether the trait is heritable, we must see whether there is resemblance between the characteristics of parents and the *characteristics* of offspring. Here the traits of the offspring must obviously be taken into account.

The examples of selectional explanation (whether they involve survival and reproduction or admission into a room) have a selectional component and a stability component. What exactly does the selectional component explain? In the schoolroom example, selection does not explain why any child can read at the third grade level. The fact that an individual now in the room can do this is accounted for by the stability assumption: That individual could read at the third grade level at some earlier time, and the trait persisted. Selection is not what does the explanatory work here.

The same holds for biological examples. Natural selection does not explain why I have an opposable thumb (rather than lack one). This fact falls under the purview of the mechanism of inheritance (Cummins 1975). There are only two sorts of individual-level facts that natural selection may explain. It may account for why particular organisms survive and why they enjoy a particular degree of reproductive success. But phenotypic and genotypic properties of individuals—properties of morphology, physiology, and behavior—fall outside of natural selection's proprietary domain.

Yet, at the population level, these limitations disappear. When conjoined with assumptions about heritability (stability assumptions), natural selection may account for why 50 percent, or all, or none of the individuals in a population have opposable thumbs. The frequency of traits in a population can be explained by natural selection, even though the possession of those traits by individuals in the population cannot. This reflects the fact noted earlier that selectional explanations, unlike developmental ones, do not explain population-level facts by aggregating individual-level ones. Selection may explain why all the individuals in the room read at the third grade level, but not by showing why each individual can do so.

The alternative to developmental explanation I have labeled "variational," following Lewontin (1983). Natural selection is the obvious prototype here, but other evolutionary forces are conceptualized in the same way. Drift may be thought of as a process of "random selection."[13] Sampling error may transform a population without any of the organisms in it changing at all. If a population begins with the gene A at 99.5 percent and a at 0.5 percent, where these are equal in fitness and no other evolutionary force impinges, it is very probable that A will go to fixation. When drift modifies the composition of the population in this way, it is not because the individual organisms change but because they vary. The grip of the variational paradigm on evolutionary thinking goes deeper than the Darwinian commitment to the historical hypothesis that natural selection is the preeminent force of evolution.

Developmental theories are familiar in the human sciences, where the entities investigated are individual organisms. Piaget's (1959) theory of cognitive development and Chomsky's (1975) theory of language acquisition seek to explain changes in human beings by postulating a mechanism that transforms them. In both cases, the organism is viewed as "internally constrained." As a result, not just any sequence of stages is to be expected.[14]

Developmental stage theories are predicated on the assumption that the sequence of states an organism occupies is not the fortuitous result of the experiences that happen to impinge. Regardless of wide possible variation in the character and order of experience, the organism will change in a certain way. The idea of a developmental pathway is precisely the idea of regularly occurring changes that are insulated from environmental influence.

So the mere fact that an organism goes through a sequence of changes is no powerful argument for the possibility of a stage theory. Radical environmentalism is also consistent with the fact of change. What is important to a stage theory is the idea that the organism's internal state approximates a sufficient parameter in the theory of transformation. The state of the environment need not be entirely irrelevant, of course. But the more irrelevant it is, the more attractive this kind of developmental theory will be. Powerful endogenous constraints make change look more like an unfolding than like a buffeting.

It now will perhaps be clearer why the Darwinian view of natural selection suggests that there can be no developmental theory of phylogeny. It isn't the mere fact that natural selection is accorded *some*

13. As long as the reader bears in mind that this is contradictory when "selection" is understood in its evolutionary rather than everyday sense!

14. Cohen (1978) interprets Marx's theory of history as a developmental hypothesis, in the sense involved here. Internal developmental constraints propel a society through a series of forms of economic organization.

role in the evolution of life. Rather, it is the idea that natural selection is the overwhelmingly most powerful force of evolution that makes the prospects dim for a developmental account of the origin of species. Natural selection stands in opposition to endogenous constraints. The Darwinian view denies that the evolution of species is "preprogrammed." It admits no inherent tendency toward complexity or anything else. In contrast with Lamarckism, there is no preordained ladder of life that living forms are inherently disposed to ascend.

It is no accident that at the end of the nineteenth century, behaviorism in psychology saw Darwinian selection as a guiding metaphor (Boring 1950). Although Darwinism concerned itself mainly with "genetically encoded" characters and behaviorism focused on "acquired" ones, both doctrines viewed the engine of change as external to the object of change. Behaviorism in ontogeny and natural selection in phylogeny leave little room for developmental stage theories.

The idea of endogenous constraints on the changes a species may undergo is hardly unknown in evolutionary theory. My claim is that *natural selection* stands in opposition to this sort of mechanism. The Soviet biologist Vavilov, a victim of Lysenkoism, suggested a "law of homologous series," according to which similar patterns of evolution are to be expected in different lineages (Gould 1983). The orthogenetic theory of a number of anti-Darwinian paleontologists early in this century likewise tried to explain changes in a given species by thinking of the species as propelled through a series of stages from childhood to maturity to senescence. Lineages, it was thought, attain a kind of evolutionary momentum so that deleterious characteristics continue to develop until, like the paradigmatic horns of the Irish elk, they drag the species down to extinction (described in Gould 1977b).[15]

More recently, Gould and Lewontin (1979) have suggested that endogenous constraints may profoundly limit the power of natural selection. To result in an improved organism, natural selection must be able to tinker[16] with one part of the organism's phenotype without disrupting the rest of its organization. Lewontin (1978) has termed this requirement "quasi-independence." The idea of biological constraints asserts that the components of an organism's phenotype are so organized and intricately coordinated that it is very difficult to manipulate one piece at a time. On this view, species are pretty much locked into their phenotypes. Significant evolutionary change does not result from the

15. Kohn (1980) describes how Darwin worked with a theory of this sort before arriving at the mechanism of natural selection.
16. This happy phrase is from Jacob (1977).

slow accumulation of piecemeal selective modifications of organisms but has a quite different cause.[17]

Thus the idea of endogenous constraints is not alien to evolutionary theory, nor is it alien to the variational paradigm. Those who now argue that populations are incapable of evolving much under the pressure of individual selection are not urging a return to Lamarck's ladder of life. Darwinism asserts that populations can deploy a significant amount of variation and that selection among organisms can thereby produce a significant amount of evolution in a population. The idea of endogenous constraints is now put forward in macro-evolutionary theory not to displace the variational paradigm but to transpose it to another level of organization. The idea is that *species* are static entities, just as Darwinism believed individual organisms to be. Whereas Darwinism thought of a single population evolving on the basis of variation among organisms, the alternative idea is that an *ensemble of species* evolves in virtue of variation among species. Hypotheses of group selection and species selection, our main concern in Part II, are perfectly at home in the variational framework.

So the variational paradigm should be understood as a very general and abstract form of explanation. One version of it is the idea of Darwinian, individual selection. However, random genetic drift on the one hand and selection among populations on the other are also in the variational mold. The distinctive feature of variational explanation is a kind of antireductionism: Change in a set of objects is not accounted for in terms of changes in those objects. A population may change although it contains entities that are always static. Formulating this new kind of explanatory model involved rewriting the propositions that require explanation. An important element in this transformation involved placing the propositions to be explained in a new contrastive context.

5.3 Population Thinking and Essentialism[18]

In addition to creating a new object of explanation (Section 5.2), the theory of natural selection also transformed the status of the concept of variability. The fact of variability, both within and between species, has been recognized as something that requires explanation for as long as there has been biological theorizing. It is characteristic of that mode of thought that Mayr (1963, 1976b) has called "typological" or "es-

17. An alternative mechanism for generating large-scale diversity (which is hotly debated at present) will be discussed in Section 9.4.
18. Some of the ideas in this section are given a fuller treatment in Sober (1980).

sentialist" that variation is viewed as the result of "interfering forces" impinging on a "natural state." "Population thinking," to use Mayr's felicitous phrase for the contrasting position, accepts variation as a fact in its own right, capable of playing a causal and explanatory role as well as being something that can itself be explained. In this section, I want to characterize these alternative "paradigms" by contrasting Aristotle's way of conceptualizing variability with the one that, since Darwin, has played a fundamental role in evolutionary theory.

Mayr's description of the difference between the typological and populational viewpoints is a useful point of departure. He describes the essentialist as holding that

> [there] are a limited number of fixed, unchangeable "ideas" under-
> lying the observed variability [in nature], with the *eidos* (idea) being
> the only thing that is fixed and real, while the observed variability
> has no more reality than the shadows of an object on a cave
> wall. . . . [In contrast], the populationist stresses the uniqueness of
> everything in the organic world. . . . All organisms and organic
> phenomena are composed of unique features and can be described
> collectively only in statistical terms. Individuals, or any kind of
> organic entities, form populations of which we can determine the
> arithmetic mean and the statistics of variation. Averages are merely
> statistical abstractions, only the individuals of which the population
> are composed have reality. The ultimate conclusions of the pop-
> ulation thinker and of the typologist are precisely the opposite.
> For the typologist the type (*eidos*) is real and the variation an
> illusion, while for the populationist, the type (average) is an ab-
> straction and only the variation is real. No two ways of looking
> at nature could be more different (Mayr 1976b, pp. 27–28).

Lewontin's (1974b, p. 5) characterization of the transformation that Darwinism effected is equally suggestive:

> In the thought of the pre-Darwinians, the Platonic and Aristotelian
> notion of the "ideal" or "type" to which actual objects were im-
> perfect approximations was a central feature. Nature, and not just
> living nature, was undestood by the pre-Darwinians only in terms
> of the ideal; and the failure of individual cases to match the ideal
> was a measure of the imperfection of nature . . . Darwin rejected
> the metaphysical object and replaced it with the material one. He
> called attention to the *actual* variation among *actual* organisms as
> the most essential and illuminating fact of nature. Rather than
> regarding the variation among members of the same species as an
> annoying distraction, as a shimmering of the air that distorts our

view of the essential object, he made that variation the cornerstone of his theory.

I want to clarify a number of the ideas in these two passages. What does it mean to say that typologists thought that variation is "unreal," a "mere abstraction"? As will become clear shortly, Aristotle did not deny that variation *exists*. Similarly, when Mayr pictures the populationist as holding that individuals are real whereas averages are "mere abstractions," we must ask what it means for averages to have this characteristic. Surely, if *height* is an objective property of each organism in a population, then *average height* is an objective property of the population. If the individual facts exist independently of human thought and inquiry, so does the population fact. This indicates that if we wish to formulate the difference between typological and population thinking as an issue about what is real, then we must not equate "reality" with "existence," or with "mind-independent existence" (Sober 1980, pp. 351–352).[19]

Aristotle's science—his physics as well as his biology—is permeated by what I will call the "natural state model." Physical objects are diverse: There are heavy and light objects; ones that move violently and ones that do not move at all. What principles of order could underlie this teeming diversity? Aristotle's hypothesis was that each kind of physical object has a *natural state*. This is the state that the object will be in if it is subject to no *interfering forces*. Sublunar objects, according to Aristotle, naturally tend to occupy a certain location—the one at which the center of the earth now happens to find itself (*On the Heavens,* ii, clr, 296b and 310b, 2–5). We observe, of course, that most objects of this kind fail to be in their natural state. The reason is that interfering forces impinge. The essence of this approach is that objects of the same kind are said to share the same natural tendencies; variability results only because of interference with what is natural.

Although the word "natural" seems to have lapsed from modern physics, the natural state model has persisted. Newtonian mechanics describes the state of rest or uniform motion as the one that a body will be in if it is not subject to forces. Relativistic physics has followed suit. The general theory of relativity, for example, describes space-time as structured by a set of geodesics, which are the paths that an object will follow when no forces impinge.

The natural state model also plays a fundamental role in evolutionary theory. Indeed, the description of evolutionary theory as a theory of

19. It follows that this dispute about "realism" differs markedly from philosophical debates about the "reality" of mathematical objects, for example. The latter controversy does concern whether certain entities exist (or whether certain propositions are true) independently of human conventions, language, and inquiry.

forces offered in Chapter 1 says as much. The Hardy-Weinberg law describes what will happen to gene and genotype frequencies in a population unaffected by any evolutionary forces. Evolutionary forces may be understood partly in terms of how they "deflect" a population from Hardy-Weinberg equilibrium. So it may seem that the Aristotelian conception is alive and well in evolutionary theory. But this appearance glosses over a fundamental conceptual shift. Zero-force states in evolutionary theory are states of *populations*, not organisms. Aristotle's natural state model, on the other hand, was intended to explain variation among *organisms*. It is this individual-level model that has lapsed from scientific discourse.

How did Aristotle characterize an organism's natural state and the interfering forces that may affect it? In *De Anima* (415a26) we find it laid down that

> [for] any living thing that has reached its normal development and which is unmutilated, and whose mode of generation is not spontaneous, the most natural act is the production of another like itself, an animal producing an animal, a plant a plant.

So aside from spontaneous generation, which obeys a different set of laws, fidelity of replication is an organism's natural state.

Aristotle's notion of natural reproductive tendencies is inseparable from his account of sexual reproduction. The main idea is that the male semen provides a set of instructions that dictates how the female matter is to be shaped into an organism.[20] Abnormality in the parents, or the fact that the parents are from different species, or trauma during fetal development may all prevent the instructions from being executed perfectly. They may, in the Aristotelian phrase, prevent the "form from completely mastering the matter." Mules (sterile hybrids) are a case in point. In fact, the existence of females also counts as a deviation from the natural state, since Aristotle held that interference-free reproduction will generate an offspring that perfectly resembles the father—i.e., a *son* (*Generation of Animals*, ii, 732a; ii, 3, 737a27; iv, 3, 767b8; iv, 6, 775a15). A further, fascinating, aspect of Aristotle's position, which I will not explore here, is that he seems to have thought that there are entire species, such as seals and lobsters, that count as monstrosities. The concept of monsters (*terrata*) was evidently not limited to an intraspecific standard of normalcy.[21]

20. Max Delbrück (1971) suggested that Aristotle's idea of semen as an information-bearing entity entitles him to a posthumous Nobel Prize for discovering DNA.
21. I have been greatly aided in this brief discussion of Aristotle's views on monsters by Preuss (1975) and Furth (1975).

Where has this amazing position gone wrong? The problem isn't the implication that natural states are rarely, if ever, occupied—that almost everything under the sun counts as a monstrosity. It is no criticism of Newtonian mechanics that objects always are acted on by some force or other and that they rarely occupy the zero-force state of rest or uniform motion. Natural states need not be statistically "normal." Rather, the problem is much more fundamental: We no longer conceptualize population variability in terms of "deviation from type." And even in the case of individual, organismic development, the idea of a natural developmental pathway has also disappeared from view.

The natural state model interprets homogeneity in a poulation as natural and variation as the result of interference. A type is postulated for all organisms of a given kind, and deviation from that type requires special explanation. Evolutionary theory, in a sense, inverts this relationship. Genetic variation in a population requires no special evolutionary forces. Even a population that arises from a single mating pair will have a great deal of genetic variability. The Mendelian mechanism can "naturally" produce offspring that are genetically diverse, even if the parents are genetically identical. Furthermore, once a population has some genetic diversity, it will persist that way if no evolutionary forces impinge. This is part of the point of the Hardy-Weinberg law. It takes no special evolutionary forces to create and maintain variability.

Uniformity, however, takes some work. Natural selection is one mechanism that can *destroy* variation. For it to act at all, there must exist variation (in fitness). But once a selection process begins, it gradually destroys the conditions needed for its continued operation. Selection eliminates variation in fitness, and thereby brings itself to a halt. This is not to say that selection always tends to eliminate *phenotypic* or *genotypic* variation, although this, too, is frequently the case.[22] Here we see the natural state model stood on its head: Variability is "natural," whereas uniformity is the result of "interfering forces."[23]

22. Recall that in Fisher's sex-ratio argument (Section 1.5), the equilibrium at which each son and daughter has an equal chance of producing offspring is the 1:1 ratio. There still is *variation* (in sex) here, but no variation in *fitness*. The same holds true for the model of heterozygote superiority described in Section 1.4: heterosis is a mechanism that preserves *genetic* variability in a population, even though it destroys variation in the *fitnesses* of genes.

23. A properly historical understanding of the replacement of the natural state model would require close examination of the development of statistical ideas in the nineteenth century. The view of variation in the writings of Laplace and Quetelet was still wedded to the idea of "deviation from type" (see, in particular, Hilts 1973 and, for further references, Sober 1980), an interpretation that seemed gratuitous by the end of the century. Quetelet, for example, argued that variation in a population—say in the girths of Scottish soldiers—results from various interferences deflecting a single developmental tendency, one that would show itself in investigations of The Average Man. It is interesting

The natural state model also has been discarded in the way evolutionary theory conceptualizes the development of organisms from genotype to phenotype. A fertilized egg, according to the Aristotelian view, naturally tends toward one possible phenotypic outcome, which may be deflected by various interfering forces. But the idea of a *privileged* phenotypic state has lapsed from the theory of the relation of genotype to phenotype. All possible phenotypes of a genotype are "natural," since all are possible. This radically non-Aristotelian idea is codified in the genotype's *norm of reaction* (Lewontin 1977), which I discussed in Section 4.1. The norm of reaction shows the different phenotypes a given genotype will produce in different environments. So, for example, we might graph the height that a corn plant with a particular genotype will attain as a function of how much water or nutrition or sunlight it receives. There is no such thing as its "natural" height. We may prefer a taller plant to a shorter one, and natural selection may, too. But the norm of reaction marks no such distinction. The idea that one phenotype will result if no interfering forces impinge while other phenotypes will result from interferences has gone up in smoke.[24]

The perspective on development offered by the norm of reaction cuts very deep. Although we may smile smugly at Aristotle's remarks about reproduction, the fact of the matter is that our commonsense ideas about development are very much rooted in the natural state model. When two human beings conceive a child, we tend to think of the fertilized egg as having certain capacities. Perhaps the child will be extremely good at mathematics, or musical, or sensitive to the feelings of others. Perhaps it will have none of these attributes. Given the child's genotype, which of these various possible outcomes occurs depends on the environment. Yet, despite this range of possibilities, we may tend to think that one developmental outcome is "natural" while the others are results of interference. Malnutrition while in the womb may reduce the child's intelligence; a well-nourished child may have its intelligence boosted by various enriching experiences. But through the complexity of these thousands of variables, we may think we discern

---

that what eventually came to be called the "normal curve" originally was known as "the law of errors." Here is one case in which ontology evidently recapitulates philology (Quine 1960, p. vii).

24. In Sober (1980), I suggested that the idea of a "privileged" (natural) phenotype of a genotype in the natural state model is like the idea of a "privileged" temporal interval between two events in classical physics. Just as the norm of reaction confers on each possible phenotype of a genotype an equal claim to being "natural," so special relativity shows that temporal intervals are relative to choices of inertial rest frames, none of which is privileged.

the child's "natural" level of intelligence—what its intelligence would be if it were neither "artificially boosted nor diminished."

It is as if common sense leads us to think of each human genotype as a bucket that has a certain "natural capacity" (Lewontin 1977). The bucket may be underfilled or overstuffed. But in addition to these two sorts of unnatural outcomes, each genotype has a phenotypic score associated with it that is its natural state. The norm of reaction, on the other hand, makes this whole system of thinking look like a delusion. The idea of a particular human being's having a "natural" level of intelligence makes no more sense than the idea of a particular corn plant's having a "natural" height.

The idea that each human being has a "potential" that he or she may realize to varying degrees has played an important role in a number of ethical and political issues, sometimes with good consequences, sometimes with bad. Oppression of various kinds looks like an underfilled bucket, unmerited privilege like an overstuffed one. For example, providing special educational opportunities for disadvantaged children may seem imperative, if the children have not reached their "full potential." But if we think of the children as already occupying their natural state, it may seem less pressing to provide them with an "artificial boost." If each individual had a natural phenotype, these normative judgments could perhaps be grounded in biology. But we must look elsewhere to find a basis for these discriminations, since the Aristotelian foundations of this idea are no longer available (Lewontin 1977).

As noted earlier, evolutionary theory has not abandoned the idea that a theory of forces begins with the identification of a zero-force state. But the theory treats *variation* as a state that requires no evolutionary forces to be brought into being;[25] in addition, the generalizations of evolutionary theory apply to population configurations, not organismic ones. Aristotle's natural state model differs on both counts, viewing uniformity as a zero-force state and the laws of variation as applying to *individuals*. The abstract structure of a theory of forces has been preserved in this transformation, although its content has been radically altered.

In addition to characterizing the idea of "interference-free reproduction," the Aristotelian natural state model also was used to characterize the notion of *species*. Each species has a unique natural state associated with it, i.e., a state that all members of the species would have if no interferences impinged. Philosophers should beware of in-

---

25. By this I mean that a single mating pair can "naturally" produce a variable population; I'm not suggesting, of course, that this fact has no evolutionary explanation.

terpreting this Aristotelian thesis so that it comes out trivially true. It *is* a necessary truth that an organism is a member of *Homo sapiens* if and only if it is a member of *Homo sapiens*. But this trivial assertion hardly vindicates the Aristotelian position. Nor does the requirement that the species definition be empirical suffice to display its intended content. If it is an empirical discovery that perpetual-motion machines are impossible, then it is a necessary empirical truth that an organism is a member of *Homo sapiens* if and only if it is a member of *Homo sapiens* or a perpetual-motion machine.

The essences posited by Aristotelian essentialism were causal mechanisms of a certain sort. The Aristotelian judgment that "man" is rational and contemplative by nature should not be understood as a piece of blind optimism about the actual state of development of every member of the species; rather, it is a claim about what human life would be like if various distortions were removed. This species concept is defective not because it rules out the possibility of variation but because it is predicated on a mistaken theory of the mechanism of variation. The theory tolerates the idea that human beings may vary in their degree of rationality. What it does not tolerate is the thought that it is meaningless to view human beings as sharing the same "natural tendency."

This discussion casts in a new light a familiar picture of how Darwinian theory undermined essentialism in biology, one energetically developed by David Hull (1965). Darwin thought of speciation as the result of a slow accumulation of modifications within a lineage. Phyletic gradualism, as it is now called, presents obvious line-drawing problems. Marking off where one species ends and another begins will be rather like determining the number of hairs that marks the difference between being bald and not. According to Hull, the Aristotelian idea that species may be informatively characterized in terms of necessary and sufficient conditions set biologists on a two-thousand-year wild-goose chase.[26] Aristotle thought of species as discrete, whereas Darwinian theory sees evolution as continuous.

I do not wish to contest the historical accuracy of this claim about where Aristotelian biologists looked, and what they failed to find. And even if Aristotle's species concept were more flexible on the subject of continuity, this would not settle the historical question of how his successors regarded him.[27] Even so, it is worth noticing that essentialism

26. The title of Hull's (1965) paper is "The effect of essentialism on taxonomy: 2000 years of stasis."
27. The following passage from *History of Animals* (5888b4 ff.) is worth considering in this connection: "Nature proceeds little by little from inanimate things to living creatures, in such a way that we are unable, in the continuous sequence to determine the boundary line between them or to say on which side an intermediate kind falls. Next, after inanimate

conflicts with evolutionary theory for reasons additional to whatever commitments essentialism may have to discontinuity.

The idea that each species has a natural state is quite consistent with the existence of phenotypic continuity between species. Two species may have different natural states, and yet a constellation of interfering forces may create variability within and continuity between them. The natural state model doesn't run afoul of the fact of variability but errs in the sort of mechanism postulated to account for it.

There is another reason to think that the clash between essentialism and evolutionary theory goes beyond the issue of gradualism. The reason is that saltational theories of evolution (i.e., ones postulating discontinuities between species) have been and continue to be critically debated *within* the evolutionary framework. Goldschmidt (1940) believed that speciation proceeds via the creation of a "hopeful monster"— a macro-mutation that creates a new living form in a single generation. His ideas were rejected by the modern synthesis, which was considerably more gradualist in orientation. Still, it is not controversial now that at least one sort of speciation occurs virtually instantaneously. This is the process in plants of *polyploidy*. By a "mistake" of meiosis, an offspring plant may contain some multiple of the number of chromosomes found in the parent. In a single generation, a viable organism is brought into being that is reproductively isolated from the parental species. When this event occurs in several plants, the polyploids may form their own breeding population and thereby count as a full-fledged sexual species.

More recently, Eldredge and Gould (1973) have argued that speciation proceeds by a process of "punctuated equilibria." According to this idea, speciation takes place very quickly compared with the average lifespan of a species. However, unlike Goldschmidt, advocates of the punctuated equilibrium view do not conceive of speciation as a one-generation affair. Rather, if a species lasts ten million years (which is frequently the case), then the speciation event that gave birth to it will be termed "geologically instantaneous" if it takes only 1 percent of that length of time. But this allows the process to go for 100,000 years! Nor is it unprecedented to suggest that evolutionary rates may vary in this way. Before discussion of "punctuated equilibria," Simpson (1953) took seriously the idea of uneven evolutionary rates and Mayr

---

things come the plants: and among the plants there are differences between one kind and another in the extent to which they seem to share in life, and the whole genus of plants appears to be alive when compared with other objects, but seems lifeless when compared with animals. The transition from them to the animals is a continuous one, as remarked before. For with some kinds of things found in the sea one would be at a loss to tell whether they are animals or plants."

(1963) proposed that "genetic revolutions" may be the sum and substance of speciation.[28]

Aristotle would have little to fear from contemporary biology if the only problem for essentialism were a model of speciation according to which evolutionary rates are perfectly constant. A modification of theory that could be accommodated comfortably within the evolutionary point of view would then seem to make the world safe for essentialism. But it is important to see that the existence of de facto phenotypic discontinuities between species would not be enough to reinstate the natural state model. Even if all the members of a species happened to share some phenotypic characteristic unique to them, this would not show that the trait in question was the "natural outcome" toward which the development of each member of the species was directed. The evolutionary theory of forces applies to populations separated by distinct phenotypic discontinuities just as much as it applies to populations that blend into one another. The Aristotelian distinctions find no articulation in this theory.

Philosophers have quite naturally wanted to consider versions of essentialism beyond those envisioned by the natural state model. A paradigm case has been the periodic table of elements, which seems to tell us that the essence of each element is its atomic number (Hull 1965). It isn't that nitrogen atoms have a "natural tendency" to attain an atomic number of 14, which may be frustrated by the impingement of various "interfering forces." Rather, the idea is that the atomic number is a characteristic that all and only the nitrogen atoms share, and that an atom *must* have this atomic number if it is to be nitrogen. Essentialism of this sort holds that it is no accident that the property of being an atom of nitrogen and that of having a given atomic number go together; indeed, they *must* covary, since the atomic number is the nature or essence of nitrogen (Enc 1975).[29] As before, we want to require that essences are necessary properties that play a certain causal (and hence explanatory) role. The fact that nitrogen has the essence it does is supposed to be an empirical discovery, not a piece of semantic legis-

---

28. To my mind, the real novelty of the punctuated equilibrium idea is not the claim that rates have varied but the explanation offered for that fact. This will be discussed in Section 9.4.

29. Apparently, there can be line-drawing problems for the concept of atomic number just as surely as there can be line-drawing problems for the species concept. How close to each other do some protons have to be to count as a single nucleus? Even if this answer had a determinate answer in classical physics, it may turn out to lack one in quantum mechanics, if the latter theory prohibits us from assuming that distances between subatomic particles have sharp values. It is interesting that this line-drawing problem seems *not* to occasion any conceptual problems for the notion of a chemical element. What this shows, I suspect, is that we want our concepts to be precise *enough*; there is no need for them to be *perfectly* precise.

lation. That an atom possesses a given atomic number *causes* it to have numerous other properties. Again, essentialism is a doctrine about causal mechanisms. Is there room in evolutionary theory for species essences in this sense?

A species essence must now be a property that *all* members of the species have; it can't be something that the species initially lacked and then acquired. Consider the parallel with nitrogen: Perhaps the essence of nitrogen is that all nitrogen atoms have atomic number 14. If so, one cannot explain why all and only the nitrogen atoms have that atomic number by saying that first there were some nitrogen atoms, then they were subject to a set of forces, and this is what endowed them with atomic number 14.[30] The strategy of explanation pursued in evolutionary theory suggests, I think, that no genetic or phenotypic characteristics are treated there as essential. Think of a species that came into existence by polyploidy. Is the species' multiple complement of chromosomes a species essence? An individual could still be a member of the species and not have this genetic structure; it would simply be "abnormal." Similarly for local populations; a deme might lack the requisite trait and still belong to the species in question.

This is not the place to undertake a full-scale positive inquiry into what view of species is appropriate in evolutionary theory. This book focuses on the ideas of fitness, selection, and adaptation, not on the species concept. I will mention, however, one suggestion that is in harmony with the negative comments just made about essentialism. If no phenotypic or genotypic characteristic of an organism can count as a species essence, what is it that makes an organism a member of one species rather than another? Rather than look at the structural properties of organisms, perhaps we should attend to the historical relationships of one organism to another. The view that species are "individuals," developed by Ghiselin (1974b) and Hull (1976b, 1978), suggests that organisms are conspecific in virtue of their actual historical relationships to each other, not in virtue of their degree of genotypic or phenotypic similarity. Species, from this perspective, are chunks of the genealogical nexus, not natural kinds.[31]

The alternative that Mayr contrasts favorably with "essentialism" is "population thinking." It is hard to get a precise fix on what Mayr takes to be distinctive of this latter point of view. At times, it seems

30. It is hard to see what *would* explain why all nitrogen has atomic number 14. Perhaps this is a proposition that can be *justified* but not explained. If so, there are some *empirical* propositions that are scientifically inexplicable, a point made in a slightly different context by Causey (1972).

31. See Kitcher (1984b) for a dissenting view, and Sober (1984d) for a dissent from his dissent.

to have a decidedly individualist cast. In the passage quoted earlier, Mayr pictures the populationist as holding that individual organisms are "real" whereas averages are "abstractions." Yet he also describes this orientation as holding that *variation* is real. But the properties used in population biology to describe variation are properties of *populations*. If population thinking amounts to the idea that a particular level of organization is "real," which level is it to be—the organism or the population?

Additionally, the word "real" remains mysterious. I noted at the outset that the issue about the reality of individuals or of averages should not be understood as the issue of whether they *exist*. When I turn to the questions surrounding the units of selection controversy in Part II, it will emerge that the realism question is best seen as a question about *causality*. There is an important sense in which the Darwinian theory of natural selection views properties of individuals as real (i.e., causally efficacious) and populational properties (like averages) as artifacts (i.e., effects). But for now, I want to set questions of causality to one side and describe how population thinking puts an emphasis on populations that is wholly different from that found in the essentialist's natural state model. At the end, I'll return to the question of causality and briefly describe what I take to be the relationship of Darwinism and population thinking.

The natural state model explains variability in a population by describing organisms one at a time in terms of their natural states and the interfering forces that affect them. Evolutionary theory describes a population's properties as the upshot of the earlier properties of that *population* and the forces that affect *it*.[32] One of the achievements of evolutionary theory was to make autonomous population-level description a scientific possibility.

Intuitively speaking, two strategies seem capable of ensuring the soundness of a scientific concept. One can legitimize a concept (1) by displaying its connections with concepts that apply at a lower level of organization or (2) by showing how it is connected with concepts that apply at the same level. Call these the vertical and horizontal methods, respectively. The standard approach to the species concept is vertical. When scientists ask for a clarification of this concept, they often want to be shown a criterion that specifies when *organisms* (or *local* populations) are members of the same *species*. A criterion of this sort would explain a higher-level concept (species) by connecting it with a lower-level one (organism or local population).

---

32. Recall the observation in Section 4.2 that the difference between deterministic and stochastic process in evolutionary theory is made at the level of populations, not organisms.

Science never has pursued the vertical strategy for all of its concepts. The vertical strategy requires that properties of wholes be understood in terms of properties of parts. Presumably, science at any one time recognizes only finitely many levels of organization. This means that at any one time there must be a lowest level at which the vertical requirement cannot be met.[33] What is more, science does not resolutely pursue this vertical strategy until rock bottom is reached. Although it has been thought important to define the species concept in terms of concepts applying to organisms, biologists have rarely, if ever, demanded that the concept of an organism be clarified in terms of the concept of a cell. The point is not that the idea of an organism is perfectly clear but that it has seemed clear *enough*.

Population biology has made it possible for population-level concepts to be treated in the same way. One talks of a population's being K-selected, or predatory, or assortatively mating. It is simply not required that such descriptions be grounded vertically. What must each organism in a predator population do for the population to count as predatory? What does each homozygote in a population have to be like for the homozygote genotype to have a given fitness value? There is no such organismic requirement. In population biology, concepts are legitimized horizontally, not vertically. One elucidates them by displaying their connections with other population-level concepts.

The vertical requirement is a kind of methodological reductionism. It says that population concepts are to be clarified in terms of organismic ones. Population thinking, in legitimizing the horizontal strategy, constitutes an alternative to this sort of reductionism.[34]

The treatment of population properties as autonomous objects of inquiry is now characteristic of evolutionary model building. A striking formulation of this methodological strategy may be found in the work of one of population biology's founders. All his life, R. A. Fisher saw a fruitful analogy between gene frequencies in a Mendelian population and molecules in motion in a chamber of gas (Provine 1984). The detailed life histories of this organism or that molecule do not materially affect the macro-description of the population or the gas. As early as 1922, Fisher wrote:

> The investigation of natural selection may be compared to the analytic treatment of the Theory of Gases, in which it is possible to make the most varied assumptions as to the accidental circum-

33. Kripke (1978) has argued that it is impossible for each commonsense concept to have a "criterion of individuation" for this very reason.
34. "Reductionism" is a term used for a variety of doctrines; this connection with population thinking reflects just one of them.

stances, and even the essential nature of the individual molecules, and yet to develop the general laws as to the behavior of gases, leaving but a few fundamental constants to be determined by experiment (Fisher 1922, pp. 321–322).

In the same vein, Fisher (1930) compared his fundamental theorem of natural selection (to be discussed in Section 6.1) with the second law of thermodynamics: Chance differences among individuals (molecules or organisms) even themselves out, leading the population to increase its value of a certain quantity (entropy or fitness).

Population thinking was the product of a century of work in which chance was understood by *macro*-analysis (Hilts 1973, Hacking 1975, Schweber 1982). An older and productive reductionist paradigm had allowed scientists to understand the properties of an object by decomposing it into parts. But probabilistic properties were rendered intelligible by other means. Evolutionary theory understood fitness, mutation, and migration—although they are properties of organisms—in terms of their roles in the dynamics of *populations*. Not that the reductionist perspective has been discarded, as the continuing successes of molecular biology attest. The conceptual change was one of supplementation rather than replacement.

Wholes are made of parts. However, this fact about the world does not dictate a strategy of investigation. Wholes may be understood by decomposing them into parts, but so may parts be understood by situating them in wholes (Wimsatt 1980). Population thinking differs from essentialism by shifting the level of theoretical description from the individual organism to a higher level of organization.

The heart of population thinking, then, is a commitment to the methodological fruitfulness of constructing theories whose parameters apply to *populations*. It is not, I would suggest, best understood as an ontological thesis about the reality—existence or causal role—of much of anything. Darwin's stress on individual selection can be construed as holding that variation among organisms is the cause of variation among populations (Lewontin 1974b, p. 4). Individual facts are causes; populational facts are effects. If population thinking and Darwinism were one and the same thing, there would be no harm in construing population thinking as holding that only individuals are "real." But population thinking is a more general perspective; it is so fundamental that it is held in common by evolutionists who disagree over the importance of individual selection. So, for example, hypotheses of group selection are examples of population thinking *par excellence*. According to such hypotheses, it is group properties that are causally efficacious, and the properties of organisms are their effects. This is a dispute over the level of organization at which the causal action takes place. Although population thinking began with the idea that individual facts are causally fundamental, it has evolved into a polymorphic point of view. Darwinism was important in the genesis of this conceptual framework; but Darwinism was only the beginning.

Chapter 6

Adaptation

The aptness of organisms—the exquisite fit of organisms to their environments—is one of the central phenomena that the theory of evolution by natural selection attempts to explain. The word "adaptation" ambiguously denotes the process of achieving this goodness of fit and also the end state of its achievement. In this chapter, I try to separate some of the conceptual threads that have entangled the apparently interchangeable concepts of adaptedness, adaptation, and adapting. As in the case of fitness and natural selection (Chapter 3), the equivalence of these ideas is more apparent than real.

In Section 6.1, I attempt to formulate and answer a question concerning the circumstances in which natural selection will improve the level of fitness found in a population. Although evolutionists often remark that their theory provides no scope for the concept of Progress, there is an interesting sense in which natural selection can be expected to be progressive. These reflections will lead to a consideration of the scope and limits of Fisher's (1930) so-called fundamental theorem of natural selection. Section 6.1 also includes a historical reflection on the role of economic ideas—due to Malthus on the one hand and the Scottish economic tradition on the other—in Darwin's thinking.

In Section 6.2, I consider another connection between the concepts of adaptation and natural selection. A number of contemporary theorists (e.g., Williams 1966, Lewontin 1978, and Gould and Vrba 1982) have stressed the importance of not confusing adaptations with *fortuitous benefits*. The latter may occur by accident, whereas the former, they insist, must have evolved by a process of natural selection. Besides making this claim precise, I also will consider its converse: If a trait evolved because there was selection for it, does this suffice for the trait to count as an adaptation? Some of the same theorists who have argued that adaptation implies selection have denied that selection implies adaptation. I will take issue with their arguments and thereby try to nail down a necessary and sufficient condition for a characteristic's being an adaptation.

6.1 Selection and Improvement

Darwin thought of organisms as being modified by their *local* environments; his theory of natural selection gives no role to the Lamarckian idea, discussed in Section 5.2, that evolution is driven by some central force that tends in all populations to produce a single sort of progressive change. Natural selection predicts a bush rather than a ladder. Opportunistic populations evolve in the various directions that environments fortuitously make available; they do not unfold in accordance with some internal dynamic.[1]

Nevertheless, it would be misleading to think of Darwin's theory or of current evolutionary biology as totally proscribing the idea that natural selection tends to increase some measurable property of organisms. There are at least two reasons why. First, although the theory of natural selection does not *demand* that selection always produce an increase in complexity, or body size, or amount of genetic information, it is *compatible* with the theory that this should be true. Darwin himself was very interested in finding a way to characterize progress (Ospovat 1981, pp. 210–228), but reflection on some obvious possibilities discouraged him from thinking that this line of inquiry would bear fruit. Increasing complexity, for example, has the evolution of parasites as an obvious counterexample. Gould (1977a) has interpreted the advice that Darwin gave himself in his notebooks "to never say higher and lower" as indicating that this distinction is rendered incoherent by the theory of evolution by natural selection. However, I think that Darwin's attitude was one of circumspection, not rejection, as the following passage from *The Origin of Species* suggests:

> There has been much discussion whether recent forms are more highly developed than ancient. I will not here enter on this subject, for naturalists have not as yet defined to each other's satisfaction what is meant by high and low forms. But in one particular sense the more recent forms must, on my theory, be higher than the more ancient; for each new species is formed by having had some advantage in the struggle for life over other and preceding forms. If under a nearly similar climate, the eocene inhabitants of one quarter of the world were put into competition with the existing inhabitants of the same or some other quarter, the eocene fauna or flora would certainly be beaten and exterminated; as would a secondary fauna by an eocene, and a palaeozoic fauna by a secondary fauna. I do not doubt that this process of improvement

1. The "opportunism" of natural selection was stressed by Simpson (1953) and, more recently, by Gould (1977a).

has affected in a marked and sensible manner the organisation of the more recent and victorious forms of life, in comparison with the ancient and beaten forms; but I can see no way of testing this sort of progress (Darwin 1859, pp. 336–337).

Although Darwin is cautious about the idea of progress, he is, in this passage, not cautious enough. It is a mistake to think of fitness relations as necessarily transitive. Pairwise competition between A, B, and C may be such that A is fitter than B, B is fitter than C, and C is fitter than A. Were the three types all present in a single panmictic population in which individuals of different types randomly interact with each other, relative fitness might be transitive. But when A and B are present in one population, and B and C in another, the relationships in these two cases do not settle what will happen in a third—namely, when A and C are the only competing types. Levene, Pavlovsky, and Dobzhansky (1954) provided a real-life example involving competition among *Drosophila* in laboratory population cages in which transitivity fails. This sort of result is really not very surprising; it is no more so than the nontransitivity of being better at chess or tennis (or of winning moves in the children's game rock-paper-scissors). Failure of transitivity simply reflects the fact that the composition of a population is itself a determiner of the fitness values of organisms in the population. Darwin's "particular sense" in which evolution by natural selection must be progressive follows neither from his own theory nor from contemporary ideas.[2]

Darwin's interest in discovering a way in which natural selection must be progressive has survived in contemporary theorizing. The idea that evolution by natural selection should increase the amount of information an organism encodes has been popular. However, if this quantity is equated with the amount of DNA an organism possesses, the proposal appears to have some jarring consequences. If a definition of progress is to place human beings at the pinnacle of evolution, then this proposed measure of it is unsatisfactory. Squid, for example, have more DNA content than any mammal (Williams 1966, p. 39).

There is a second reason that it is misleading to banish the idea of progress from evolutionary theory. True, the *supervenience* of fitness implies that no single *physical* quantity will always increase as a result of natural selection. And the *opportunistic* character of selection suggests

2. Of course, if we consider only single, pairwise competitions, then Darwin's "rematch criterion" of progress is unobjectionable: If A supplanted B because the former was the fitter, we might expect a rematch under identical conditions to produce the same result. It is obscure why one should wonder about the possibility of testing this criterion of progress.

that there may be no interesting *abstract* quantity (such as an increase in genetic information) that allows one to characterize evolution by natural selection as "progressive." Still, there is the concept of fitness itself. If the fitter supplant the less fit, doesn't it follow automatically that selection is a device for improving fitness? It is possible to think that selection is progressive in this sense, even after one has rejected an overarching Lamarckian idea of Progress with a capital P. We will see in this section that this idea does have some justification in contemporary theory, though one that is much more limited than is often supposed.

Let us make the question at issue more precise. A population evolving under the influence of natural selection may be thought of as "tracking" its environment (Lewontin 1978). Selection modifies the distribution of traits that determine how well organisms fit into the world in which they live. But the information the population obtains is always somewhat out of date. As the population is modified in response to a selection pressure created by the environment, the environment may change. So the population may be thought of as adapted to a past environment, rather than to the present one.

When the environment changes slowly, this time lag may be minimal or nonexistent. But when the environment changes quickly, the evolving population may lag farther and farther behind the environment it is "attempting" to track.[3] In the latter case, the process of natural selection will not improve the fit of organism to environment. The environment must be reasonably stable, if natural selection is to be an improver.

Other requirements must be met before natural selection can be relied upon to increase the level of adaptedness.[4] For example, one cannot simply assume that a trait that would be advantageous will necessarily evolve. There is no guarantee that the advantageous phenotype will appear in the population. If it does, there is no guarantee that it will be heritable. And even if these conditions are met, there is no guarantee that it will not be lost in virtue of the activity of other evolutionary forces. For example, genetic drift may confound the de-

---

3. Van Valen (1973) has advanced what is in effect an intermediate conjecture—the so-called Red Queen hypothesis—according to which the environment changes just fast enough to prevent organisms from gaining on it, while adaptation serves to prevent the population from falling too far behind. The hypothesis is named for the Red Queen in Lewis Carroll's *Through the Looking Glass*, who had to move as fast as possible just to stay in the same place. See also Fisher's (1930, pp. 44–45) discussion of the "deterioration of the environment."

4. I use the term "adaptedness" interchangeably with "fitness." The level of adaptedness of an organism is simply how fit it is. The term "adaptation," however, requires some special attention, to be provided in Section 6.2.

terministic influence of natural selection. And even in an infinite population, a favorable (point) mutation may still be lost by accident in reproduction. The chance of its traversing the path from introduction to fixation is only twice its selection coefficient.[5] Since selective values frequently are on the order of fractions of a percent, the probability of fixation is quite small. Still, if mutants of the advantageous type are introduced again and again into the population, chances are that sooner or later one will go to fixation. But this, again, requires that selection continue to favor the characteristic; the environment must remain constant, if the favorable type is to be given sufficiently many opportunities of going to fixation, so that it is likely that sooner or later this will occur.

The prediction that enough monkeys equipped with enough typewriters will eventually type out *Hamlet* is true enough, as long as we don't forget the importance of the parameters involved. We may think of the constraints listed above in the same way: natural selection will be able to increase adaptedness only if certain conditions obtain.

Another clarification of the idea that natural selection is an improver is required. This is the idea of trade-offs. Darwin recognized that a characteristic may have multiple effects on fitness (Ghiselin 1969, chap. 9). The large horns on a stag may be advantageous if females prefer to mate with such males. On the other hand, large horns may be somewhat deleterious when it comes to escaping from predators. It would be absurd to assume that natural selection must perfect a characteristic relative to each of its functions. Rather, the idea is that large horns will be selectively advantageous if the *overall* benefits outweigh the costs, compared with the other characteristics that happen to be present in the population.

A related point is that natural selection does not guarantee that the adaptations found in nature are perfect. Darwin (1859) concluded *The Origin of Species* by pointing out that the theory of natural selection makes the existence of *imperfect*, though serviceable, adaptations comprehensible in a way that the doctrine of special, divine creation does not. Gould (1980d) has recently offered the panda's "thumb" as a nice example of this kind. Pandas are part of the order *Carnivora* which, as a group, does not have an opposable first digit. For a complex set of reasons, pandas have evolved to a vegetarian diet, subsisting almost entirely on bamboo. The animal strips the bamboo by running branches between its wrist and a spur of bone that looks like a thumb until the anatomy is examined more closely. The bone is basically a bulge in the panda's wrist. If God had designed pandas, he presumably would

5. A result due to Haldane (1932); see Crow and Kimura (1970), pp. 418–422 for discussion.

have given them a more efficient way of preparing bamboo shoots. The method is clumsy but works tolerably well. Herbert Spencer's slogan of "the survival of the fittest" would have been less misleading, although less striking, if it had been "the survival of the fitter."

Having said all this, I now can isolate the main question of interest. Granted, *other* evolutionary forces may block natural selection from improving the level of adaptedness found in a population. Furthermore, the overall level of adaptedness of an organism is itself a compromise between potentially conflicting design features. But imagine a population that exists in a constant physical environment that is subject to no forces besides selection. In this circumstance, can selection be counted upon to improve the level of overall fitness in the population? Here I am asking a question about the nature of selection, considered in itself.

We must be clear on what a "constant environment" means here. If natural selection is to improve the level of adaptedness, it must change the composition of the population. If "constant environment" is taken to imply that the frequencies of traits in the population do not change, then the question we are asking has a trivial answer: natural selection cannot improve anything unless it changes *something*. So the question we want to ask is whether natural selection will improve adaptedness, when the only change that occurs is the change in trait frequencies that natural selection itself produces.

I will approach this issue first by considering two traits (X and Y). The fitness of each characteristic (i.e., the average fitness of the organisms possessing it) may be represented on a graph that shows how it is influenced by the frequencies of the two traits in the population. Although this device may be used to represent differences in viabilities or fertilities, I will interpret the fitness values as taking account of both: An organism's expected number of offspring reflects its probability of reaching reproductive age as well as its prospects for reproductive success. Figure 5 shows the simplest situation, in which the fitness values of the two traits are the same regardless of their frequencies.

This graphical representation allows us to describe what will happen to a population of Y individuals when an X mutant is introduced. Since X is the fitter type, X individuals will, on average, outreproduce Y individuals. What will this mean for the relative representation of the two types in the population? To answer this question, we need some assumptions about heritability (Section 5.2): How do the characteristics of parents predict the characteristics of offspring here? To simplify discussion, I'll assume that the two traits are perfectly transmitted from parent to offspring in a uniparental, asexual mode of reproduction. They may therefore be single alleles, diploid genotypes, complexes of genes scattered through an organism's entire genome, or phenotypic

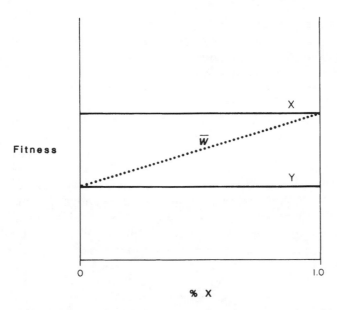

Figure 5. Selection for X and against Y, where the two traits have frequency-independent fitness values. $\overline{w}$, the average fitness of organisms in the population, increases as X goes to fixation.

characters. In any case, the greater fitness of the X type results in an increase in the representation of X individuals. Once the two types have moved to this new frequency (one "notch" to the right on the x axis), we may consult the graph again to see which trait is fitter. Since the fitness values are not affected by the change in frequency, the process is repeated; again, X increases. Selection will eventually take X to 100 percent representation (fixation). Having reached that point, the introduction of a Y mutant would have no effect, since selection would return the population to 100 percent X.

What happens to the level of adaptedness in the population in this process? The dotted line shows the change in the average fitness of organisms in the population ($\overline{w}$—pronounced "W-bar"). It increases steadily under selection, reaching its maximum when X has gone to fixation. In this circumstance, selection raises the level of adaptedness found in the population.

I should stress at the outset that although the quantity $\overline{w}$ is sometimes referred to as "the fitness of the population," this description is misleading. Suppose in a given circumstance that a population's fitness is related to its size. Perhaps the larger of two populations is fitter because it is farther away from the zero census size that spells e-x-t-i-n-c-t-i-o-n. Or perhaps the smaller of the two is more secure because it is consuming its nonrenewable resources at a lower rate. Whichever of these situations obtains, the populations may nevertheless have equal values of $\overline{w}$, if the organisms in each are simply reproducing at replacement numbers. $\overline{w}$ measures fitness *in* populations, not fitness *of* populations.

The fact that $\overline{w}$ increases in this simple model of asexual reproduction has within it the core idea behind Fisher's (1930) fundamental theorem of natural selection. The theorem, naturally enough, is usually stated without mentioning the assumptions it depends upon: *The rate of increase in fitness in a population at a time equals the additive genetic variance in fitness at that time.* Note that as long as there is (additive genetic) variation in fitness, the level of fitness found in the population will increase. I now want to consider the constant-viability model of selection discussed in Section 1.4, with an eye to revealing some of the assumptions on which the theorem depends.

In the model discussed in Section 1.4, constant fitness values (i.e., ones that are independent of trait frequencies in the population) were assigned to genotypes in a randomly mating sexual population. This departure from the asexual situation described in Figure 5 leaves it open whether $\overline{w}$ increases under the influence of natural selection in the present case as well. We now will see that in that model $\overline{w}$ increases. Three possible fitness relations must be considered: (1) the heterozygote is intermediate in fitness; (2) the heterozygote is fittest; (3) the heterozygote is least fit.

Figure 6 represents the first case: Each genotype has a constant (frequency-independent) fitness value, with AA greater in fitness than Aa, which in turn is fitter than aa (i.e., $w_1 > w_2 > w_3$). The two dashed lines show that the two alleles—A and a—have frequency-*dependent* fitness values. It may seem odd that alleles should change in fitness, even though genotypes do not, but it is easy to understand why this should be so. When a is very rare, it almost always occurs in heterozygote form. However, when it is at fixation, it must be found entirely in homozygotes. So in the former case, the fitness of a will be virtually the same as the fitness of Aa; in the latter, the fitness of the allele will be the same as the fitness of aa. The allele changes its fitness as the population evolves, because the allelic fitness is just a weighted average of the fitness values of the two genotypes in which it may occur (Section 1.4). The weighting simply reflects how often the allele is present in heterozygote or homozygote condition, and this fact, of course, is influenced by how much heterozygosity there is in the population. A similar explanation shows why the fitness of A is frequency-dependent.

When AA is fitter than Aa, which in turn is fitter than aa, selection

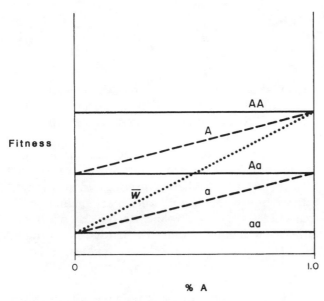

Figure 6. The constant viability model for one locus with two alleles, where the heterozygote is intermediate in fitness. Allelic fitnesses are frequency dependent. $\overline{w}$ increases as A goes to fixation.

drives A to fixation. Note that at every moment in the evolution of the population—from the introduction of A into a population of a's all the way to the end point at which A attains 100 percent—A is fitter than a. Although both alleles increase in fitness in the process, A is always superior. This suggests a handy rule of thumb: *If you want to know what will happen to gene frequencies under selection, look at the allelic fitness values.*[6] This may take a bit of doing, since models do not always explicitly assign fitness values to single genes.

The dotted line in Figure 6 shows the average fitness ($\overline{w}$) of organisms in the population. Note that it increases from a minimum when a is universal to a maximum when A has reached fixation. In Section 1.4, I defined $\overline{w}$ as a weighted average of the three genotypic fitness values w_1, w_2, w_3:
$$\overline{w} = p^2w_1 + 2pqw_2 + q^2w_3.$$
But this expression can be rearranged to show it as a weighted average of the allelic fitness values,
$$\overline{w} = p(pw_1 + qw_2) + q(qw_3 + pw_2),$$
where the allelic fitnesses are, respectively,

6. This is Fisher's (1930, p. 30) idea of the *average excess* of a gene.

$$W_A = pw_1 + qw_2$$

and

$$W_a = qw_3 + pw_2.$$

$\overline{w}$ increases as the population evolves; in this model, selection is an improver.

The same holds true for the second case—heterozygote superiority with constant genotypic fitness values, shown in Figure 7. Note that there is a frequency at which the two alleles are equal in fitness. This is the stable equilibrium point ($\hat{p}$), described in Section 1.4, that enables heterotic systems to maintain a balanced polymorphism. In the example of the sickle-cell trait discussed there, both alleles are maintained in the population because heterozygotes have increased resistance to malaria, but do not suffer from anemia. Again, the handy rule of thumb applies: *If you want to know when, if ever, selection will stop changing gene frequencies, see if there is a gene frequency at which the alleles are equal in fitness.*

The dotted line in Figure 7 represents $\overline{w}$. I pointed out in Section 1.4 that the equilibrium $\hat{p}$ of this system is stable. Not only will selection fail to modify the population if it is at this equilibrium point, but also,

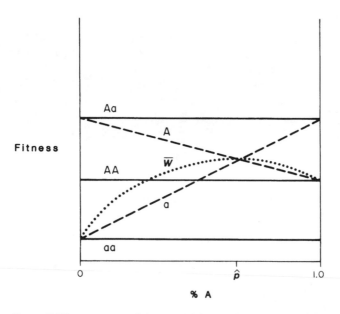

Figure 7. The constant viability model for one locus with two alleles, with heterozygote superiority. $\overline{w}$ reaches its maximum as selection drives the population to its global gene frequency equilibrium ($\hat{p}$).

if the population is not at equilibrium, selection will drive it there. In this case, p̂ is a *global* equilibrium: No matter what gene frequency you begin with (as long as the two alleles have some representation), the population will evolve to p̂. The point of interest here is that moving to p̂ always increases $\overline{w}$. In this model as well, selection is an improver.

The third and final situation follows suit. Figure 8 represents *underdominance* with constant genotypic fitness values. The heterozygote is the least fit genotype. As before, constant genotypic fitness values give rise to frequency-dependent allelic fitnesses. As in the case of heterozygote superiority shown in Figure 7, there is an equilibrium point. The two alleles are equally fit at p̂. When the population has this gene frequency, there will be no selection. But heterozygote inferiority is special, because the equilibrium point here is *un*stable. Note that if the population is to the right of p̂, *A* will have the higher fitness value, and so selection will take *A* to fixation. On the other hand, if the population is anywhere to the left of p̂, *a* will be fitter, and so *a* will be driven to fixation.

The dotted line in Figure 8 represents $\overline{w}$, the average fitness of organisms in the population. Note that the unstable equilibrium p̂ is the point at which $\overline{w}$ has its *lowest* value. But when a population is perturbed

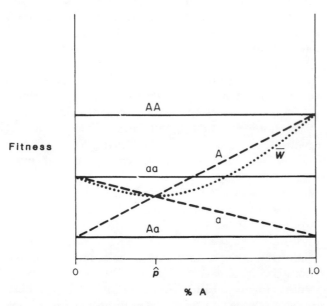

Figure 8. The constant viability model for one locus with two alleles, with heterozygote inferiority. The equilibrium point is *un*stable. If the initial allele frequency is p > p̂, then *A* will go to fixation. If the frequency is p < p̂, then *a* will go to fixation. In each case, selection increases $\overline{w}$.

from this unstable equilibrium, natural selection will push either A or a to fixation. In either case, $\overline{w}$ will increase. In this third and final circumstance, natural selection is once again an improver.

Heterozygote superiority (Figure 7) and heterozygote inferiority (Figure 8) both have their equilibria. To see the difference between the stable equilibrium that obtains when the heterozygote is most fit and the unstable equilibrium that obtains when the heterozygote is least fit, let us invert the metaphor. Balls situated on hillsides or mountaintops have a "tendency to fall." A stable equilibrium (heterozygote superiority) is like a ball at the bottom of a valley; an unstable equilibrium (heterozygote inferiority) is like a ball balanced precariously at the top of a mountain. The ball in the valley is at a point of stable equilibrium because a push will cause the ball to roll back and forth until it returns to its original position. But a little nudge to the ball on the mountaintop will have the opposite effect. Its equilibrium is unstable; a small perturbation drives it even farther from its equilibrium point.

The three cases just considered exhaust the possibilities for a one-locus two-allele system with constant fitness values. The result—that selection increases $\overline{w}$—remains unchanged when the argument is generalized in two directions. First, a system in which there are more than two alleles at the locus still conforms to the requirement that natural selection increases $\overline{w}$ (Roughgarden 1978, pp. 101–110). The situation becomes more complicated when one attempts to model a selection process acting on several loci simultaneously. Where those loci are independent of each other (i.e., one assumes that there is no linkage between loci and free recombination occurs), selection will again serve to maximize $\overline{w}$ at each locus. This result does not mean that natural selection in the real world succeeds in optimizing, since, in the real world, various loci are nonindependent. But, recall, the question at issue here is what natural selection is like in itself. And the answer to this question obtained so far is that natural selection is an improver, when it acts alone.

The main limitation of this result is the assumption that genotypic fitness values are frequency-independent. It is easy to show this graphically. Figures 5 through 8 can each be tilted so that the genotypic fitnesses run downhill from left to right. Since tilting the whole graph leaves all of the relationships unchanged, the population will evolve to precisely the same gene frequency it goes to when the genotypic fitnesses have zero slope. However, tilting affects the fate of $\overline{w}$. Anything can happen when one is allowed to fiddle with the fitness functions in this way. Selection may increase or decrease the value of $\overline{w}$. Once frequency-dependent selection is taken into account, no general statement can be made as to whether selection tends to improve.

Perhaps the most familiar case of frequency-dependent selection is described in the theory of mimicry (Wickler 1968; Owen 1982). Monarch butterflies (*Danaus plexippus*) taste terrible to blue jays. Another butterfly species, *Limenitis archippus*, has evolved the characteristic visual appearance of monarchs, but not their bad flavor. *Limenitis* is the mimic and the monarch is the model. To understand the evolution of the mimetic trait, we must imagine an earlier population of *Limenitis* in which the trait is absent. Mimicry—the novel mutant—is then introduced; it has a fitness advantage, and thereby increases in frequency.

The special wrinkle in this process is that the fitness of the mimic declines as it becomes more common. Blue jays learn from experience (Brower 1969; Brower, Pough, and Meck 1970). If the nasty-tasting monarchs are common, and the tasty mimics are rare, blue jays learn to avoid eating butterflies with that characteristic appearance. But as the mimics become more common in the *Limenitis* population, their representation relative to the monarch models also increases. A blue jay eating a butterfly with the characteristic appearance is now more likely to find it tasty.

At each point in this process, we may imagine that mimicking monarchs are fitter than nonmimicking monarchs. Mimicry thereby goes to fixation in the *Limenitis* population, but the absolute fitness of mimics is frequency-dependent. Figure 9 is a guess at the shape of the characteristic fitness functions of mimics (M) and nonmimics (N) in the monarch population. Notice that the average fitness of organisms in the population increases. The dotted line shows the increase in $\overline{w}$. A population of 100 percent mimics has a higher average fitness than a population of 100 percent nonmimics. Again, selection is an improver.

This type of graphical representation allows us to identify cases in which the average level of adaptedness, $\overline{w}$, shows no net change as well as cases in which it actually declines. A slight modification of the mimicry example, shown in Figure 10, provides a case in which $\overline{w}$ initially increases during the selection process but then declines so that no net change results.

As a final and even more depressing hypothetical fitness function, consider the one shown in Figure 11. Selection drives the trait S to fixation when it is introduced into a population of A individuals, ever reducing the average fitness of individuals in the process. Selection, in this case, is the very opposite of an improver.

Does the structure illustrated in Figure 11 correspond to anything in natural populations, or is it merely a perverse theoretical possibility, invented to chide those optimists who rely on natural selection to build a bigger and better world? One possible interpretation of Figure 11 is especially germane to recent evolutionary investigations. Suppose that

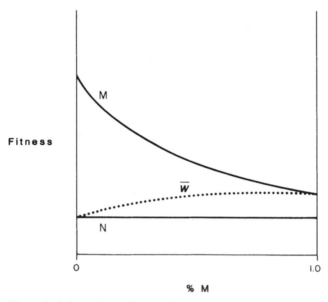

Figure 9. A hypothetical fitness function comparing mimics *M* and nonmimics *N* in a population of monarch butterflies. The greater fitness of mimicry leads it to increase in frequency relative to nonmimicking monarchs. In the process mimicking monarchs also become more frequent compared to models in the species *Limenitis*. Hence the advantage of mimicry declines as the trait becomes more common. Mimicry nonetheless goes to fixation and $\overline{w}$ shows a net increase.

A is an *altruistic* trait—one that causes individuals with the trait to benefit others at their own expense. *S* is the selfish alternative; it receives benefits, but never reciprocates. A population composed entirely of altruists will be vulnerable to "invasion." A selfish mutant will be at a reproductive advantage and eventually go to fixation. In the process, however, the average fitness of organisms in the population will be reduced; the individuals in a population of 100 percent altruists do better than the individuals in a population of 100 percent selfish types. The many biologists who have argued that altruism is a biologically unstable property—subject to "subversion from within"—have imagined a fitness function of precisely the sort shown in Figure 11.[7]

---

7. In Part II, I explore the suggestion that group selection may favor groups that contain altruists over groups that contain only selfish individuals. For now, the point is to show how selection among the organisms in a single population may fail to improve average fitness.

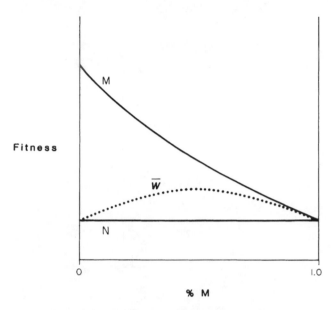

Figure 10. A slight modification of Figure 9. As before, mimicry M goes to fixation and nonmimicry N in the same population is eliminated. In this case, however, the average fitness of organisms in the population ($\overline{w}$) is the same after the process is over as it was before it began.

There are other possible interpretations of A and S in Figure 11. Think of the A individuals as themselves quite selfishly involved in their own survival and reproduction. Think of the S individuals as *spoilers*. They are fitter than the resident A types, but in the process of reaching fixation, S individuals spoil the environment for themselves and for the alternative type as well. Spoilers reach fixation in part by causing a decline in everyone's level of adaptedness.[8]

Taking these various graphical representations together, I hope it is clear why Fisher's theorem fails to apply when fitness values are frequency-dependent. Selection may be relied upon to be a "progressive" force of evolution in the sense discussed here only when fitnesses are frequency-*in*dependent.

The condition needed for selection to drive a novel mutant to fixation is precisely that at every point in the selection process, the mutant must be fitter than the trait it displaces. This constraint involves relative fitness values. It thereby leaves open what happens to the absolute

8. Haldane (1932) gives a nice example: Imagine a rabbit genotype that causes its bearer to feed on baby rabbits that are not its own progeny. Although the trait would initially be selectively advantageous, it would be disastrous if it reached fixation.

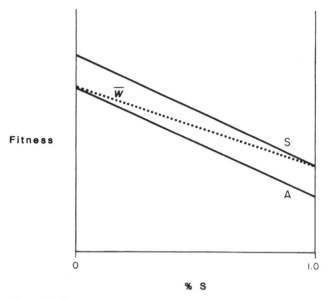

Figure 11. Frequency-dependent selection in which trait S, once introduced into a population of A individuals, goes to fixation. $\overline{w}$ declines along the way. This illustrates how a selfish trait can displace an altruistic one, even though a population of altruists does better than a population of selfish individuals.

fitness of the mutant as it increases in frequency. Selection as a cause of evolution in a population depends on *relative* adaptedness; selection as a cause of improvement depends on what happens to *absolute* fitness. This is why evolution by natural selection can fail to improve.

Hence, natural selection is no more intrinsically inclined to improve the level of fitness found in a population than it is to grind the population into the dust. Each is possible, and which occurs depends on the shape of the fitness function. Having settled the issue of principle, I now want to briefly turn to the question of relative frequency: How common is it for selection to have one effect rather than another? In general, should we expect natural selection to increase $\overline{w}$, or not?

We can answer this question by thinking about the value of $\overline{w}$ before a novel mutant is introduced, and the value it will attain after selection has run its course. Many populations maintain themselves at something close to a steady state. They reproduce neither very much above nor very much below replacement numbers. If a new mutant is to increase in frequency owing to natural selection, it must do better than the organisms already on the scene. Its fitness, we may suppose, will therefore allow it to reproduce above replacement levels. As it sweeps through the population, however, the increase in population size must sooner or later come to a halt. The reason is that the universe is a finite place and can sustain only so many living things. So sooner or later the population must return to its steady state, in which organisms reproduce at replacement numbers.

If this rough picture has anything like general (not to say universal) application, we have an answer to our puzzle about selection and improvement. $\overline{w}$ may increase temporarily during periods of expansion in population size. But what goes up must come down. If it rises above replacement numbers, $\overline{w}$ must sooner or later decline to that earlier value. If $\overline{w}$ is a measure of progress, we can see why evolution by natural selection will often not be progressive. This will be true when evolution by natural selection has a particular frequency-dependent structure.[9]

This picture does not prohibit selection processes from exhibiting a somewhat attenuated sort of optimizing property. We may time-index the trajectory of the population, and consider progressive those moments in evolution when $\overline{w}$ rises above unity and *nondegenerative* those times when $\overline{w}$ is at or above replacement values. Natural selection will improve a population when the population is still in the expansion state—for example, when it invades a new niche. But unoccupied niches fill up, and so the population must eventually return to a steady state.

This melancholy result may encourage one to look for a different measure of progress. Surely a population that lives at a carrying capacity of 500 progresses when a novel mutant allows that population to eventually reach a census size of 500,000. Individual selection will favor mutants that can exploit new resources or invade as yet unoccupied niches. Perhaps it should be census size or total biomass, not average reproductive rates, that defines what is progressive.

I have already mentioned one problem with this proposal. A bigger population may be more precarious than a smaller one, if the former overexploits its environment. Selection among organisms may nevertheless increase population size and lead the population straight to disaster.[10] But an additional objection is worth registering. Even if increase in census size were accepted as a criterion of progress, there is no guarantee that natural selection will always push in that direction.

9. The reader may wish to compare Kojima's (1971) discussion of the issue of frequency dependence in selection experiments. I can't resist quoting the emphatic title he gave this work: "Is there a constant fitness value for a given genotype? No!"
10. As we will see in greater detail in the next part of this book, Wynne-Edwards (1962) argued that a hypothesis of *group* selection is needed to explain why populations do not typically overexploit their resources. The present point, however, is that there is nothing in the idea of individual selection that implies that this cannot occur.

This is another lesson that can be learned from Figure 11. Selection for S may reduce population size, even carrying it to zero. This revised notion of progress provides little support for the idea that natural selection is inherently an instrument of improvement.

When a population evolves in compliance with any of the constant viability fitness functions shown in Figures 6–8, $\overline{w}$ will increase. Hence, if you want to know whether an allele will go up or down in frequency at a given time, you need only find out which direction will increase $\overline{w}$. This raises two questions: Can one *explain* the change in frequency of the gene by saying that it *had* to move in a certain direction in order to increase $\overline{w}$? Is this property of the system a *cause* of the change in frequency? If we analogize this situation to the barometer case discussed in Section 5.1, the answer will be *no* on both counts. The barometer reading allows us to predict tomorrow's rain, but the reading neither explains nor causes the storm. $\overline{w}$ is a barometer, we might say; it reflects the causal facts that guide selection and allows one to predict change in gene frequency. But the increase in $\overline{w}$ and the change in gene frequencies are joint effects of a common cause; neither causes the other.

Although this resolution of the causal question is, I think, clear-cut, the query about explanation is more difficult. Another analogy suggests why. Fermat's law asserts that light passing from a point A in one medium (e.g., air) to a point C in another medium (e.g., water) will pursue a path of least time. Can this law of refraction be used to explain why the light passed through point B, a point on the surface of the water? Is it explanatory to point out that point B was on the path of least time between points A and C? I am inclined to say that it is. Fermat's law describes a property of the causal structure of the system in question, even though reaching point C is obviously not a cause of the light's passing through B. In just the same way, remarks about $\overline{w}$ may help us understand why gene frequencies changed in one direction rather than another, even though increase in $\overline{w}$ is not the cause of the change. Here again we see that explanations need not be causal (Section 5.1).

The intellectual sources of Darwin's ideas on natural selection are brought into focus by considering the role played by the idea of progress in his thought. The passage quoted earlier from *The Origin* is in some ways typical in its stress on the way natural selection perfects and improves. Yet, at the same time, there is in Darwin's work an acknowledgment of the sobering lesson that we now can clarify in terms of the idea of frequency-dependent selection. Not that Darwin had at his fingertips the quantitative formulation that now shows how each of these two contradictory perspectives has its own domain of applicability. In his case, we see the clash while it was still less than fully explicit. Each of these conflicting biological ideas had its analogue in the human sciences; the Scottish economists on the one hand, and Malthus on the other, were powerful influences on Darwin's thought (Schweber 1983). Their intersecting penumbras created the problematic that contemporary evolutionary theory is able to resolve.

The Scottish economists furnished Darwin with a model in which individuals interact in a self-interested way and yet augment the national wealth as an unintended consequence. Markets were open and wealth could be increased, indeed *would* be increased, if competition were unconstrained. In *The Wealth of Nations*, Adam Smith summarized the idea that selfish competition produces beneficient consequences as a fortuitous by-product in his striking metaphor of the invisible hand:

> Every individual is continually exciting himself to find out the most advantageous employment for whatever capital he can command. It is his own advantage, indeed, and not that of the society, which he has in view. But the study of his own advantage naturally, or rather necessarily, leads him to prefer that employment which is most advantageous to the society (Smith 1776, p. 421).

> ... every individual necessarily labours to render the annual revenue of the society as great as he can. He generally, indeed, neither intends to promote the public interest, nor knows how much he is promoting it ... in many other cases, led by an invisible hand to promote an end which was not part of his intention.... By pursuing his own interest he frequently promotes that of the society more effectually than when he really intends to promote it (Smith 1776, p. 423).

One important ingredient in Darwin's idea of natural selection is almost a substitution instance of this logical schema. Instead of entrepreneurs choosing lines of work of their own free will, organisms randomly vary in the adaptive strategies they deploy. Instead of economic competition motivated by self-interest increasing the national wealth, the struggle for existence among organisms favors individually adaptive characteristics and thereby increases the level of adaptedness in the population. $\bar{w}$, as if by an invisible hand, is made to grow.

Darwin also absorbed from the economists the central principle of the division of labor. How could natural selection account for organic diversity? In the same way that economic competition accounts for diversification and specialization. Organisms reduce the deleterious effects of competition by moving into hitherto unoccupied niches. They discover new ways of making a living and thereby prosper. The invasion of new habitats provides the occasion for drastic increases in population numbers. In the process, $\overline{w}$ rises above its previous steady state. The principle of the division of labor became, for Darwin, the principle of divergence (Kohn 1980).

The exponential phase of population growth, with its progressive increase in the level of adaptedness, must sooner or later run into the melancholy lesson of Malthus. Population cannot increase indefinitely, and so average reproductive rates cannot indefinitely remain above unity. A trait that facilitates greater reproductive success than its competition will prosper, but its prosperity can only be temporary. Malthusianism is density-dependent selection by another name.[11]

Silvain Schweber (1980) describes a particularly instructive episode in Darwin's reflections in which these two impulses were brought face to face. We may perhaps gain a sense of which was the dominant mode of thought by seeing how Darwin evaluated the pros and cons. Darwin kept records of what he read and took copious notes. The depth of his reading in political economy, though perhaps not surprising in a Victorian intellectual, would be quite remarkable in a contemporary biologist. Besides steeping himself in Hume's works, Darwin also read Adam Smith, Thomas Reid, and Dugald Stewart (Gruber 1974). In 1840, he read J. R. McCulloch's *Principles of Political Economy* and found there yet another statement of the classical picture of the relation between competition and improvement:

> Every individual is constantly exerting himself to find out the most advantageous methods of employing his capital and labor. It is true, that it is his own advantage, and not that of society, which he has in view; but a society being nothing more than a collection of individuals, it is plain that each, in steadily pursuing his own aggrandizement, is following that precise line of conduct which is most for the public advantage (quoted in Schweber 1980, p. 268).

McCulloch draws a conclusion from this supposed connection between individual competition and collective advantage that is now an old standby in the defense of *laissez-faire* capitalism:

> The true line of policy is to leave individuals to pursue their own interest in their own way, and never to lose sight of the maxim *pas trop gouverner*. It is by the spontaneous and unconstrained ... efforts of individuals to improve their condi-

---

11. I am grateful to William Provine for driving home to me the connection between frequency-dependent selection and Malthusianism.

tions . . . and by them only, that nations become rich and powerful (quoted in Schweber 1980, p. 268).

Although McCulloch's conclusion was a standard one, Darwin was exposed to at least one dissenting voice. Jean Charles Sismondi was an eminent Swiss economist who had married Emma Darwin's favorite aunt. Sismondi died in 1842, and the *Quarterly Review* published an extended commentary on his views the following year. Sismondi was described as holding an unusually pessimistic view of the consequences of competition:

> Division of labour, according to Adam Smith, is the great source of national [wealth], or "general plenty, diffusing itself through all the different ranks of society." Sismondi says, "No." . . .
>
> Unlimited competition according to the popular theory, is the great source of national riches. Sismondi says "no"—Unlimited competition renders the whole system of commerce a vast game of "beggar-my-neighbor." . . .
>
> "Permit each person"—quoth the political economist call him Adam Smith, call him McCulloch, call him Chalmers, it is all the same— "to seek his own interest in the way which suits him best, and you must be, since society consists only of individuals, promoting the general interest of society." Sismondi contradicts this doctrine (quoted in Schweber 1980, p. 269).

In 1847, Darwin read Sismondi's *Political Economy* and labeled the work "poor" in his reading notebook (Schweber 1980, p. 270).

In the age of capitalist expansion, economists usually (though not universally, as the example of Sismondi attests) thought of the free market as having benign consequences. Now that resources are much more fixed and the prospects for expansion are more constrained, it is not uncommon for the deleterious consequences of competition to leap to mind. One standard way to illustrate how competition can make everyone worse off is the metaphor of the "tragedy of the commons." Suppose we are dairy farmers with separate herds who share a field for grazing. At present, we each sell cream and butter and earn $10,000 per year. Each of us has a certain surplus of funds, which makes it feasible for each to consider adding another cow to his herd. You realize—indeed we all do—that if each of us adds another cow, the yield per cow will decline to $9000 because the supply of grass will deteriorate. But if you add a cow and no one else does, your income will go up to $11,000. And if everyone adds a cow except you, your income will decline to $8000. The decision problem you face may be represented as a matrix, in which the different payoffs are a function both of what you do and of what everyone else does:

| | | Possible states of the world | |
		Others add	Others do not add
Possible actions	You add	$9,000	$11,000
	You do not add	$8,000	$10,000

The decision you would make if you wanted to maximize your income would be to add a cow. This action is "best" in the technical sense that no matter what the others do, adding yields more cash than not adding. But, of course, everyone is in the same position, and so everyone adds a cow, if he is rational. The outcome is that everyone's income drops from $10,000 to $9,000. The result is not "Pareto optimal," in that everyone would be better off if no one added a cow.

One way to avoid the tragedy of the commons[12] is for the farmers to voluntarily reach a binding arrangement under which some add a cow and pay compensation to those who do not (Hirshliefer 1982). Or the state may intervene and require each farmer to refrain from adding another cow. Human beings have the capacity, at least in principle, to escape from tragedies of the commons. When a process is about to begin that rational reflection predicts will harm everyone, we can restructure the decision problem. But organisms caught in a frequency-dependent selection process are usually not so lucky. There is no way to opt out of a selection process in which all the organisms at the end of the process are worse off than all the organisms at the beginning.

It is important to be clear on the analogy with natural selection. In the tragedy of the commons problem, the farmers do not survive or perish in consequence of their decision. Organisms in a selection process, of course, do. The point is that the processes are guided by a similar dynamic and generate an end result that is measurable in a certain similar way. Farmers select actions whose "fitness" is measured in dollars; natural selection favors traits according to which is better in terms of the currency of survival and reproductive success. The result

12. This problem is also called the Prisoner's Dilemma: you and an accomplice are arrested. Each of you has to decide whether to remain silent or confess to the police. The payoff matrix is structured in such a way that both you and your accomplice choose to confess and thereby go to jail for longer than either would have if both had remained silent.

of the selection process may be characterized in terms of the average situation of the players—income earned or fitness.[13]

The economists of Darwin's time tended to think that since a society is "nothing but" a collection of individuals, the society will maximize its well-being if each individual endeavors to maximize his welfare. A similar elision is found in Darwin's (1859, p. 489) concluding remarks in *The Origin*:

> And as natural selection works solely by and for the good of each being, all corporeal and mental endowments will tend to progress towards perfection.

Before one understands the structure of the tragedy of the commons, it perhaps sounds like a contradiction to be told that each individual, acting rationally and correctly foreseeing the consequences of his or her actions, will make a decision that leaves everyone worse off than he would be if some "irrational" choice had been made. How could every individual suffer, if each individual acts in his or her best interests with full knowledge of the consequences?

Darwin had a low opinion of Sismondi's suggestion that such deleterious effects can be the natural outcome of unrestrained competition. What may have been unexpected at the time Darwin wrote is now a routine observation of elementary decision theory. It is difficult to establish the precise influence that political economy exerted on Darwin's view of natural selection. It is more difficult still to show how both the economics and the biology flowed from mistaking a contingent property of some competitive interactions with a necessary structural property of all competition.

With the wisdom of hindsight, we can see that selection need not improve and that rational decision making can leave everyone worse off. It is an attractive hypothesis that the incomplete views found in earlier biology and economics were a consequence of ideological influence. However, it is impossible to dismiss the suggestion that the more complete picture we now possess was inaccessible to earlier theorists for purely technical reasons. Perhaps it takes a rigorous decision theory on the one hand and the abstract picture of natural selection afforded by Fisher's theorem on the other to bring the issues into focus. It will be a subtle piece of historical reasoning indeed that can disentangle these explanatory strands.

---

13. Notice the structural similarity between the Prisoner's Dilemma and the kind of frequency-dependent selection process described in Figure 11 (p. 186). Two strategies S and A are such that S is better than A, but if everyone does S, everyone is worse off than they would be if everyone did A.

The account I have given of Malthus' contribution to our present understanding of natural selection differs markedly from the standard one. Malthus thought he had shown how the geometrical increase in population and the arithmetic increase in food would inevitably produce mortality. Although his primary interest was the state of the poor in England, Malthus thought that he had discovered a principle of universal scope. It is not uncommon for contemporary writers on evolution to think of this Malthusian idea as a central axiom of the theory of natural selection.[14] This is, I think, a considerable distortion of our present understanding of natural selection; I even would venture to say that Darwin's own thought is not best represented as beginning with this principle.[15]

Malthusianism is not a proper starting point for the theory of natural selection for reasons made abundantly clear by Fisher (1930, pp. 46–47). It is not to be denied that if populations reproduced at above the replacement level, and if this excess reproduction were not offset by mortality, the population would increase without limit.[16] But this describes a counterfactual situation, and so does not by itself establish that natural selection in fact operates in the real world. If the Malthusian idea is stated as the factual claim that populations *do* reproduce more offspring than can survive, this is sufficient to guarantee the existence of selection pressures, provided survivorship is nonrandom. Fisher wryly identifies what is wrong with making this idea an axiom of the theory of natural selection:

> There is something like a relic of creationist philosophy in arguing from the observation, let us say, that a cod spawns a million eggs, that *therefore* its offspring are subject to Natural Selection; and it has the disadvantage of excluding fecundity from the class of characteristics of which we may attempt to appreciate the aptitude. . . . The historical fact that both Darwin and Wallace were led through reading Malthus's essay on population to appreciate the efficacy of selection, thought extremely instructive as to the philosophy of

14. See, for example, Mary Williams' (1970, 1973) axiomatization of the Darwinian theory of evolution and Sober (1984c) for criticisms.

15. Especially if Darwinian natural selection is construed as including what Darwin called sexual selection. See below for details.

16. Fisher (1930, p. 47) stresses that the idea that population "tends" to increase *geometrically* and food only *arithmetically* is irrelevant to this point. In addition, Richard Lewontin (personal communication) has remarked that this quantitative principle, besides being extraneous to the Malthusian point, also has the property of being contradictory when stated generally. Most populations are also food for other populations; so is their "natural tendency of increase" arithmetic or geometric?

their age, should no longer constrain us to confuse the consequences
of that principle with its foundations.

Fisher's point is that the theory of natural selection does not take
reproductive rates as an unalterable given, from which mortality is
computed as a consequence. Reproductive rates can themselves be
manipulated by selection. In addition, selection, as we now conceive
it, can take place even in the absence of overreproduction. Adjust the
tilt of the fitness functions displayed in Figure 11 so that $\overline{w}$ has zero
slope. Selection will still drive trait S to fixation, but we might imagine
that $\overline{w}$ remains at the replacement level throughout the selection process.
A steady-state population can evolve by natural selection just as much
as one that shows excess reproduction. Symmetrically, selection can
modify a population even when all offspring survive.

Fisher calls Malthusianism a consequence, not a foundation, of the
theory of natural selection. Excess reproduction with a finite carrying
capacity is a special case, although arguably one of great importance.
Not that Fisher's "fundamental theorem of natural selection" deserves
pride of place; frequency-independent fitness values are, I have argued,
a *very* special case. The foundational idea common to both approaches
is that evolution by natural selection depends only on *relative* fitness
values. The fundamental theorem shows how the additional assumption
of frequency independence implies that selection is progressive; the
Malthusian idea that a trait's spread through a population undermines
the conditions that make it advantageous shows that the additional
assumption and its implications are not inevitable.

Malthusianism survives in contemporary selection theory, but not
because Malthus identified natural tendencies of population and food
that guarantee that the grim reaper of mortality selection will always
be with us. Malthus stressed the importance of *intra*specific competition,
an idea that arguably was crucial in Darwin's shift from thinking of
competition as occurring among groups to seeing it occur at the individual level (Herbert 1971). If the members of different species make
a living in very different ways, they are not apt to be at each other's
throats; but members of the same species, according to the Malthusian
idea, compete for the same resources, and so competition between them
must be more intense. The logical extension of this idea that similarity
increases competition is that a characteristic's own worse enemy is
itself. As natural selection destroys the variation found in a population
by driving the fittest trait to fixation, it increases the problems that the
new trait faces. Immediately after a trait is introduced as a novel mutant,
the organisms possessing the trait compete mainly with organisms that
do not have it. But as the trait becomes common, organisms with the

trait frequently interact with other organisms that also have it. In this way, success spoils part of the advantageousness of the trait, since part of its advantage consisted in its rarity.[17] This melancholy Malthusian idea has an importance to the theory of natural selection far beyond his more familiar principles of geometrical and arithmetic increase.

6.2 Retrospect and Prospect

"Adaptedness," as I have used the term, is simply fitness by another name. Whether selection in a population will tend to increase the level of adaptedness is, as we have just seen, a delicate question. Some clarification is now required for the idea of *adaptation*. Like many terms in ordinary language (e.g., "marriage" or "selection"), it may denote both a process and a product of a process. "Adaptation" may describe what an organism or a population *does*; it also may label those traits of an organism or population that result. Although the process/product ambiguity occasions no great confusion in the case of "marriage," it has been less than harmless in the case of interest here. The process of getting married necessarily issues in the end state of being married; and people who occupy that conjugal state must have gone through that process. Here, process and product necessarily go together. The analysis I intend to recommend here does not quite follow this quotidian pattern. It has a rather paradoxical result: Although "adaptedness" and "adaptation" may appear to be terminological variants for the same thing, they really denote aspects of the evolutionary process that are logically independent. A trait can be an adaptation without marking an improvement in adaptedness; conversely, it can enhance its bearer's level of adaptedness without being an adaptation.

Current evolutionary usage is fairly consistent in applying the term "adaptation" only to those traits that have emerged by a process of natural selection (Williams 1966, Lewontin 1978, Gould and Vrba 1982). More specifically, the requirement is that if a trait is an adaptation for performing a particular task, then it must have evolved because there was selection for the trait, where the trait's selective advantage was due to its helping perform the task. Consider Lewontin's (1978, 1980) example of dorsal fins on the dinosaur *Stegosaurus*. If the dorsal fins are adaptations, they must have evolved by natural selection. And if they are adaptations for defense against predators, they must have evolved because there was selection for possessing dorsal fins, where

17. Not that there are no cases in which the advantage goes to traits that are common rather than rare. The point is that there is a quite general reason to think that rare traits experience less competition than common ones.

the selection for the trait was due to the fact that dorsal fins contributed to defense against predators.

It follows that the concept of adaptation needs to be understood in terms of the idea of *selection for properties*; the idea of *selection of objects* will not suffice (Section 3.2). If a neutral trait is pleiotropically linked to an advantageous one, it may emerge because of a process of natural selection. It was selected, but this doesn't mean that it is an adaptation. The reason is that although it was selected, there was no selection *for* that trait.

The main idea here is to contrast adaptations with fortuitous benefits. After dorsal fins became prevalent in the species, they may have performed any number of functions. Perhaps they served as cooling systems. Perhaps they helped stegosaurs attract mates. But if these additional benefits did not play a causal role in the emergence of the trait, it will be a mistake to describe the fins as adaptations for temperature control or mating. The concept of adaptation describes why a trait became prevalent, not why it remained so.

It is natural to inquire what this claim about adaptation implies about the first ancestor of *Stegosaurus* to have the trait. Suppose the animal had the trait because of a mutation, rather than by selection. Can we say that the trait was an adaptation *in the case of that single organism*? Here are some options: (1) apply the concept of adaptation to historically persisting populations, not single organisms; (2) allow that dorsal fins were an adaptation for the original organism because of what happened later; (3) deny that dorsal fins are adaptations for the initial organism but are adaptations when they occur in subsequent organisms. My inclination is to prefer choice 3. The first two suggestions prohibit certain discriminations; each denies that different organisms or groups of organisms in the same lineage may differ with respect to the question of whether a particular trait counts as an adaptation. We will see later in this section an example in which this implication is counterintuitive. A characteristic that initially is present fortuitously can be selected and thereby become established in a population. Initially, it is not an adaptation for doing anything; later on, it is.

There is no assumption here that traits go to fixation because they perform precisely one function. If dorsal fins were multiply useful air-conditioning units that also helped attract mates and repel predators, and if these three characteristics of the fins each contributed to its fixation in the population, then there would be nothing amiss in claiming that the fins, once evolved, were adaptations for thermal control, sexual display, and self-defense. What is excluded are those functional benefits that arise *ex post facto* as new uses for old structures.

What would this necessary condition on a trait's being an adaptation imply about characteristics that do not go to fixation but are driven to a stable intermediate frequency? Presumably, we would want to say that the heterozygote genotype in the sickle-cell system is an adaptation for malaria resistance. But, according to the model developed in Section 1.4, it never reaches fixation. This suggests that the requirement on adaptation should be weakened. Perhaps we should say that if a trait is an adaptation, then selection for the trait after it was introduced into a population as a novel mutant increased its frequency. However, the ideas to be developed next do not depend on attending to this nicety; so let's simply imagine that the characteristics considered are driven to fixation by natural selection.

The point thus far is that a trait is an adaptation only if it evolved because there was selection for it. But is the converse true? Will every trait that has become prevalent because there was selection for it thereby count as an adaptation? Lewontin (1978) and Williams (1966) have both answered this question in the negative, but for very different reasons.

Lewontin (1978) defends this conclusion by appeal to an example in which frequency-dependent selection drives a characteristic to fixation, with $\bar{w}$ having the same value after the process is over as it had before it began. Suppose a trait doubles an organism's fecundity but leaves everything else about the organism unchanged. The trait will then sweep through the population. But if the population has a fixed carrying capacity, the population will reproduce at replacement numbers after the process is over, just as it did before. I do not contest Lewontin's claim that the evolution of the trait does not improve the level of *adaptedness* (i.e., fitness) found in the population. But Lewontin additionally concludes that the trait, once it has evolved, is not an adaptation. His idea is that an adaptation must not only arise by natural selection but also must bring with it an increase in adaptedness (see also Gould and Lewontin 1979, p. 592).[18]

We may bring this issue into clearer focus by considering the example of mimicry discussed before (Section 6.1). The absolute fitness of the mimicking *Limenitis* declines as it becomes more common. Suppose, just for the sake of argument, that the fitness function of mimicry and nonmimicry is the one shown in Figure 10 (p. 185). Here selection takes the mimetic trait to fixation, even though $\bar{w}$ is the same after the process as it was before. Does it follow that mimetic appearance is not

---

18. Lewontin also mentions that the emergence of the trait may create problems for the individuals in the population that their ancestors never had. However, this does not suffice to show that the trait fails to be an adaptation, since many unproblematic cases of adaptation presumably have this sort of by-product.

an adaptation? This sounds like a mistake. The visual appearance of the mimic is an adaptation for avoiding being eaten by blue jays. No matter that it is an adaptation whose evolution brings with it no net improvement. Mimicry is an adaptation; it "solves problems" and becomes prevalent as a result. But, as this example of frequency-dependent selection makes clear, its evolution may undermine the very conditions that made it advantageous in the first place.

What of Lewontin's own example of the trait that doubles fecundity? The situation is described so meagerly that we have no idea how the trait manages to do this. We therefore find ourselves at a loss when it comes to saying what the trait is an adaptation *for*. However, suppose we flesh out the story. Imagine that the trait increases fecundity by allowing energy to be extracted more efficiently from food, thus providing a parent with a greater quantity of the raw materials out of which offspring may be built. If so, the trait is an adaptation for energy extraction. A consequence of possessing the trait is that fecundity is doubled, and this is why the trait is selected. The trajectory of its evolution will have the same shape as the mimicry case just described. The trait is an *adaptation* that brings no improvement in the level of *adaptedness* in the population. If this line of reasoning is correct, we must reject Lewontin's argument that selection for a trait can drive the characteristic to fixation without the trait's counting as an adaptation.

Adaptation is a historical concept. To call a characteristic an adaptation is to say something about its origin. It is for this reason that adaptations stand in contrast to fortuitous benefits. The lesson to be learned from Lewontin's example might be put by saying that it is a *purely* historical concept. The benefits that account for a trait's fixation need not persist into the present. A trait can be an adaptation for performing some task, even though performing that task in the present environment confers no benefit. The point is that it *was* beneficial.

Williams' (1966) argument that selection is not sufficient for adaptation is also based on an example but is more complicated. The issue arises in the context of his discussion of C. H. Waddington's (1957) ideas on *genetic assimilation*. A superficial look at Waddington's experimental work might suggest that he found a mode of evolution that involves Lamarckian inheritance of acquired characteristics. The discussion of Lamarckism in Section 4.1 helps show why this is not so: the experimental result exhibits the *pattern* but not the *process* that Lamarckism demands.

Williams (1966, pp. 70–81) discusses a striking experiment of Waddington's that involved the genetic assimilation of the bithorax phenotype in *Drosophila*. When eggs from normal (single-thorax) parents were exposed to a near lethal dose of ether, a few of them developed into flies with the abnormal bithorax phenotype. These individuals were then selected as the parents of the next generation, and the fertilized eggs they produced were themselves subjected to an ether shock. Again, a certain number of offspring developed bithorax. When Waddington repeated this procedure for about thirty generations, he obtained an interesting result. Eventually, fertilized eggs were produced that developed the bithorax phenotype *without* the application of the ether treatment. What began as an environmentally induced acquired characteristic seems to have evolved into a genetically preprogrammed one. Williams reports that the experimental verification of this interesting outcome is beyond question.

The model of the relationship between genotype and phenotype that explains this result is also something on which Waddington and Williams agree. Before the treatment began, all the flies developed normal thoraxes in the normal environment. However, some of these flies had genes that would have produced the bithorax phenotype, were it not for the fact that they also had suppressor genes at other loci. So before the treatment began, there was latent, unexpressed genetic variability. The phenotype was, in the terminology of Waddington (1957) and Rendel (1967), *canalized*. Once the experiment was underway, the ether shock prevented the suppressor genes from blocking the development of the bithorax form. It allowed the cryptic variation to express itself.

It is additionally assumed that the occurrence of bithorax in a fly is influenced by the number of "plus" alleles that are found at various loci. At the beginning of the experiment, the organisms had, on average, relatively few of these—so few that suppressor genes managed to prevent bithorax from developing in an ether-free environment. But by the end of the selection process, the concentration of plus alleles had been increased in the population. By then, the organisms had, on average, so many plus alleles that the genes that succeeded in suppressing bithorax at the beginning were no longer able to do so.

We may use the distinction between pattern and causal mechanism developed in Section 4.1 to pinpoint what is and is not Lamarckian about this experimental result. If one compares the norm of reaction of a fruit fly with relatively few plus alleles with the norm of reaction of a fly with a great many plus alleles, where ether is the environmental variable, the graph will have the same shape as the giraffe example of Figure 3(a) (p. 107). The *pattern* described there is one in which the mother genotype is labile (it produces a long neck when the mother stretches and a short neck when she doesn't) whereas the offspring genotype is not (it produces a long neck whether or not the offspring stretches). Waddington's experiment exhibits the same pattern. The flies at the start of the experiment have labile genotypes; they produce bithorax in the presence of ether and a single, normal thorax in its absence. At the conclusion of the experiment, however, the genotypes have lost this lability, producing a bithorax phenotype whether or not there is an ether shock. The *pattern* of change that Waddington produced is in the Lamarckian mold.

But just as clearly, there is nothing Lamarckian about the *process*— the causal mechanism—underlying this sequence of changes. In addition to the parent and offspring norms of reaction just described, a Lamarckian mode of inheritance requires a causal connection: The offspring giraffe must have the genotype it does *because* its mother acquired a longer neck. The heart of Lamarckism is the idea that organisms come equipped with detectors; somehow they have ways of monitoring which acquired characteristics are beneficial and then directing mutations to produce them. It is this idea about process that is wholly absent in Waddington's experiment. The genetic variability that Waddington selected on did not come into existence because it would be beneficial. Waddington could have obtained his results without the occurrence of a single mutation. The crucial fact about the bithorax experiment is that latent genetic variability was made to express itself phenotypically because of an environmental cause. The frequencies of preexisting genes were then modified by natural selection.

Is the bithorax condition at the end of the experiment an adaptation? Williams and Waddington disagree. Waddington viewed genetic assimilation as an adaptive response of the organisms in the population to an environmental stress. Williams sees the emergence of bithorax as a failure of the organism to respond. It is for him "a simple kind of degenerative evolution" (p. 75)—"a disruption . . . of development" (p. 78). Williams' analysis thereby implies that traits can evolve because there is selection for them without those traits counting as adaptations. Waddington's selection regimen brought the bithorax form to fixation; but, says Williams, bithorax is not an adaptation.

There is a very simple argument against thinking of bithorax as an adaptation, one which Williams, quite correctly, does not deploy. This line of reasoning maintains that bithorax is not an adaptation because it would not evolve or remain advantageous in the wild. It is only the "artificial," "unnatural" environment of Waddington's laboratory that permitted the aberrant structure to thrive.

I have already made some skeptical remarks about the idea of an "unnatural phenotype" (Section 5.3). The present point is just that this argument implicitly assumes that human beings are outside of nature. Waddington's laboratory was one of the niches in which *Drosophila* lived. The selection pressures there differed from those found elsewhere, but there is no rule that says that a species must live in a uniform environment. Nor should one stipulate that a trait in a local population is an adaptation only if that trait is or would be an adaptation in the species' full range. What Waddington did to his flies is comparable with what a predator might do to its prey. Of course, Waddington and his collaborators did not *eat* the flies that were selected against. The point is that the experiment involved the interaction of members of one species with members of another. The fact that one of those species happens to have been *Homo sapiens sapiens* does not affect the point. Artificial selection is a variety of natural selection; the relation is one of set inclusion, not disjointness.

Williams (p. 75) begins his argument that bithorax is not an adaptation by introducing the distinction between an adaptive response and a failure to respond. He proposes a neat analogy:

> The term "response" usually connotes an adaptive adjustment of some sort, and would not be used for disruptive effects. It would be normal usage to say that some Frenchmen responded to the Terror by conforming to the Jacobin demands and that others responded by fleeing the country. It would not be normal usage to say that some responded by losing their heads. Decapitation was the result, not of a response, but of a failure to respond soon enough or in an effective manner. Similarly we may say that some of the flies responded adequately to the ether and produced a normal phenotype in spite of the difficulty presented by the treatment. Others showed an inadequte response. They were able to survive in the protected confines of the culture bottle but could only produce a grossly imperfect phenotype, bithorax.

Williams then (p. 76) generalizes the fundamental idea behind his distinction:

> For a response to occur, there must be sensory mechanisms that perceive particular aspects of the environmental situation and activate effectors that efficiently prevent the same, or other, correlated environmental factors from producing a certain undesirable effect. Susceptibility results from the environmental factor's getting through and producing an effect in spite of any responses that may be activated.

Williams then offers another example. A man's heartbeat remains constant in a room that gets hot; this counts as a response to the environment. But a lizard's heart will beat faster when the environment gets hotter; this does not count as a response to the environment. The man's heartbeat is held constant by *devices*—"sensors" and "effectors"—like the sweat glands in the skin. The lizard's increased heartbeat, on the other hand, is not a response to the rise in ambient temperature, any more than a rock's getting hotter would be.

Williams then shifts gears and tries to justify the conclusion that bithorax is not an adaptation by considering the idea of *genetic information*. The idea is to explain the intuitive distinction between degeneration and adaptation in terms of the difference between losses and gains in information. Again, he begins with an example. If the human skin is subject to frequent friction, it will develop a callus. But the sole of the foot begins developing a callus *in utero*, before any friction impinges. Williams asserts that the unconditional development of calluses on the feet (i.e., whether or not friction occurs) is not an adaptation at all.

The evolution of callused feet, says Williams (p. 80), began with the germ plasm encoding the facultative instruction "thicken the sole if it is mechanically stimulated; do not thicken it if this stimulus is absent." The process ended with an obligate instruction that reads: "thicken the soles." Williams' conclusion is this:

> I fail to see how anyone could regard this as the origin of an evolutionary adaptation. It represents merely a degeneration of a part of an original adaptation. . . . It must, as a general rule, take more information to specify a facultative adaptation than a fixed one.

I interpret Williams as arguing that if an obligate ("Do X") instruction evolved from an earlier facultative ("Do X if Y") instruction, the obligate instruction is not an adaptation. Granted, it may have emerged by natural selection. However, Williams' position is that it takes more than this for a trait to be an adaptation.

We may begin sorting out these interesting ideas by distinguishing *evolutionary* from *ontogenetic* adaptation. An adaptation, in the sense that matters to us here, is the result of natural selection. The process of adapting that produces it is an *evolutionary* one—it must involve change in gene frequencies (Section 1.2). But notice that neither the lizard's increasing heartbeat nor the man's sweating and constant heartbeat are events in which something adapts in the *evolutionary* sense. Gene frequencies do not instantaneously change because I begin to sweat on a summer afternoon. Perspiring may help me to "adapt" to my humid study, but what I did was not a case of evolutionary adaptation. This, of course, is not to deny that the inclination to perspire when it gets hot is an evolutionary adaptation. But this capacity was long since fixed in our species before I sat down to write this page.

It may be true, as Williams asserts, that *ontogenetic* adaptation requires fancy equipment. Perhaps we may say that the man's maintaining his constant heartbeat counts as an adaptive response only because he has sensors and effectors. It would follow that a fruit fly that develops the bithorax phenotype is not in the business of *ontogenetic* adaptation to its environment. After all, it is implausible to think that a fruit fly is equipped with ether sensors that tell it to build a bithorax phenotype when conditions are right. But this has nothing to do with whether the bithorax phenotype is an *evolutionary* adaptation.

Take any example of an evolutionary adaptation you please. To choose an old standby, think of the melanic (dark) coloration that the peppered moth (*Biston betularia*) evolved because predators had trouble seeing dark moths on soot-covered trees in certain areas in the north of England (Kettlewell 1973). Industrial melanism is an adaptation. Yet never in the evolution of this trait did individual moths have sensors that told them whether there was soot and effectors that turned on the protective coloration. Peppered moths are not chameleons. Individual moths were not engaged in ontogenetic adaptation, but industrial melanism is an evolutionary adaptation for all that.

This reflects a point I developed in Section 5.2: Natural selection requires *stasis*, not change, in organisms, if it is to issue in evolution. Variational explanations require a kind of stability assumption. Although no moth alters its coloration, the coloration of the population evolves; evolutionary change in the population is the result of selection on *un*changing organisms.

Ontogenetic adaptation, as it were, has an agent and a beneficiary. When an organism begins to sweat, it is the organism that is adapting to the heat. Moreover, the organism's sweating serves to advance its own prospects for survival. The thing that does the adapting is the thing that receives the benefit of adapting. On the other hand, when a population evolves by natural selection, what, if anything, is the entity that does the adapting? No single organism is in this business, since mutations occur at random and in a multigeneration selection process, no single organism even survives from start to finish. Indeed, as just noted, evolutionary adaptation can occur even when no organism changes at all: No organism adapts itself to its environment, because no organism makes any adjustment in its own morphology, physiology, or behavior at all. Nor is it especially satisfying to say that in evolutionary adaptation, the *population* adapts to environmental changes. Typically, populations have sensors and effectors no more than individual organisms do. Furthermore, the idea of a population's adapting may misleadingly suggest that it is populations that compete against each other and necessarily benefit from the process of adapting. We have just seen (Section 6.1) that if the good of the population is measured by the average fitness of organisms in it (i.e., by $\bar{w}$), then evolution by natural selection need not benefit the population.[19] Even if we move to some other measure of population fitness, it need not be true that individual selection among organisms must benefit the population. Organisms are not the agents of evolutionary adaptation; populations need not be the beneficiaries of the process of adapting. In ontogenetic adaptation, the agent of adaptation and the beneficiary of adaptation are standardly one and the same entity—an organism. But this simple arrangement fails to transfer in any simple way to the case of evolutionary adaptation.

I have argued that the response/failure to respond distinction is relevant to the issue of ontogenetic adaptation, not to the question of evolutionary adaptation. Evolutionary adaptations can arise by natural selection even when no organism does anything at all by way of adapting. The organisms may be totally static, but as long as they live and die differentially, the net result may be an evolutionary adaptation.

I now want to raise some questions about Williams' information argument and its bearing on the bithorax case. The flies at the start of the treatment produced the bithorax form conditionally—depending on whether ether was present or absent. At the end of the experiment, they produced the bithorax phenotype unconditionally. Yet I doubt that the genomes of the flies first encoded a conditional instruction which then was supplanted by an unconditional one.

There are lots of things that an organism does (and lots of things that happen to it) that are not caused by behavioral instructions written down in the genome or anywhere else. A fly that is smashed by a fly swatter is not executing a conditional program that says "Die if hit by a large object." Nor is it following any other sort of program either. Likewise, I see no reason to think that flies at the beginning of the experiment encoded the instruction "develop bithorax if and only if an ether stimulus is present."

We have here a distinction that is quite familiar from the philosophy of mind—the difference between acting in accordance with a rule and following a rule. The planets act in accordance with a rule that dictates how their orbits are to be shaped. But the planets do not follow any rule. There is no "program" inside them that they execute. In philosophy of mind, just as much as in philosophy of biology, it is a challenging problem to say which behaviors of an organism result from the organism's following a program. The question of saying what the program

---

19. Though the idea that populations benefit from natural selection is a natural one for advocates of group-selection hypotheses, it does not follow simply from the fact that evolutionists wish to talk about adapting and adaptation.

is that an organism follows—describing what its content is—is similarly central.[20]

I have already mentioned that Williams sees the development of bithorax at the initial stage of the experiment as a *failure* of flies to respond. This would suggest that the ether shock *prevented* the developmental program from being executed correctly. But if this is right, then the program cannot say "develop bithorax precisely when an ether stimulus is present," since the phenotypically abnormal flies have not developed in a way that violates *this* instruction. So, for this reason as well, I find it implausible to attribute this conditional (facultative) instruction to the flies at the beginning of Waddington's experiment.

Even if we grant that the organisms began with a facultative instruction and then evolved to an obligate one, it is not entirely clear that this change must involve a *loss* of information. This all depends on the way the instructions are encoded. Suppose, for example, that the genetic system works as described above. A fly's genome has a certain number of plus alleles (which code for bithorax) and a certain number of suppressor alleles. At the start of the experiment the number of plus alleles is low enough for the suppressors to prevent bithorax, except when there is an ether shock. By the end of the experiment, the number of plus alleles has risen to the point where the suppressor alleles cannot prevent bithorax, even when there is no ether. That is, we are imagining the experiment beginning with m alleles encoding the instruction "build bithorax" and ending with n alleles encoding that instruction, where $n > m$. Here we have an *increase* in the informational burden to the genome.[21] Not that I place much credence in this picture of how the organism's developmental program is organized, although as indicated above, something like it is precisely what Williams assumes. My point here is just that it is far from inevitable that conditional and unconditional developmental regularities are always underwritten by genomic instructions of greater and lesser informational content.

A final objection is worth registering. Let us suppose for the sake of argument that the flies began with a conditional (facultative) instruction and evolved to an unconditional (obligate) instruction. Suppose further that this transition marked a decline in informational content. Williams concludes that such a reduction should be viewed as a degeneration, not as an adaptation. I now want to suggest that a trait may be an adaptation even if it means a decline in genetic information.

20. On the question of when such attributions are reasonable, see Dennett (1971). On the question of what makes it true that the program reads one way rather than another, see Stampe (1977) and Dretske (1982).
21. I owe this observation to Jean Murphy.

Immediately after his discussion of bithorax, Williams (p. 82) reminds the reader of "the principle of the economy of information" introduced earlier in the book. This idea flows from the fact that the fitness of a developmental or behavioral instruction is not to be gauged merely by the relation it effects between the organism and the external, physical environment. Also relevant is the amount of storage space the genome uses up in encoding the instruction:

> genetic information is limited in amount and must be utilized as economically as possible (Williams 1966, pp. 41–42).

Imagine an organism that received wounds all of which happened to be shaped like fleur-de-lis. It would end up behaving the same whether it followed the instruction "initiate tissue growth if a wound occurs" or the instruction "initiate tissue growth if a fleur-de-lis-shaped wound occurs." But the latter, says Williams, would take more genetic information to encode, not to mention the fact that implementing the instruction would require fitting out the organisms with a shape detector. Although the two instructions might be behaviorally equivalent, they differ in what we might call their informational fitness.

Williams (pp. 82–83) holds that the principle of informational economy has a bearing on the transition from facultative to obligate instructions in the bithorax experiment:

> Whenever a given character would be more or less universally adaptive in a population, it can be expected to be obligate. Only when adaptive adjustment to uncertain conditions would be important would one expect facultative control. Since the obligate is more economical of information, it can always be expected in situations in which a facultative response would not be significantly more effective.[22]

The principle of informational economy faces two questions. First, it is unclear how much of a premium the genome places on economy of information. The hypothesis of junk DNA (to be discussed in Section

---

22. Williams (p. 82) suggests that the idea of informational economy bears on the nativist/empiricist controversy: "All elements that can be instinctive . . . will be instinctive. Instinct costs less than learned behavior, in the currency of genetic information." Much later in the book (pp. 203, 263), he uses this idea to explain "imprecisions" of adaptive behavior, such as the existence of sexuality in children and homosexual behavior. Also, Williams deploys this idea in his argument against group-selection hypotheses, which I will discuss in Part II. A trait selected because it confers individual advantage may occasionally lead an organism to help the group at its own expense. This will not threaten the individual selection point of view, however, if fine-tuning the organism's behavioral strategy so that it took care only of itself would involve huge informational costs.

9.1) asserts that large portions of the genome perform no function at all in the construction of the organism's phenotype. No one now knows for certain how close this idea is to the truth. If it is on the right track, then storage capacity may not be as precious a resource as Williams assumes.

In addition, if economy of information is adaptive, why should a reduction in "information load" automatically count as a degeneration rather than as an adaptation? Williams' own principle suggests that *simplification* can serve an adaptive function. Discarding useless baggage can be an adaptation that benefits overburdened travelers. This is another reason to reject Williams' informational argument that the emergence of obligate bithorax cannot be an adaptation.

I conclude that neither Lewontin's nor Williams' arguments show that selection is insufficient for adaptation. They agree that adaptation implies selection but reject the converse. I, on the other hand, see no problem with the following characterization of adaptation:

> A is an adaptation for task T in population P if and only if A became prevalent in P because there was selection for A, where the selective advantage of A was due to the fact that A helped perform task T.

Notice that this definition characterizes the idea of a trait's being an *adaptation for* a given task, not the idea of adaptation *simpliciter*. But the extension to the latter notion is obvious enough: A is an adaptation precisely when there is some task such that A is an adaptation for performing it. Note that the causal notion of *selection for properties* occupies center stage in this fundamental evolutionary concept.

Although I have been critical of Williams' arguments for thinking that bithorax is not an adaptation, there is a way of making sense of his conclusion in terms of this characterization of adaptation. The fruit flies at the beginning of the experiment were *labile*; they had genomes that generate normal thoraxes *or* bithorax, depending on the presence or absence of ether. That is, it was the fact that these flies had nonflat norms of reaction that permitted the selection regimen to go forward. Presumably, this property of the genome is not an adaptation: There never was selection for genomes that were capable of generating bithorax phenotypes when exposed to ether and normal phenotypes when not. The fact of canalization should receive precisely the opposite interpretation. Selection favored those genotypes that produced normal thoraxes; it is the merest accident that some such genotypes yield bithorax in an environment in which there are ether shocks. The point is that the bithorax phenotype, as well as the properties of the genome that caused flies to produce bithorax, were not adaptations at the *beginning* of the experiment.

Not that the opposite interpretation is *impossible*; it's just implausible. There are cases, after all, in which the lability of the genome may be viewed as the product of natural selection. For example, in many species, sex is environmentally determined, and it is not incoherent to think that the existence of this facultative response exists in some cases because being male was advantageous in some environments and being female was in others (Bull 1983).

This way of formulating the thesis that nonflat norms of reaction (in the case of the facultative bithorax flies at the start of the experiment) are not adaptations makes sense only on a certain presupposition. It is the view of disjunctive causes discussed in Section 3.1. When a particular *Drosophila* genome, at the start of the experiment, produces a bithorax phenotype, we do not view that phenotype as an adaptation. The genome that produced it was, in the distant past of the lineage, selected for its ability to generate normal thoraxes; the bithorax in the ether environment is fortuitous. It happened that in the peculiar environment of Waddington's laboratory, this phenotype was advantageous.[23] But even so, we classify the initial appearance of bithorax in the laboratory population as an accidental benefit, not an adaptation.

Let's compare this example with another. A *Drosophila* population is subject to high-temperature stress. It gradually evolves a thicker skin, which is selectively advantageous because it insulates. Later on, the environment is changed and the temperatures are now very low. The thickened skin it evolved before remains advantageous. Is the thickened skin an adaptation merely for *high* temperature tolerance, or is it an adaptation for *generalized* temperature tolerance? If the former, then the benefit conferred by the thickened skin at low temperatures counts as fortuitous—the trait is not an adaptation for coping with low temperatures. If the latter, then the thickened skin is an adaptation for coping with the environmental stress that obtains, and so we describe its ability to withstand low temperatures as an adaptation. My own intuitions here are that there is nothing amiss in viewing thickened skin as an adaptation for generalized temperature tolerance, even though it evolved only in the setting of high-temperature stress.

The problem is to find a difference between these two cases. It is perverse, I take it, to think of the "lability" of the fruit fly genome in the bithorax case as an adaptation. There was selection for normal thoraxes, not selection for "lability." But why are we less inclined to put our foot down in the same way in the temperature tolerance case?

23. And by the *end* of the experiment, bithorax had become an adaptation.

Why not insist that there was selection for high-temperature tolerance, not selection for general temperature tolerance?

I suggest that the analysis developed in Section 3.1 helps explain this difference. The causal processes that lead to a normal thorax in an ether-free environment are quite different from the causal processes that produce bithorax under ether shock. On the other hand, in the second example, the way in which the organism withstands high temperatures and the way it tolerates low ones are the same—added insulation. Here we have a recapitulation of the comparison of the examples of the ice cube melting and the toy's being crushed. I will not repeat the theses about disjunction and causation developed there but will only stress their application to the present context. The biologist's distinction between adaptation and fortuitous benefit requires that certain sorts of comparisons make sense. We must be able to look at what a trait does in the present environment and what it did in the past and say whether these are "the same." If contrived disjunctive predicates were allowed to describe selection processes, every present benefit of a trait would count as "the same" as the historical advantages that made the trait prevalent in the first place. We would be able to say that the bithorax phenotype produced under ether shock at the beginning of the experiment and the normal thorax in ancestral populations are manifestations of the same underlying adaptation—the "lability" of the genome. The effect would be that the biological distinction collapses; every present advantage now counts as an adaptation.[24] The conclusion is not that the biologist's distinction is untenable but that it has an interesting metaphysical presupposition.[25]

Adaptation and fitness (adaptedness) are complementary concepts. The former looks to the past, reflecting the kind of history that a trait has had. The latter looks to the future, indicating the chances that organisms have for survival and reproductive success. These retrospective and prospective concepts are mutually independent. An adaptation may cause problems for the organisms that have it; a changed environment may mean that an adaptation is no longer advantageous. Similarly, an adaptive trait may arise by chance; mutation may "at

24. In more detail, here is how the fallacious inference would run: Ancestrally, there was selection for normal thoraxes in ether-free environments. Hence, there was selection for being a normal (single-thorax) fly in an ether-free environment or a bithorax fly in an ether shock environment. Hence, producing a normal thorax in ether-free environments or bithorax under ether shock was an adaptation. In Waddington's laboratory, the "labile" flies at the start of the experiment had this characteristic. Hence their producing bithorax under ether shock was an adaptation.

25. This idea was suggested to me by an argument advanced in Dretske (1984) concerning the relationship between learning and the attribution of content to psychological states.

random" produce a novel character that enhances fitness. That is, a trait can be an adaptation at a given time without being adaptive at that time; and it can be adaptive at a time without being an adaptation then. But, as I understand these concepts, adaptations cannot reach fixation by a process of random drift; nor can an adaptive trait be deleterious. To say that a trait is an adaptation is to make a claim about the cause of its presence; to say that it is adaptive is to comment on its consequences for survival and reproduction.

These distinctions would be easier to grasp if "adaptation," "adaptedness," and "adapting" did not sound like different terms for the same thing. But it is no accident that these very different concepts are given names that sound so similar. In a simpler world, the process that produces adaptation would be certain to increase the level of adaptedness. In a simpler world, an organism's "adapting" to its environment by modifying its behavior and physiology would be the same kind of process as a population's "adapting" to its environment by modifying its genetic composition. If matters were simple, frequency-dependent selection would not occur and evolutionary adaptation would have the same structure as ontogenetic adaptation. The vocabulary we have inherited was perhaps designed with these simpler assumptions in mind. It is a sign of the conceptual progress made by evolutionary theory that we now are forced to recognize distinctions that our commonsense vocabulary makes less than transparent.

PART II
The Group Above and the Gene Below

Chapter 7
Beginnings

In the next three chapters, the units of selection problem will be set forth as a *conceptual* puzzle. The goal is not to decide what the unit of selection in fact is (or what the units are) but to understand what it means for something to be a unit of selection. Chapter 7 plays an almost totally negative role in this effort. After briefly surveying the historical background of the problem in Section 7.1, I describe a series of puzzles and try to expose a number of fallacies. The cumulative effect will, I hope, be one of constructive confusion. The confusion should arise from seeing that biologists have often talked at cross purposes about the problem of the units of selection. The constructive contribution should come from assembling a number of requirements that an adequate view of the biological problem should obey.

I should say at the outset that my criticisms are aimed all over the map. In Sections 7.3 to 7.5, I challenge several arguments that have been made in favor of the view that the single gene is the unit of selection. However, in Sections 7.2 and 7.8, I criticize certain conceptions of what a unit of selection is that lend support to the idea that higher-level units (like whole organisms or groups) are units of selection as well. Since no positive thesis will be advanced in this chapter, I will be able to alienate reductionists and antireductionists alike.

The critique provided in this chapter clears the ground, so that a positive foundation can be laid in the next. In Chapter 8, I try to clarify the concept of cause and to show how this idea can be used to make sense of the units of selection controversy. Chapter 9 continues this project, with detailed application of my causal analysis to several theoretical proposals that are currently attracting considerable biological attention.

7.1 Historical Background

So far in this book, I have given natural selection its classic Darwinian construal. Organisms within a single breeding population differ with respect to their abilities to survive and reproduce. Heritable characteristics that make for a fitter organism will tend to spread through the population. It is essential to realize that Darwinian (i.e., individual) selection favors organisms that are best at facilitating their own survival and reproductive success. This sort of selection helps those who help themselves. Organisms that are "altruistic," in the sense of helping others survive and reproduce at their own expense, will be at a selective disadvantage. Darwinian selection works for the good of the individual organism, not for the good of the group or the species.

Notice that the abstract structure of the concept of natural selection does not dictate that it must be *organisms* that are the units of selection. Natural selection impinges on a set of *objects* if there are (heritable) differences in fitness among them (Lewontin 1970). However, this abstract requirement does not say what sorts of things those objects might be. A stunning variety of units is in principle available, ranging from macro-molecules, genes, and cells on the one hand up through organisms, populations, species, and communities on the other. The bulk of Darwin's thinking on natural selection involves the substantive assumption that *organisms* are the units of selection.[1]

What would it be to choose another unit—for example, groups—as the objects that undergo a selection process? As we will see in what follows, there are a number of conceptual problems that must be solved before this question can be answered precisely. For the moment, we may note that the phenomenon of *altruism* must be a central puzzle for the Darwinian paradigm. Individuals who donate reproductive advantages to others at their own expense will do less well than individuals who never make such donations but benefit from the donations of others. Under Darwinian selection, it is better to receive than to give. However, if groups containing altruists found more colonies and go extinct less often than groups composed just of selfish individuals, altruism might become established. Group selection appears to be an alternative principle that makes the existence of altruism intelligible.

Although Darwin was extremely scrupulous about treading the Darwinian straight and narrow, at least once in his career he advanced a group selection hypothesis. In *The Descent of Man*, Darwin considers altruism in human moral behavior. First comes the statement of the problem:

1. I therefore use the term "Darwinian selection" to denote a form of individual selection in which an organism's fitness reflects just its chances of surviving and reproducing. As we shall see, Darwin occasionally described selection in other terms and more than occasionally wrote in such a way as to blur the distinction between different units of selection. "Darwinian" is used here in neither of two ways that have become all too prevalent—first, as a synonym for "true" and second, as a synonym for "stodgy."

It is extremely doubtful whether the offspring of the more sym-
pathetic and benevolent parents, or of those which were the most
faithful to their comrades, would be reared in greater number than
the children of selfish and treacherous parents of the same tribe.
He who was ready to sacrifice his life, as many a savage has been,
rather than betray his comrades, would often leave no offspring
to inherit his noble nature. The bravest men, who were always
willing to come to the front in war, and who freely risked their
lives for others would on average perish in larger numbers than
other men (Darwin 1871, p. 163).

Then Darwin gives his selectionist explanation:

It must not be forgotten that although a high standard of morality
gives but a slight or no advantage to each individual man and his
children over the other men of the same tribe, yet that an advancement in the standard of morality and an increase in the number
of well-endowed men will certainly give an immense advantage
to one tribe over another (Darwin 1871, p. 166).

We see here a group selection approach to the puzzle of altruism.
Although individual selection favors the selfish behavior of shunning
risks and letting others assume them, *groups* of selfish individuals would
be at a competitive disadvantage compared with groups that contain
altruists. Hence altruism would have a chance of spreading, opposed
by Darwinian individual selection but propelled by the force of group
selection.

Darwin and Wallace corresponded at length about the feasibility of
a group selectionist explanation of another phenomenon, quite unrelated
to human morality (Darwin and Seward 1903, pp. 288–297; discussed
in Ruse 1980). Why are hybrids—the result of crosses between organisms from different species—so frequently sterile? Darwin observed
that if hybrids were bound to be sterile or inviable, individual selection
would lead organisms to avoid mating with members of other species,
because interspecific mating would be a poor investment of an organism's reproductive effort. But this does not show why the hybrids
should be sterile in the first place. In fact, individual selection seems
to predict that hybrids should *not* be sterile, just as it does for crosses
between members of the same species.

Wallace offered, as he frequently did for other phenomena, a group
selection hypothesis:

It appears to me that, given a differentiation of a species into two
forms, each of which was adapted to a special sphere of existence,
every slight degree of sterility would be a positive advantage, not

> to the individuals who were sterile, but to each form. If you work it out, and suppose the two incipient species $a \ldots b$ to be divided into two groups, one of which contains those which are fertile when the two are crossed, the other being slightly sterile, you will find that the latter will certainly supplant the former in the struggle for existence; remembering that you have shown that in such a cross the offspring would be more vigorous that the pure breed, and therefore would certainly soon supplant them, and as these would not be so well adapted to any special sphere of existence as the pure species a and b, they would certainly in their turn give way to a and b (Darwin and Seward 1903, p. 268).

Wallace reasoned that if hybrids are more vigorous than pure breeds, pure breeds will tend to disappear if the hybrids are not sterile. Hence a species that is not cross-fertile with another will have a better chance of surviving than a species that is. Wallace does not say why a hybrid species should not swallow up nonoutcrossing species as well as its own outcrossing parents; there are ecological assumptions here that have not been made explicit.[2] In any event, the point is that Wallace conjectured that hybrid sterility evolved for the good of the species.

Darwin had an alternative explanation. Neither individual nor group selection is the reason for hybrid sterility. Rather, Darwin thought the characteristic evolved as a pleiotropic (Sections 1.2 and 3.2) consequence of selection for other traits. The process of speciation leads populations to diverge from each other. Each species adapts to its own way of life. A consequence of these physical differences between species is that crosses between species are sterile. But nowhere in the separation process was there selection for hybrid sterility.

Darwin and Wallace went back and forth on what Darwin called "this terrible problem." Darwin held constant. Wallace grew more uncertain, but his confidence returned many years later when he repeated the group selectionist argument in his book *Darwinism* (1905).

There is another place where Darwin seems to come close to offering a group selection account. In *The Origin*, he takes up the question of how the existence of sterile castes in the social insects is to be explained:

> How the workers have been rendered sterile is a difficulty; but not much greater than that of any striking modification of structure; for it can be shown that some insects and other articulate animals in a state of nature occasionally become sterile; and if such insects had been social, and it had been profitable *to the community* that a number should have been annually born capable of work, but

2. I owe this observation to David Wilson.

incapable of procreation, I can see no very great difficulty in this being effected by natural selection (Darwin 1859, p. 236; emphasis mine).

Darwin's subsequent discussion, however, suggests that the group selectionist phrase "profitable to the community" may have been a verbal slip. Darwin points out that a parent may benefit by producing some sterile offspring, if the sterile offspring help their fertile siblings. This is individual selection, in which there is selection for a certain reproductive strategy. An individual selection account does not impose the preposterous requirement that organisms gain a reproductive advantage by being sterile. Sterility is explained as a consequence of selection for having some mix of fertile and sterile offspring.[3]

One might propose to explain the origin of sterile castes in terms of an individual parent's being selected in virtue of its use of a given reproductive strategy. Could one also describe the process in terms of a *kin group*'s being selected in virtue of its having a certain mix of sterile and fertile individuals? Is there any real difference between these two construals of what the unit of selection is in this case? Perhaps the difference between *individual* parent and kin *group* is merely terminological. The same question may be raised about Darwin's remarks on human morality. Darwin suggested a group selectionist explanation. But if the tribal groups were composed of genetically related individuals, could one redescribe the group selection process in terms of individual selection? Are there objective differences between different sorts of selection processes or do the differences merely arise from the different perspectives that investigators may choose to assume? One purpose of this chapter and the ones that follow will be to provide this question with a precise answer.

If we leap ahead half a century to the Modern Synthesis of Darwinism and Mendelism in the 1930s and 1940s, what has become of the issue of the units of selection? The situation is complicated by the fact that a schism between mathematical theoreticians and field naturalists had developed. The split was not first and foremost one of doctrine, but of outlook. The highly abstract mathematical character of population genetics models made them either unpalatable or indigestible to many naturalists. A gap was widening; some biologists thought of simple

3. Compare Fisher's sex-ratio argument (Section 1.5). On logical grounds alone, a sex ratio (other than 1:0 or 0:1!) is a property not of an organism but of a group. Yet the property of producing offspring with a given sex ratio is a property of an organism. Darwin's ideas about sterile castes, like the argument about the sex ratio that he presented in the first edition of *The Descent of Man*, only to withdraw it in the second, involves taking the production of grandoffspring into account in computing an organism's fitness.

models as a source of insight into a complex world, while others were suspicious of the models on the grounds that they were oversimple.

Not that the mathematical theoreticians were of one mind on the question of group selection. But Fisher (1930), Haldane (1932), and Wright (1945) clearly recognized that standard Darwinian selection within a single breeding population would not work in favor of altruistic characteristics. The idea that altruism required a novel evolutionary mechanism was much less obvious to naturalists, however, who often used the jargon of "group adaptation" and "evolution for the good of the species" without realizing the theoretical consequences of such turns of phrase.

Fisher (1930, pp. 49–50), in a section of his book called "the benefit of the species," is quite explicit about how alien the idea of group selection is to his view of evolution:

> It will be observed that the principle of Natural Selection, in the form in which it has been stated in this chapter, refers only to the variation among individuals (or co-operative communities) and to the progressive modification of structure or function only in so far as variations in these are of advantage to the individual, in respect to his chance of death or reproduction. It thus affords a rational explanation of structures, reactions and instincts which can be recognized as profitable for their individual possessors. It affords no corresponding explanation for any properties of animals or plants which, without being individually advantageous, are supposed to be of service to the species to which they belong.
>
> This distinction was unknown to the early speculations to which the perfection of adaptive contrivances naturally gave rise. For the interpretation that these were due to the particular intention of the Creator would be equally appropriate whether the profit of the individual or of the species were the objective in view. The phrases and arguments of this pre-Darwinian viewpoint have, however, long outlived the philosophy to which they belong. It would be easy to find among modern writers many parallels to the thought expressed in the following quotation:—'Of what advantage could it be to any species for the males to struggle for the females and for the females to struggle for the males?'
>
> This sort of question might appropriately be put to an opponent who claimed that the instincts of animals were in each case due to the direct contrivance of the Creator. As a means of progressive change, on the contrary, Natural Selection can only explain these instincts in so far as they are individually beneficial, and leaves entirely open the question as to whether in the aggregate they are

a benefit or an injury to the species.

There would, however, be some warrant on historical grounds for saying that the term Natural Selection should include not only the selective survival of individuals of the same species, but of mutually competing species of the same genus or family. The relative unimportance of this as an evolutionary factor would seem to follow decisively from the small *number* of closely related species which in fact do come into competition, as compared to the number of individuals in the same species; and from the vastly greater *duration* of the species compared to the individual. Any characters ascribed to interspecific selection should of course characterize, not species, but whole genera or families, and it may be doubted if it would be plausible to point to any such character, with the possible exception . . . of sexuality itself, which could be interpreted as evolved for the specific rather than for the individual advantage.[4]

Sewall Wright's position on group selection and altruism was different. Wright devoted much of his productive career to arguing for the importance of a model he called *interdemic selection* or the *shifting balance theory*, but this idea has little per se to do with altruism (as Wright 1980 himself has recently emphasized). In fact, it is a real question whether Wright's "interdemic selection" is correctly treated as a case of group selection or as a process that simply combines the effects of individual selection, migration, and drift. This question will occupy our attention in Section 9.2.[5]

Wright's best-known discussion of group selection for altruism was in a book review of George Gaylord Simpson's *Tempo and Mode in Evolution* (Wright 1945). He sketches a model in the space of about one page. There are numerous semi-isolated populations. In each, the altruistic characteristic is less fit than the selfish one. But populations of altruists are better able to send out migrants to colonize other groups. Wright's model thus involved two forces, one opposing the evolution of altruism, the other favoring it. However, since the model fails to specify the relative values of certain population parameters, it is impossible to tell which push is stronger. Wright simply observes that "it is . . . difficult to see how socially advantageous but individually dis-

4. Fisher's two arguments for thinking that group selection is a comparatively *weak force* of evolution—the one appealing to numbers, the other to time—will be discussed in Sections 9.2 and 9.3, respectively.
5. Whether or not Wright's model ought to be thought of as defining what group selection is, it is clear that many biologists who discussed group selection in the period following the publication of Williams' (1966) book interpreted Wright's models in this way (Wilson 1983).

advantageous mutations can be fixed without some form of intergroup selection" (Wright 1945, p. 417).

Haldane (1932) also deemphasizes the evolution of "socially valuable but individually disadvantageous characters," relegating it to a few pages in the Appendix (pp. 207–210). He argues that the evolution of altruism depends on small group size and the relatedness of individuals in the same group. Again, evolution "for the good of the species" is recognized to require a special mechanism, one that did not deserve center stage in a book called *The Causes of Evolution*.

I have already mentioned the schism that arose within the Modern Synthesis between naturalists and mathematical model builders. Members of the latter group, I have suggested, were not much taken with the idea of group selection; they either explicitly rejected it or viewed it as a rather minor wrinkle. But, as Fisher mentions in the above quotation, there were "modern writers," mainly found within the ranks of naturalists, who had failed to completely assimilate this Darwinian point of view. Not that all naturalists thought in terms of the good of the species. But those who thought this way were mostly naturalists.

David Wilson's (1983) review article on the concept of group selection cites some instructive examples from the influential ecology textbook by Allee et al. (1949):

> If whole populations are adaptive, it seems possible that adaptations producing beneficial death of an individual—death for the benefit of the population—might evolve (p. 692).

> It may be concluded from these data that . . . natural selection operates upon the whole interspecies system, resulting in the slow evolution of adaptive integration and balance. Division of labor, integration, and homeostasis characterize the organism and the supraorganismic intraspecies population. The interspecies system has also evolved these characteristics of the organism and may thus be called an ecological supraorganism (p. 728).

> . . . it may be expected that surviving populations will be coordinated under the formula "one for all and all for one" (p. 694).

A similar approach is at work in this passage from Odum's ecology textbook (quoted in Wilson 1980, p. 2):

> although the anaerobic saprophages (both obligate and facultative) are minority components of the community, they are none-the-less important in the ecosystem because they alone can respire in the dark, oxygenless recesses of the system. By occupying this inhospitable habitat they "rescue" energy and materials for the majority of aerobes. What would seem to be an "inefficient" method

of respiration, then, turns out to contribute to the "efficient" exploitation of energy and materials in the ecosystem as a whole (Odum 1963, p. 28).

Evolution for the good of the species also was a persistent theme in the ethology literature (see, for example, Lorenz 1966).

The division of labor between mathematical theorizing and natural observation had permitted a split to develop. Division of labor, whether it occurs in insect colonies, automobile factories, or scientific communities, has certain familiar advantages. But advantageous arrangements can be accompanied by deleterious (pleiotropic) effects. In the case before us here, it isn't the consideration of group selection hypotheses that was regrettable. Rather, what was unsatisfactory was the repeated appeal to hypotheses that theoreticians had claimed to be implausible *without this occasioning any substantial foundational debate* between the two camps. Whatever may be said of trees falling in the forest, problems may exist even when no one is trying to solve them.

Perhaps it is unrealistic to expect such a heterogenous, multidisciplinary scientific community to have immediately recognized and come to grips with this issue. Foundational crises take a while to become visible. And, of course, there were other pressing problems that evolutionary biology addressed in this period. Whatever the correct historical explanantion of this time lag, the issue finally surfaced during the 1960s.

In 1962, V. C. Wynne-Edwards published a book called *Animal Dispersion in Relation to Social Behavior*. This book systematized and consolidated the idea that Darwinian, individual selection is not a sufficient mechanism to account for the adaptations that are observed in nature. Wynne-Edwards realized that the heart of the Darwinian theory counterpredicts adaptations that exist for the good of the group. But since he thought that careful natural observation shows that altruism is widespread, he concluded that a hypothesis of group selection was required to account for the data.

The main kind of adaptation that interested Wynne-Edwards was the restraint a population might show in reproducing and in exploiting food supplies. Wynne-Edwards thought that populations often live below the maximum "carrying capacity" that the environment permits. A "selfish individual" that eats more and reproduces more than its prudent colleagues will gain a reproductive advantage. But populations made up of such "over-exploiters" would be more likely to crash, a victim of the famine they helped to cause.[6]

6. Gilpin (1975) and Maynard Smith (1976) have pointed out that just as Darwin had his Malthus, so the idea of group selection à la Wynne-Edwards was anticipated in the

D. S. Wilson (1983, pp. 165–166) has pointed out that although Wynne-Edwards appealed to Wright's work on interdemic selection, Wynne-Edwards and Wright were thinking about kinds of processes that differed in two ways:

> First, Wynne-Edwards clearly emphasized highly persistent groups (such as breeding colonies of shorebirds), although he also considered more ephemeral groups (such as wood lice occupying separate tree trunks. . .). Wright also envisioned multigenerational groups, but his emphasis was rather the reverse. Second, although both acknowledged many kinds of group-beneficial traits, Wynne-Edwards focused almost exclusively on population regulation. He imagined animals reproducing below their potential, so that the population does not go extinct. By contrast, the socially beneficial trait in Wright's model causes the population to grow faster, sending out more dispersers to colonize new groups. Wright therefore emphasized the differential productivity of groups, while Wynne-Edwards emphasized differential extinction. In retrospect these two approaches can easily be united; groups that grow below their potential can ultimately produce more dispersers in the same way that birds who reduce their clutch size ultimately can produce more offspring. At the time, however, the difference may have appeared more substantial.

Wynne-Edwards' book was historically significant in three ways. First, as I've mentioned, group selectionist ideas were implicit in a good deal of evolutionary theorizing, often without precise recognition of their departure from the Darwinian paradigm. Wynne-Edwards brought the novel theoretical status of the group selection hypothesis into focus and argued that group selection must have played a major role in evolution. Second, Wynne-Edwards typified one reaction of naturalists to the genetical models in the Modern Synthesis that said that group selection—evolution for the good of the group—would not work. Naturalists not infrequently reacted to mathematical models by thinking that if they conflicted with natural observations, there must be something wrong with the models. For reasons independent of the group selection issue, population genetics had been criticized for taking seriously models

---

theory of Carr-Saunders (1922). Carr-Saunders posited a competition between nations, in which nations of cooperative individuals would do better than nations all of whose citizens competed with each other. Gilpin (1975) and Maynard Smith (1976) do not indicate whether this idea actually influenced group selectionist thinking in this century or merely represents an anticipation of it. Wynne-Edwards (1962) mentions that he read Carr-Saunders only after he wrote *Animal Dispersion*.

that were known to be oversimplified.[7] Wynne-Edwards' nonquantitative approach was hardly without precedent in evolutionary biology.

The third salient fact about Wynne-Edwards' book was that it was one of the principal targets of George C. Williams' immensely influential *Adaptation and Natural Selection*, published in 1966. Before Williams' book, evolutionary biology was able to persist in the sort of schizophrenia just mentioned. Even if mathematical models predict that altruism is extremely improbable if not impossible, there is still room for disagreement. One may reject the mathematical models as unrealistic and persist in the claim that altruism exists. Or one may insist on the approximate accuracy of the models and conclude that altruism is an illusion. Williams reached across the disciplinary divide. Although he spoke from the point of view of population genetics theory, his arguments are almost entirely nonmathematical. They address the phenomena that naturalists had focused upon directly. His goal was to show that hypotheses of group selection rest on a variety of fallacies. Since the role of mathematical models was almost completely de-emphasized in his presentation, it was no longer so easy to dismiss the criticism as based on an oversimplified understanding of evolutionary theory. Rather, the book mainly reads as if it is the core of evolutionary theory itself, and not the contingent assumptions of some special mathematical model, that militates against the group selection idea.

Williams advocated a return to the Darwinian fold, but with a new wrinkle. The Modern Synthesis had made the field of population genetics possible; the synthesis of Darwinism and Mendelism had provided the conceptual equipment needed for understanding evolutionary processes in terms of changes in gene frequencies. This perspective was the one that Williams brought to bear in his own positive proposal for what the unit of selection is. Selection does not occur at the level of species or groups, or even, strictly speaking, at the level of organisms. Rather, it is the single gene that is the unit of selection.

Several lines of thinking evolved in response to Williams' challenge. Mathematical theoreticians took up the idea of group selection on its own terms, and by modeling the process tried to find out what natural circumstances would make group selection a significant cause of evolutionary change. These models generally came up with negative results, indicating that the natural conditions required for group selection to work are so extreme or severely constrained that group selection explanations of real-world phenomena are to be viewed as implausible. But, in an important review article (to be discussed in Section 9.2), Michael Wade (1978) argued that each of these models contains several unrealistic assumptions that a priori bias the case against group selection.

7. An example is Mayr's (1963) antipathy to "beanbag genetics," to be discussed in Section 7.2.

During the same time, David S. Wilson (1975, 1980) offered arguments for the plausibility of postulating selection at the level of groups and communities. In addition, a number of paleobiologists have argued that certain patterns of diversity in the fossil record are best explained via a hypothesis of species selection (Section 9.4). Although Williams (1966) gave the group selection concept a good scorching, it now appears to be rising, phoenix-like, from its own ashes.

Besides being a possible explanation of altruism, group selection has been discussed as a candidate account for the existence of sex. As the passage cited above from Fisher (1930) suggests, sexuality is puzzling from the perspective of individual selection, since it seems to be a waste of reproductive effort for a parent to produce sons.[8] If selection favors organisms that maximize their genetic representation in the next generation, asexual reproduction would seem to be favored. An asexual female has 100 percent of her genes in each offspring; a sexual female can expect to have only 50 percent of her genes in each offspring. Sexuality seems to pose a "twofold" disadvantage for the organism. On the other hand, it may conceivably be true that asexual groups are more likely to go extinct than sexual ones, since sexual groups will harbor a greater store of variability. This problem is as difficult as it is central. Sexuality provides no straightforward argument for the effectiveness of group selection; but accounting for sex is a problem that continues to keep group selection a subject of controversy.

Williams' (1966) positive proposal—genic selectionism (i.e., the view that the single gene is the unit of selection)—has also produced a lively debate.[9] But the character of the dissension is worthy of note. Those who deny that the single gene is the unit of selection have not produced any novel experiment or unexpected natural observation in reply. Rather, they have simply mustered a small number of fairly familiar facts. As in the dispute about higher-order selection processes, there is an important component of the problem that is conceptual, one that involves clarifying what the claim of genic selectionism amounts to and determining what sorts of facts would count as evidence in its favor.

7.2 Transitivity and Context Dependence

Williams' (1966) positive proposal—that the unit of selection is the single gene—contradicts some core ideas in mainstream Darwinism.

8. See Williams (1975) and Maynard Smith (1978) for discussion of this central puzzle.
9. I should stress that "genic" should not be confused with "genetic." Genic selectionism is the view that all selection is selection for and against *single* genes. The idea that gene complexes—i.e., ensembles of genes—are the units of selection is thus contrary to the thesis of genic selectionism.

For example, Ernst Mayr (1963, p. 184) asserted that "natural selection favors (or discriminates against) phenotypes, not genes or genotypes." The idea here is that organisms survive and reproduce in virtue of their phenotypes—for example, because they are resistant to malaria (Section 1.4), or have protective coloration, or produce offspring with a given sex ratio (Section 1.5). To be sure, if phenotypic characteristics are correlated with genotypic ones, a *consequence* of natural selection may be that gene frequencies change. But the unit of selection, according to this line of argument, is the organism's phenotype, not the single gene.

Gould (1980a, p. 90) gives voice to this line of thinking when he observes:

> Selection simply cannot see genes and pick among them directly. It must use bodies as an intermediary. A gene is a bit of DNA hidden within a cell. Selection views bodies.

I will call this argument against the thesis of genic selectionism *the directness objection*.

A second Darwinian idea has been mustered against genic selectionism. It appeals to a fact that has been well known since the demise of beanbag genetics, as it has been pejoratively called.[10] Mayr (1963, p. 296) has called it the "genetic theory of relativity":

> No gene has a fixed selective value; the same gene may confer high fitness on one genetic background and be virtually lethal on another.

Gould (1980a, p. 91) makes use of the same line of argument:

> There is no gene "for" such unambiguous bits of morphology as your left kneecap or your fingernail. Bodies cannot be atomized into parts, each contructed by an individual gene. Hundreds of genes contribute to the building of most body parts and their action is channeled through a kaleidoscopic series of environmental influences: embryonic and postnatal, internal and external. Parts are not translated genes, and selection doesn't even work directly on parts. It accepts or rejects entire organisms because suites of parts, interacting in complex ways, confer advantages.

This argument against genic selectionism, based on the fact that the gene/phenotype relation is many-many rather than one-one, I will call *the context dependence objection*.

---

10. But see Haldane (1964) for a defense of this approach on the grounds that it is methodologically valuable, if not totally realistic.

Notice that these two objections to the thesis that the single gene is the unit of selection lead in slightly different directions. The directness objection says that phenotypes—not single genes, not gene complexes—are the level to focus upon. The context dependence objection, on the other hand, apparently allows that *gene complexes* are selected for and against. Its animus is the *single gene*, not the genetic level as a whole.

Both these objections base their case on certain *causal* facts concerning the relationship of genotype and phenotype. Selection does not work for and against single genes, they say. Instead, the correct view, according to the directness objection, is that selection works for and against whole phenotypes. According to the context dependence objection, selection works for and against ensembles of genes. As we will now see, advocates of genic selectionism have a very different picture of the causal relation of genotype, phenotype, and natural selection, one which, if true, refutes both of these objections.

Those who argue that the single gene is the unit of selection often seem to think of genes as the deeper cause of evolution by natural selection. A phenotype, on this view, is just the rather superficial face that genes show to the world. Dawkins (1976, p. 21) provides a metaphor that nicely illustrates the thesis that genes are the causal heart of the matter. Organic evolution began four billion years ago with replicator molecules constructing "survival machines" around them as a "protective coat":

> But making a living got steadily harder as new rivals arose with better and more effective survival machines. Survival machines got bigger and more elaborate and the process was cumulative and progressive. . . . Was there to be any end to the gradual improvement in the techniques and artifices used by the replicators to insure their own continuance in the world? . . . They did not die out, for they are past masters of the survival arts. But do not look for them floating loose in the sea; they gave up that cavalier freedom long ago. Now they swarm in huge colonies, safe inside gigantic lumbering robots, sealed off from the outside world, communicating with it by tortuous indirect routes, manipulating it by remote control. They are in you and in me; they created us, body and mind; and their preservation is the ultimate rationale for our existence. They have come a long way, those replicators. Now they go by the name of genes, and we are their survival machines.

The picture given here is that genes cause phenotypes, and phenotypes then determine survival and reproductive success. Assuming that causality is *transitive*, it will follow from this that genes cause survival and reproductive success. Genes, phenotypes, and evolution are links on a causal chain. Genes cause evolution, because they cause something else (i.e., phenotypes) which themselves cause evolution.

So the directness objection will have no force, if causality is a transitive relation. The idea of transitivity is a familiar one. For example, "being taller than" is a transitive relation. Consider any three individuals x, y, and z. If x is taller than y, and y is taller than z, then x is taller than z. "Feeling affection for" is not a transitive relation, however, as the experience of introducing one's friends to each other can reveal.

But *is* causality transitive? It is easy to think of examples that suggest that it is. If I dial your telephone number, this will cause your telephone to ring. If your telephone rings, this will cause you to lift the receiver. Does my dialing cause you to lift the receiver? Intuitively, the answer seems to be *yes*.

However, the question of the transitivity of causality is less straightforward than this little example might suggest. The wrinkles will be ironed out in Chapter 8. For now, it should be clear how the assumption of transitivity may be exploited by advocates of genic selectionism. They may grant that selection acts "directly" on an organism's whole phenotype. Similarly, the "direct" cause of your picking up your telephone is that it rang. But this hardly shows that my dialing your number was causally inert. If causation is transitive, the "directness" issue matters very little. If genes cause phenotypes and phenotypes cause survival and reproductive success, then genes *cause* (and are not merely correlated with) survival and reproductive success. Genic explanations of evolution will be "deeper" than phenotypic ones, in the sense that ultimate causation goes to the heart of the matter in a way that proximate causation does not. As Dennis Stampe has observed, the selfish gene idea makes genes look like generals who send their troops into battle. The fact that generals seldom die in wars, but allow their soldiers to do this for them, hardly shows that generals fail to powerfully influence the outcome. Their lack of direct involvement is, if anything, a sign of their power, not their impotence.[11]

11. Brandon (1982) has used the vocabulary of Salmon's (1971) "screening off" idea to characterize a sense in which phenotypes are units of selection. An organism's phenotype *screens off* its genotype from its subsequent survival and reproductive success. This means that an organism's probability of surviving and reproducing, given just the information about its phenotype, is not altered by including further facts about its genotype. But, says Brandon, an organism's genotype does not screen off its phenotype from its reproductive success. Although genotypic facts about an organism may confer on it a probability of survival and reproduction, that probability would be revised by taking the organism's phenotype into account. It is not surprising that this asymmetry between genotype and phenotype obtains, in view of the fact that if G causes P and P causes R, it is frequently the case that the intermediate link in the chain, P, will screen off G from R. Think of the relation of my dialing, your phone's ringing, and your answering. But screening off

So the directness objection to the view that the single gene is the unit of selection is refuted, if causality is transitive. However, the context dependence objection is less vulnerable. It may grant the transitivity of causality but still dig in its heels over the thesis that the single gene is the unit of selection. That is, both these objections find fault with the following "genic syllogism":

Genes cause phenotypes.
Phenotypes cause reproductive success.
Causality is transitive.

Therefore, genes cause reproductive success.

The directness objection is committed to challenging the third premise. The context dependence objection rejects the first premise (and perhaps others besides).

Whether a gene helps or hurts an organism depends on what other genes occur in the organism's genome. This is the phenomenon of "polygenic effects"—Mayr's "genetic theory of relativity." The context dependence objection rejects genic selectionism in favor of the view that causal efficacy is a property of *ensembles* of genes. The fact of context dependence is supposed to show that selection acts for and against gene complexes, not for and against single genes. However, for this objection to go through, a puzzle about causality must be solved.

For one event to cause another, it is not required that the former be sufficient for the occurrence of the latter. I strike a match and this causes it to light. But other things had to be true for the striking to have this effect. The match had to be dry and oxygen had to be present. The causal efficacy of the striking depended on the context. But we don't conclude from this that the striking didn't cause the lighting. How, then, can the context dependence of a gene's impact on the

---

is a three-place relation; whether phenotypic information screens off genotypic information, or vice versa, depends on the third item considered. Brandon chose an organism's reproductive success. But suppose we choose *change in gene frequencies*. Then the screening-off relation is inverted. Gene frequencies and genic selection coefficients *determine* change in gene frequencies, if the population is infinitely large, and confer a probability distribution on future gene frequencies, if drift is taken into account. In either case, adding information about phenotypes does not make any difference.

My point here is not to argue that there are no asymmetries between genotype and phenotype but that the screening-off relation is a very coarse-grained instrument to use in describing them.

phenotype disqualify the gene from being the unit of selection? This question, like the earlier one about the transitivity of causality, will be given careful attention in Chapter 8.

Williams (1966, pp. 56–57) anticipated this reply to the context dependence objection. He grants that the genotype/phenotype relation is complex but denies that this counts against the idea of genic selectionism:

> Obviously, it is unrealistic to believe that a gene actually exists in its own world with no complications other than abstract selection coefficients and mutation rates. The unity of the genotype and the functional subordination of the individual genes to each other and to their surroundings would seem at first sight, to invalidate the one-locus model of natural selection. Actually these considerations do not bear on the basic postulates of the theory. No matter how functionally dependent a gene may be, and no matter how complicated its interactions with other genes and environmental factors, it must always be true that a given gene substitution will have an arithmetic mean effect on fitness in any population. One allele can always be regarded as having a certain selection coefficient relative to another at the same locus at any given point in time. Such coefficients are numbers that can be treated algebraically, and conclusions inferred for one locus can be iterated over all loci. Adaptation can thus be attributed to the effect of selection acting independently at each locus.

The last phrase in this quotation suggests an ambiguity. Is Williams arguing that it is one-locus genotypes that are the units of selection, or is he maintaining that the unit is the single gene? That is, is the designated unit the genotypes AA, Aa, and aa, or the alleles A and a? The pages immediately following the quoted passage make it clear that Williams has the latter in mind. Williams there describes how the fitness of an *allele* may be defined algebraically and how the fitness of an allele depends on other alleles that may occur at the same locus as well as at other loci.[12]

Dawkins (1976, p. 40), in his readable popularization of the genic selectionist point of view entitled *The Selfish Gene*, approached the same problem: How can the single gene be the unit of selection, if genes build organisms only in elaborate collaboration with each other? He answers by way of a clever analogy:

---

12. In any event, even if we reformulated Williams' position so that the commitment were to diploid genotypes at a locus, rather than single alleles, it would fare no better. The arguments he advances are subject to the same line of criticism, as we shall see.

An oarsman on his own cannot win the Oxford and Cambridge boat race. He needs eight colleagues. Each one is a specialist who always sits in a particular part of the boat—bow or stroke or cox, etc. Rowing the boat is a cooperative venture, but some men are nevertheless better at it than others. Suppose a coach has to choose his ideal crew from a pool of candidates, some specializing in the bow position, others specializing as cox, and so on. Suppose that he makes his selection as follows. Every day he puts together three new trial crews, by random shuffling of the candidates, for each position, and he makes the three crews race against each other. After some weeks of this it will start to emerge that the winning boat often tends to contain the same individual men. These are marked up as good oarsmen. Other individuals seem consistently to be found in slower crews, and these are eventually rejected. But even an outstandingly good oarsman might sometimes be a member of a slow crew, either because of the inferiority of the other members, or because of bad luck—say a strong adverse wind. It is only *on average* that the best men tend to be in the winning boat.

The oarsmen are genes. The rivals for each seat in the boat are alleles potentially capable of occupying the same slot along the length of a chromosome. Rowing fast corresponds to building a body which is successful at surviving. The wind is the external environment. The pool of alternative candidates is the gene pool. As far as the survival of any one body is concerned, all its genes are in the same boat. Many a good gene gets into bad company, and finds itself sharing a body with a lethal gene, which kills the body off in childhood. Then the good gene is destroyed along with the rest. But this is only one body, and replicas of the same good gene live on in other bodies which lack the lethal gene. Many copies of good genes are dragged under because they happen to share a body with bad genes, and many perish through other forms of ill luck, say when their body is struck by lightning. But by definition luck, good or bad, strikes at random, and a gene which is consistently on the losing side is not unlucky; it is a bad gene.

So the coach's selection, although it may appear to be a case of group selection, is really individual selection. And natural selection, although it may appear to act on the whole organism's phenotype (or on its entire genome) is really selection for and against single genes. Later in the book, Dawkins considers a situation that might seem to require interpretation in terms of group selection. Suppose the ability of a crew to win a race were strongly influenced by communication back and forth within the team. Suppose further that half the candidate rowers speak only English, and the other half speak only German:

What will emerge as the overall best crew will be one of the two stable states—pure English or pure German, but not mixed. Superficially it looks as though the coach is selecting whole language groups as *units*. This is not what he is doing. He is selecting individual oarsmen for their apparent ability to win races. It so happens that the tendency for an individual to win races depends on which other individuals are present in the pool of candidates (Dawkins 1976, pp. 91–92).

There are two salient features of this reply to the context dependence objection. The first is the idea that context dependence is not a problem for genic selectionism. Whether a single gene will produce a given phenotype depends on what genes it is with; whether a single oarsman will win a race depends on what other rowers are on his team. But, according to Williams and Dawkins, this does not prevent us from attributing causal efficacy to the single gene or oarsman. To see whether this is correct, we need a clearer idea of what the concept of cause amounts to. Granted, striking the match was able to make it light only because certain background conditions were right, and this does not prevent us from saying that striking the match caused it to light. But, as with the question of transitivity, we cannot content ourselves with simple plausible examples. In Chapter 8, we will see that context dependence does indeed present serious difficulties for genic selectionism.

The other salient feature of the passages from Williams and Dawkins is that they show, at best, that selection *can* be interpreted as operating at the level of the single gene. But why *should* we opt for the lower-level description? This is especially striking in Dawkins' crew analogy. Perhaps we *can* describe the coach as selecting single oarsmen. But no argument has been given to show that the coach is not selecting whole teams. Apparently, we have come a long way from the problem of altruism, in which group selection hypotheses and individual selection hypotheses make genuinely different predictions. But what difference does this distinction make in the crew case or the genetic case? The suspicion that we are dealing here only with an aesthetic or utilitarian preference, not with a matter of fact, has not been allayed.

Williams and Dawkins have an answer; they do say why we should prefer the lower-level description. They want to show that genic selectionism is the *correct* view to hold, not just that it is one among many possible descriptions. In the next section, we will examine one of their main lines of reasoning for the view that the single gene is the unit of selection.

7.3 Parsimony

Several lines of argument interweave in Williams' book. One that is philosophically salient involves the idea of *parsimony*. Williams scrutinized a number of natural observations that had been cited as evidence favoring group selection and invented alternative explanations that appeal to individual selection alone. In some cases, Williams was able to produce no decisive independent evidence favoring his alternative story. Yet he did not conclude that we should remain agnostic on the question of group selection. Rather, he offered the greater parsimony of lower-level selection hypotheses as a reason for preferring them.

Dawkins also places heavy weight on the idea of parsimony. After considering the version of the crew team example in which rowers communicate with each other (Section 7.2), he remarks that it is better to describe the situation in terms of individual oarsmen being selected. The reason this is better, he says, is that it is more parsimonious.

There are two assumptions here that need to be scrutinized. The first is that the lower-level description is more parsimonious than the higher-level one. The second is that parsimony is a reason for preferring one hypothesis over another. Philosophers have labored long and hard over the latter idea, which is a foundational one for the question of non-deductive inference.[13] Although I will have something to say about the justification of parsimony, let's begin with the more specifically biological problem: Why think that lower-level selection hypotheses are more parsimonious than higher-level ones?

Williams (pp. 17–18) takes the view that natural selection is an "onerous principle." We should account for characteristics as adaptations only if no "physicalistic" explanation is available; and, if we have to advance an adaptationist account, a lower-level selection hypothesis is superior to a higher-level one. So we have here a simplicity ordering: a physicalistic explanation is more parsimonious than an organismic selection account, which in turn is more parsimonious than one that invokes group selection. If the available evidence is consistent with more than one of these, we should tentatively adopt the one that is simplest.

In defending the idea that physicalistic explanations are preferable to selectional ones, Williams discusses an example about flying fish (p. 11). It is a good thing for flying fish that they return to the water

---

13. Hesse (1969) summarizes the main extant proposals for explicating and justifying simplicity. I tried my hand at the general problem, without any special reference to biological issues, in Sober (1975). In Sober (1983d and 1984a), I describe some of the philosophical issues in connection with the use of parsimony as an inference procedure in phylogenetic reconstruction.

after their flight. However, no special biological explanation is required here, since it is a matter of physics that what goes up must come down. One doesn't need to bring natural selection into the story, since the laws of physics suffice to explain. In just the same way, an earthworm's digestive system benefits the other organisms that share its habitat, but there is no reason to think that the gut evolved for the good of the community (p. 18). Group selection should not be invoked, if individual selection suffices.

Let's think more carefully about the flying fish. What characteristic are we trying to explain? Williams says that the fish's return to the water is "physically inevitable." Now, not every organism is bound by the laws of physics to return to the earth's surface, so this is perhaps not the characteristic that Williams has in mind. On the other hand, there seems to be a sense in which every event is in principle explicable by the laws of physics; organisms are physical objects, and what they do is caused by the physical properties of their physical parts. Williams is not toying with vitalism when he says that some, but not all, characteristics of organisms are physically inevitable.

Let us take a characteristic that is, in a very strong sense, physically inevitable. It is the characteristic of *having mass*. It would be preposterous to account for why the flying fish has this trait by postulating a selection process. However, no principle of parsimony is needed to show why this explanatory strategy is absurd. Selection requires variation. And even the most tolerant inventor of "just-so stories" will draw back from imagining that the ancestral population of flying fish contained some organisms with mass and others without. A trait that must be universal in a population cannot evolve by natural selection among the organisms in that population. This is one clear sense in which "physically inevitable" characteristics cannot be adaptations.[14]

So let us take another characteristic of the flying fish, one that may conceivably have been subject to variation in the ancestral population. Flying fish have gills, not lungs. Given this, they must return to the water if they are to stay alive. If we think of this gill system as fixed, then there may have been selection against flying fish that remained aloft for a very long period. There is no particular reason, as far as I know, to think that the ancestral population was variable with respect to "flying time." But if it were, the flying fish's returning to the water sooner rather than later may be the result of selection, and may therefore count as an adaptation.

A third possibility may be considered briefly. Suppose we do not

---

14. The idea that a trait can be an adaptation only if it emerged via a selection process was discussed in Section 6.1.

think of the gill system as an unalterable given in the lineage. Consider the rather remote possibility that some ancestral flying fish had lungs as well as gills. Then, the fact that flying fish return to the water once in flight starts to sound like something that may be due to a selection process. Conversely, to the degree that we think of this idea as biologically preposterous, we will deny that the fish's returning to the water is an adaptation.

My point here is not to argue that these scenarios are biologically plausible but to make a point about parsimony. When we reject the selectional account of the flying fish's behavior, it is because of some biological assumption about the fish, not because physicalistic explanations are more parsimonious than selectional ones. For parsimony to even be a candidate for deciding matters here, there would have to be two explanations that are both consistent with everything we believe about the phenomenon. This is simply not the case here.

So the implausibility of a selectional account of the flying fish's behavior should not convince us that *parsimony* favors physicalistic over selectional explanation. Is there any independent reason for thinking that it does? And, more directly relevant to our current concerns, is there any general reason to think that parsimony favors lower-level selection hypotheses over higher-level ones?

If a characteristic is "physically inevitable," in the sense that no organism in a population can lack it, selection is ruled out.[15] I do not see why it is more parsimonious to regard a characteristic as physically inevitable in this sense than it is to regard it as the result of a selection process. If the characteristic is sufficiently far out (e.g., having mass), we may reasonably doubt that there is any variation for selection to work upon. And if the characteristic is like others that are known to vary in the population, it may be plausible to expect the requisite variation to be present. Parsimony does not say which of these is more plausible.

Of course, if we knew that the trait in question had emerged by selection in another, similar population, this might provide some basis for thinking that the requisite variation obtained in the case of the flying fish. Conversely, if the trait showed no variation in the one case, this would offer some support for thinking that it exhibited no variation in the other. These inferences are *inductive extrapolations* of sorts. We reason that the mechanism that accounts for one phenomenon also accounts for another, similar one. This variety of inference does have

15. Note that this sense of physically inevitable is different from the one that simply means that the trait has some physicalistic explanation. Presumably, every trait is "physically inevitable" in this latter sense.

an interesting connection with the idea of parsimony. However, the point so far is simply that it does not *on its own* say that nonselectional accounts are superior to selectional ones. This conclusion is reached only with the addition of substantive biological assumptions.

In comparing physicalistic and selectional explanations, we must distinguish between two sorts of problems. We might simply want to know whether there is a physical cause for a given biological characteristic's presence in a population. If we are not vitalists, this question has an easy answer, regardless of the characteristic considered. Alternatively, we might be interested in whether the trait is present for *nonselective* physical reasons. Here the problem is more difficult and we presumably will want to explain different traits in different ways. In the first sort of question, "physicalistic" is perfectly compatible with, and is even required by, "selectional," and the problem is trivial. In the second, "physicalistic" and "selectional" are mutually incompatible, and the question is substantive.

The substantive question about higher- and lower-level selection processes construes each as a distinct possible cause of evolution. When group and individual selection accounts are formulated in such a way that each rules out the other, what role might parsimony play in deciding between them?

The inductive extrapolation from known to unknown cases discussed above is one arena in which the principle of parsimony finds an application. Occam's razor says that new causal mechanisms should not be postulated when old, familiar ones do the trick (Sober 1981c). If we know that individual selection is the cause of a variety of familiar phenomena, better to think that it also is the explanation of the cases under contention than to make up a novel mechanism to account for them. The principle of parsimony is, in part, a principle of explanatory conservatism.

Even if parsimony sometimes works in this way, the question remains of why parsimony is a reason to regard some hypotheses as true and others as false. Philosophers have frequently answered this question by pointing out that rejecting parsimony leads to skepticism. *Every* causal hypothesis goes beyond the evidence at hand. "Parsimony," or "simplicity," or "a priori plausibility" must be brought in, if any unique choice of an explanation is to be made. Even those hypotheses that we think are extremely well supported have competitors that are equally consistent with the evidence. If we fail to bring in parsimony as a reason, we will have to concede that we have little in the way of evidence favoring even those hypotheses that we think are well confirmed.

This is not so much a justification of parsimony as a warning about the consequences of thinking that parsimony is not justified. I want to suggest a modest rationale that is more positive. Parsimony, as a principle of explanatory conservatism, is a form of inductive inference. Put crudely, it asserts that the best explanation of a new phenomenon is the explanation that has worked for old ones. It is structurally similar to the sorts of inferences one makes about the unobserved members of a population based on information about an observed sample. Such inductive inferences may be strong or weak and so, too, may appeals to parsimony. My point here is not to provide any blanket justification for all parsimony arguments but just to suggest that they represent no *special* epistemological problem.[16]

One familiar reason for objecting to an inductive extrapolation is the belief that the new items one makes a judgment about are not relevantly similar to the old, familiar ones. When a biologist replicates an experiment that has been replicated a hundred times before, there is some reason to think that the new outcome will resemble the old ones. But if the experimenter doesn't know that the new laboratory setup matches the standard ones in ways that may be causally relevant, there is more room for skepticism. Just this sort of suspicion may weaken the appeal to parsimony as a defense of lower-level selection hypotheses. Perhaps familiar cases of selection are correctly explained by individual selection. Why think that the new and controversial cases ought to be handled in the same way? Unless one has some confidence that the relevant biological parameters that obtain in the former cases also hold in the latter, the appeal to parsimony will not be decisive. Here, the idea that parsimony is sometimes a species of inductive argument helps explain why parsimony arguments can be strong in some contexts and weak in others.

There is another reason for thinking of parsimony as a form of inductive argument. Even if parsimony were a reason for preferring individual selection hypotheses, this would not explain why they are true. Hence a biologist will quite rightly regard the appeal to parsimony as in principle insufficient to settle the theoretical matter. What is needed is a specification of the empirical conditions that group and individual selection processes each require, and some general argument about how often these conditions are satisfied in nature. This, it will be recalled, is what Fisher tried to do in the brief passage about group selection cited earlier.[17]

16. I therefore reject a fairly common position among scientists and philosophers alike: Inductive inference (i.e., an inference from known to unknown cases, or from known cases to generalizations) is viewed as a reasonable guide to belief, whereas simplicity or parsimony is taken to be a purely aesthetic matter, having no bearing on what is true. This is, I think, an untenable methodological dualism.

17. I do not use "parsimony" as a synonym for plausibility. A number of arguments

Once we see the importance of this explanatory problem, which appeals to parsimony cannot begin to solve, it becomes unclear whether the parsimony argument can carry much weight in the units of selection controversy. Perhaps individual selection is needed to explain familiar phenomena. Why extend it to cover new and controversial ones, rather than apply a group selection hypothesis to them? Let us grant that it would be more parsimonious to explain all adaptations as due to individual selection, rather than say that some are due to individual selection while others were produced by higher-level selection processes. This hardly settles matters, since the most important theoretical question has yet to be addressed. We would like to know whether the empirical conditions that make the individual selection hypothesis correct in the familiar cases are also satisfied in the controversial ones. If we knew that, then the inductive extrapolation from old phenomena to new ones would be stronger. Without such information, it may be reasonable to simply remain agnostic; we may be unwilling to let parsimony *on its own* decide the question.

I do not pretend that this rationale for parsimony is complete. A second sort of analysis is worth considering. According to the suggested justification, to see which explanation of a phenomenon is most parsimonious, one must have an opinion concerning how *other* phenomena are to be explained. This does not permit one to tell which of two competing hypotheses is simpler on purely "intrinsic" grounds. Now it may be suggested that there must be more to parsimony than this; one might claim that even if we had no knowledge of how other adaptations are to be explained, there is something intrinsically unparsimonious about group selection. Consider a typical example— warning cries in crows. One might describe this as a form of altruistic behavior: the sentinel who issues the warning cry benefits the group but increases his risk of being caught by the approaching predator. Alternatively, one might argue that the cry benefits the sentinel's offspring and is also hard for the approaching predator to detect, and is therefore individually advantageous. The altruistic interpretation requires two sorts of evolutionary processes. Altruists do less well than selfish individuals within each group, but groups of altruists do better than groups of selfish individuals; altruism requires the simultaneous

---

investigate quantitative models with an eye to seeing what range of parameter values permit group selection to play a serious role in evolution. These arguments may conclude that group selection rarely occurs, but they do not do so on the basis of the claim that lower-level selection hypotheses are more parsimonious than higher-level ones. Such arguments produce biological reasons, not purely formal ones. A number of these models are reviewed in Wade (1978); arguments of this sort will be discussed in Chapter 9.

occurrence of individual and group selection.[18] On the other hand, reinterpreting the trait so that it no longer appears altruistic may simplify the evolutionary explanation. If the trait is individually advantageous, then a single process is able to drive it to fixation. The problem of assessing the role of parsimony would be circumvented, if some independent argument were produced concerning the relatedness of members of the flock or the acoustical properties of warning cries. But suppose we have no such information. Is the individual selection hypothesis preferable because it is more parsimonious? If so, it seems that the internal structure of the explanation, and not its relationship to other explanations we accept, is what makes it more or less parsimonious. Two processes are less parsimonious than one, and that is the end of it.

I do not deny the appeal of this line of reasoning. But it requires amplification in several directions. First, one must be careful not to sacrifice plausibility for the sake of parsimony. A story can always be invented wherein a trait is shown to be individually advantageous. The more strained this conjecture is, the less impressed we are with the greater parsimony of the individual selection account. This point is hardly limited to the units of selection debate. A monolithic panselectionism—the view that every trait is the product of individual selection—has a certain pleasing simplicity. But those who have stressed the importance of chance or architectural constraints in evolution have been reluctant to allow parsimony to decide matters (Section 1.2). Other considerations enter into evaluating hypotheses, ones with which parsimony can conflict. Second, although altruism implies that two sorts of selection processes have been at work, this complexity should not be automatically assumed to attach to all group selection hypotheses. There can be group selection for nonaltruistic characteristics (Sections 7.7 and 9.3). Finally, my point about the incompleteness of parsimony arguments still stands. Even if parsimony justifies a preference for lower-level selection hypotheses, a central theoretical question remains unanswered. Parsimony may be evidence that a hypothesis is true, but the simplicity of the hypothesis does not explain why it is true. A full resolution of the units of selection controversy must tell us not only *whether* but *why* group selection is common or rare.

Williams (1966) and Dawkins (1976) both appeal to parsimony but do not make it entirely clear what they have in mind by this idea. In this section, I have presented two formulations of their parsimony argument against group selection. According to the first, there is nothing inherently more parsimonious about lower-level accounts; if we had

18. The idea of altruism will be clarified in Section 9.3.

habitually explained uncontroversial phenomena by hypotheses of group selection, then parsimony would advise us to explain controversial phenomena in the same way. If we construe parsimony simply as a principle of conservatism, it says nothing directly on the subject of higher versus lower.

The second construal of parsimony is more partisan. If a characteristic is prevalent in a population, labeling it altruistic rather than selfish makes a difference in the parsimony of the selection story one is obliged to tell. Altruism implies group selection and individual selection, whereas an individually advantageous trait can become prevalent by individual selection alone. Although two processes are more complicated than one, this does not imply that one of the processes is more complicated than the other. We still lack a perfectly general argument about the relative simplicity of higher and lower.

There is another version of the argument from parsimony that is more intimately connected with the structure of population genetics models of selection processes. It is suggested by Williams' and Dawkins' remarks and, moreover, makes direct contact with the distinction between higher and lower. The argument begins with the assertion that all cases of evolution by natural selection can be "represented" in terms of selection acting at the level of single genes. The argument goes on to say that this mode of representation is more "parsimonious" and that this, in turn, is a reason for thinking that it is true.

According to this line of reasoning, genic selectionism provides a mode of representation that works for all cases of natural selection. The same cannot be said of the group selection idea, as we will see. If parsimony implied that modes of representation are not to be multiplied beyond necessity, the connection of *representability* and *simplicity* would be established. Our next order of business is to clarify what it means to say that all forms of selection can be "represented" from the gene's eye point of view.

7.4 Representability

In Section 4.3, I described Laplace's dream of a being who could predict all future events—both great and small—from physical laws and initial conditions. One salient fact about this demon is that it is science *fiction*. Even if nature were deterministic, our calculating powers would still be too limited for us to represent and predict in the idiom of particle physics. Our imprecise commonsense lore about macro-physical objects allows us to get around in the environment in a way that more precise laws about ensembles of microscopic particles would not.

Despite its uselessness in practice, Laplace's demon has retained its grip on our theoretical conception of ourselves. The idea that we are physical objects and that the interaction of physical parts determines what happens to the physical whole has remained immensely attractive. The idea of genic selection has benefited from the appeal of this reductionist perspective. Where Laplace thought that macroscopic events are simply consequences of the interaction of elementary particles, Dawkins holds that organisms are just "survival machines" that genes construct to advance their own selfish interests.

However, two differences between the Laplacean view and the idea of genic selectionism should be noted. First of all, Laplace thought that *all* events could be predicted from the point of view of the physics of elementary particles. The position of Williams and Dawkins, on the other hand, is much more circumspect. Genic selectionism holds that all *selection* occurs at the level of the single gene.[19] The view does not imply that selection has mattered more to evolution than drift or mutation, although, as it happens, advocates of genic selectionism often emphasize the importance of selective over nonselective forces. Nor does the view deny that characteristics of organisms are influenced by forces that are not covered by the theory of evolution, properly so-called. So, for example, there is nothing in the idea of genic selectionism that prevents cultural evolution from playing a role that is entirely autonomous from the "gavotte of the genes." In the last chapter of *The Selfish Gene*, Dawkins considers the idea of a "meme," a unit of culture that propagates itself from mind to mind for reasons that may be wholly independent of whether it helps organisms to survive and reproduce. Perhaps some thoughts are simply more contagious than others, and their contagiousness has little or nothing to do with biological fitness. This idea, whatever its plausibility, is entirely consistent with the thesis that the single gene is the unit of selection.[20]

The second difference between Laplacean determinism and genic selectionism is more relevant to our present purpose. Laplace's demon was a *dream*. No one, then or now, had the slightest idea how to represent typical macroscopic phenomena in terms of elementary particles and then do the calculations required to predict the system's change of state. But the idea that selection can be represented in terms of selection coefficients attaching to single genes is the very opposite

19. The doctrine might be weakened slightly to read that genic selection has been the form of natural selection that has played by far the most powerful role in evolution.
20. Although it must be said that Dawkins' pluralistic attitude that allows for the importance of nongenetic factors in cultural change makes it rather misleading for him to write (1976, p. 21) that we are "lumbering robots" manipulated by "remote control" by replicators, which "are in you and in me," having "created us, body and mind, their preservation [being] . . . the ultimate rationale for our existence."

of a dream. It is something that students of elementary genetics routinely learn to do. We are not talking about science fiction here, but about scientific fact. Laying out exactly how selection may be routinely represented from the gene's eye point of view will help us understand exactly what sort of a claim genic selectionism is.

In Section 1.4, I discussed a simple constant-viability model of heterozygote superiority. In that model, each genotype had a fixed fitness value that represents the "mortality tax" that it could expect to pay in its passage from egg to adult. Recall that the model assigned fitness values to genotypes (i.e., to *pairs* of genes at a single locus). The model did not assign fitness values to single genes. However, it is a trivial matter to take the fitnesses assigned to the three genotypes AA, Aa, and aa (namely, w_1, w_2, w_3, respectively) and compute from these the fitnesses of the two genes A and a.

The fitness of a gene should mimic the mathematical role of the fitness of a genotype. Recall that the genotypic fitness is a number that transforms the genotypic frequency at the egg stage in a generation into the genotypic frequency at the adult stage in that generation, the quantity being normalized by $\overline{w}$. So, for example, w_1 plays an algebraic role constrained by the following condition:

w_1 × frequency of AA before selection $=$
   frequency of AA after selection × $\overline{w}$.

The allelic fitness W_A should therefore obey the following condition:

W_A × frequency of A before selection $=$
   frequency of A after selection × $\overline{w}$.

Since the frequency of A before selection is p and after selection is

$(w_1p^2 + w_2pq) / \overline{w}$,

it follows that

$W_A = w_1p + w_2q$.

By parity of reasoning,

$W_a = w_3q + w_2p$.[21]

There is no difficulty in representing heterozygote superiority from the point of view of the single gene. A gene's fitness is simply a weighted average of the fitness values of the genotypes in which it appears. When a gene is rare, it will most often occur next to copies of its alternative allele; when it is common, it will most frequently occur next to copies of itself. Since the heterozygote is the fittest of the three genotypes, it follows that each gene is fittest when it is rare. The reader

---

21. Another way to see how to define allelic fitnesses is simply to decompose the average fitness of organisms in the population ($\overline{w}$) into a weighted average of the two allelic fitnesses. $\overline{w}$ equals $p^2w_1 + 2pqw_2 + q^2w_3$. A occurs with frequency p and a occurs with frequency q; so we can reexpress $\overline{w}$ as $p(w_1p + w_2q) + q(w_3q + w_2p)$.

may wish to consult Figure 7 (p. 180) to see how frequency-*in*dependent genotypic fitness values give rise to frequency-*dependent* allelic fitnesses.

The fitnesses of the two genes have the same values where the lines that represent their fitness functions cross. This is the equilibrium point (p̂) defined in Section 6.1. At this gene frequency, there is no difference in fitness between the two alleles, so selection ceases to bring about any further change in gene frequency. The equilibrium at this point is stable, in that departures from the point will create a selection pressure that returns the population to its equilibrium value.[22] So evolution by natural selection that occurs in accordance with this model will carry the population to this equilibrium gene frequency, at which it will remain.

The mathematical representation of this process in terms of allelic fitness values is derivable from the standard model in which *pairs* of genes—genotypes—are treated as units of selection. If the genotypic model correctly identifies the trajectory of gene frequencies, so does the single-gene model. This may strengthen the suspicion that the reader perhaps felt when considering Dawkins' racing crews (Section 7.2): What difference can there be between alternative claims about "units of selection"? Perhaps one formulation is simply a notational variant of the other.

There is one difference that ought to be noted, however (Sober and Lewontin 1982). Let's consider the special case of heterozygote superiority mentioned in Section 1.4—the balanced lethal system in which the two homozygotes have fitnesses of 0 and the heterozygote has a fitness of 1. This means that in each generation, all the homozygotes are killed off and all the heterozygotes survive the passage from egg to adult. Whatever the starting frequencies of the three genotypes, by the end of the first generation the population will have evolved to a 0-1-0 configuration. When the heterozygotes reproduce, the next generation of fertilized eggs will shift to a 1/4 - 1/2 - 1/4 arrangement; selection will then bring the population to the 0-1-0 configuration. The population will zigzag between these two configurations of *genotype* frequencies.

I have described two forces that modify the frequencies of genotypes: Mendelism (the formation of haploid gametes and the pairing of these into diploid fertilized eggs) takes the population from 0-1-0 to 1/4 - 1/2 - 1/4; then selection takes the population from 1/4 - 1/2 - 1/4

---

22. In Section 6.1, we also saw that the simple one-locus two-allele model for heterozygote *inferiority* (underdominance) implies an *un*stable equilibrium at p̂. Notice that from the point of view of allelic fitnesses, the difference between Figures 7 and 8 basically boils down to the question of whether a gene is fitter when it is rare or when it is common.

back to 0-1-0. But genic selectionism—the view that all selection is selection for or against single genes—cannot give this natural description. The zigzag of genotype frequencies is not accompanied by any change in gene frequencies. Throughout, the two alleles occur in equal frequencies, and their fitnesses are likewise identical. There is no selection at the level of single genes here, since there is no difference in the fitness values of single genes. This endless zigzag occurs at the gene frequency equilibrium $(\hat{p})$.

So the first reason for denying that genic selectionism is a correct account of this phenomenon is that it makes certain causal processes invisible. It is quite possible for selection to occur, even when there is no difference in the (overall) fitnesses of alternative alleles. Selection for the heterozygote may be balanced by selection against the homozygotes, with the result that gene frequencies do not change at all. However, we should not assume that the absence of change indicates the absence of a selective force.

The mistake here is the same as one engendered by a misinterpretation of the notion of physical force. Suppose you observe a billiard ball that is at rest. From this fact, can you conclude that no forces are acting on it? Newton's law—$F = ma$—is sometimes erroneously thought to say that the answer must be *yes*. After all, if there is zero acceleration, doesn't a little multiplication show that there must be zero force?

This mistake arises from a failure to distinguish *component* and *net* forces. $F = ma$ describes *net* forces, not *component* ones.[23] When an object fails to accelerate, there may be many (component) forces acting on it that nullify each other. The billiard ball in question may be at the center of a powerful struggle; you are pushing it due north while I push it due south. Of course, if the ball doesn't move, then the *net* force must have a value of zero. But the point remains that zero acceleration is, in itself, consistent with two possibilities: (1) there are no component forces acting on the object; (2) there are component forces acting, but they cancel out each other.

What is true for billiard balls is also true for genes. Zero change in gene frequency at gene frequency equilibrium $(\hat{p})$ in the model of heterozygote superiority implies that the *net* selective force at the genic level is zero; it does not imply that no *component* selective forces act. However, genic selectionism says that all selection is selection for and against single genes. It says that there can *be* no selection if the two alleles are equal in fitness. But this is clearly false, since heterozygotes

---

23. This is not to deny that $F = ma$ can be used to compute the "component accelerations" due to various component forces. My point is that you can't use the equation to deduce component forces from net accelerations.

survive and homozygotes are killed off even at the equilibrium gene frequency. Selective forces are at work here as surely as physical forces are at work in our struggle over the billiard ball. In evolutionary biology as well as in physics, the distinction between net and component force is indispensable.

In this example of selection at gene frequency equilibrium ($\hat{p}$), there is no *evolution* by natural selection. The example does not contradict the fundamental fact that evolution by natural selection will occur in a population when and only when there are fitness differences between genes.[24] Why not insist that genic selectionism is a thesis about *evolution* by natural selection, and simply discard this example as so much irrelevancy?

Let us grant that evolution by natural selection occurs when and only when there is variation in the fitness values attaching to single genes. This follows trivially from the definition of evolution as change in gene frequencies. If genic selectionism were simply pointing out a consequence of the definition of evolution, it would be trivially true. But this truism should not be taken to establish the *causal* claim that all selection is selection for and against single genes.

The sorting toy in Section 3.2 helps clarify the distinction involved here. In evolution by natural selection, there must be selection *of* single genes. But this doesn't mean that evolution by natural selection always involves selection *for* single genes. Selection of doesn't imply selection for: In the toy there was selection of green balls but no selection for greenness. The concept of *selection of* records the effects of selection; in Wimsatt's (1980) felicitous phrase, it is one of evolutionary theory's "bookkeeping" concepts. But the causal question of which properties are selected for and against is left open by this decision about how the books are to be kept.

Dawkin's persuasive analogy of the racing crew (Section 7.2) may be analyzed with this distinction in mind. It is uncontroversial that the selection process resulted in the selection *of* individual oarsmen. It also should be obvious that the process resulted in the selection *of* racing teams. What is more interesting is the question of causal mechanism: Was the coach selecting for winning teams or for individually skillful oarsmen? The question might be reformulated slightly: When an oarsman ended up on the winning team, was he selected for his individual skills or for belonging to a skillful team? This causal question is rather

---

24. In Section 1.2, we saw some reasons to reject this formula as a hard-and-fast definition. However, the objections I want to register against genic selection—the thesis that *single* genes are the units of selection—are ones that apply even within a resolutely genetic view of natural selection. It is for this reason that I assume here that evolution is change in gene frequency.

more difficult to answer than the straightforward question concerning selection *of*. It is obvious that there was selection *of* individual oarsmen, but how we should describe the causal process leading up to that result is less clear-cut.[25]

I have granted that genic selectionism is a truism if it is construed simply as a thesis about selection *of*. Should the dispute about the units of selection be interpreted so that it is settled by pointing out an obvious consequence of the definition of evolution as change in gene frequency? I think not. The trouble with this construal is that it proves far too much. *Group* selection processes can be represented in the idiom of gene frequencies just as much as any other evolutionary process can. If evolution by group selection occurs, there must be differences in the fitness values attaching to single genes. Does it follow that group selection is just a form of genic selection?

Williams' (1966) attack on group selection hypotheses concedes that one real case of group selection has been empirically confirmed. It is the dynamics of the *t* allele in the house mouse *Mus musculus*, studied by Lewontin and Dunn (1960); this will be discussed in Section 7.7. Williams argues that the empirical conditions that make group selection possible in this case will very rarely be satisfied in nature. In his later book on the evolution of sex, Williams (1975) considers the group selection explanation of the origin and maintenance of sexual reproduction as a genuine empirical possibility, one that he sought to disconfirm. My point here is not that Williams was right about how sex or the *t* allele are to be explained, but rather to stress the methodological stance he took: Group selection is an empirical possibility that can only be confirmed or disconfirmed by specifically biological evidence about the populations involved.

The units of selection problem would be an easy one to resolve if appeal to a definition sufficed. However, it seems clear that the important biological question about whether the group, the organism, or the single gene are units of selection is an empirical one. If this is right, then any simple resolution of the units of selection controversy by appeal to the fact that all evolutionary process may be "represented" in the language of gene frequencies must be misguided.

Williams' book contains a variety of conceptual and empirical arguments. These two lines of attack, I suggest, are at odds with each other. If the dynamics of the *t* allele involves group selection, and if group selection and genic selection are genuinely different mechanisms, then the "representation argument" in favor of genic selection is misguided.

---

25. The account of causality developed in Chapter 8 is applied to the crew analogy in Section 9.1.

Dawkins' (1976) book is more monolithic. It does not cite a single case in which group selection is empirically plausible. In fact, one gets the distinct impression from Dawkins that considerations of the most elementary kind show that group selection cannot occur in nature. There thus appears to be no internal inconsistency in Dawkins' emphasis on the parsimony of genic representations as a reason for favoring genic selectionism. Indeed, his amplification of the selfish gene perspective in *The Extended Phenotype* (1982) begins with the claim that the position he advocates is not empirically testable but rather constitutes a novel and useful way of looking at familiar facts. And in his article "Replicator selection and the extended phenotype" (1978, pp. 126–127), he explicitly embraces the conclusion that group selection is an instance of genic selection, not a process that can oppose it.[26]

My own construal of the units of selection problem is quite different: I will interpret it as a problem about evolutionary causation that is entirely empirical. It concerns selection *for*, not selection *of*. Although I have not yet presented my positive account of what this set of causal questions amounts to, this much should be clear: *If the units of selection controversy is an empirical one, then the parsimony argument that appeals to the fact that natural selection can always be represented in terms of genic selection coefficients must be entirely irrelevant.*

No matter what the mechanism of selection might be in a particular evolutionary process, its evolutionary *effects* may be represented in the idiom of the single gene. And if evolution is defined as change in gene frequency, then we have here an absolute consistency check for evolutionary model builders to use in their proposed selectional explanations: Regardless of the causal mechanism invoked, be certain that there are net differences between genic selection coefficients. To fail to meet this requirement is to fail to give a model that predicts evolution at all (Wilson 1980; Dawkins 1982). But, as we will see, there is no a priori reason to think that models of group selection must fail to satisfy this condition.

In Section 7.2, I noted an ambiguity in Williams' (1966, pp. 56–57) remarks on how selection can always be "represented" in terms of allelic fitness values. Although his comments were mainly in defense of genic representation, he occasionally asserts the adequacy of single-locus models. This suggests the idea that it is not the single gene but

---

26. I should point out that Dawkins (1982, p. 82) claims that the important questions in the dispute between group and organismic selection are different from those that matter to the choice between organismic and genic selection. He thinks of the former as a factual matter concerning what he calls "vehicle selection," whereas the latter is terminological. He favors the genic terminology because it is, he says, less likely to lead biologists into fallacious arguments.

the one-locus genotype that is the unit of selection. This reformulation, I want to emphasize, does not improve the representability argument. Just as allelic fitness values average over the genotypic contexts in which they occur, so genotypic fitness values average over the multilocus contexts in which they may occur. A one-locus model of heterozygote superiority provides an example in which a gene can be very fit in one context but disastrous in another. In the sickle-cell model discussed in Section 1.4, the allele S is maximally fit when it is next to a copy of A but least fit of all when it is next to a copy of itself. What can happen within a locus can be true between loci. Consider the three genotypes that can occur at one locus with two alleles. Their fitness ordering can depend on the genotype present at another locus. An empirical example due to Lewontin and White (1960) of this situation, known as *epistasis*, will be discussed in Section 9.1. Averaging is a universal tool, regardless of how far down you go—to single genes or only to single-locus genotypes. Changes produced by group selection can always be represented in terms of allelic fitness values; they also can be represented in terms of the fitness values of single-locus genotypes. The representability argument is vacuous, whether it is used in defense of the single gene or the one-locus genotype as the unit of selection.

7.5 The Unit of Replication

Another argument that has, I think, helped confuse the units of selection controversy appeals to the idea that it is single genes, not gene complexes, organismic phenotypes, or group phenotypes that are passed down from generation to generation. Genes, it is argued, are the units of *replication*, and this fact is supposed to help settle the question of what the unit of selection is.

Williams (1966) advances this line of argument early in his book. He begins (pp. 22–23) by asserting that

> [t]he essence of the genetical theory of natural selection is a statistical bias in the relative rates of survival of alternatives (genes, individuals, etc.). The effectiveness of such bias in producing adaptation is contingent on the maintenance of certain quantitative relationships among the operative factors. One necessary condition is that the selected entity must have a high degree of permanence, and a low rate of endogenous change, relative to the degree of bias (differences in selection coefficients).

Dawkins (1976) gives this idea a Jacobin formulation: longevity, fidelity, and fecundity are prerequisites for something to be a unit of selection.

Williams then argues (pp. 23–24) that this requirement disqualifies a number of candidates from being units of selection:

> The natural selection of phenotypes cannot in itself produce cumulative change, because phenotypes are extremely temporary manifestations. They are the result of an interaction between genotype and environment that produces what we recognize as an individual. Such an individual consists of genotypic information and information recorded since conception. Socrates consisted of the genes his parents gave him, the experiences they and his environment later provided, and a growth and development mediated by numerous meals. For all I know, he may have been very successful in the evolutionary sense of leaving numerous offspring. His phenotype, nevertheless, was utterly destroyed by the hemlock and has never since been duplicated. If the hemlock had not killed him, something else soon would have. So however natural selection may have been acting on Greek phenotypes in the fourth century B.C., it did not of itself produce any cumulative effect.
>
> The same argument also holds for genotypes. With Socrates' death, not only did his phenotype disappear, but also his genotype. Only in species that can maintain unlimited clonal reproduction is it theoretically possible for the selection of genotypes to be an important evolutionary factor. The possibility is not likely to be realized very often, because only rarely would individual clones persist for the immensities of time that are important in evolution. The loss of Socrates' genotype is not assuaged by any consideration of how prolifically he may have reproduced. Socrates' genes may be with us yet, but not his genotype, because meiosis and recombination destroy genotypes as surely as death.

Williams concludes his argument with the assertion (p. 24) that

> It is only the meiotically dissociated fragments of the genotype that are transmitted in sexual reproduction, and these fragments are further fragmented by meiosis in the next generation. If there is an ultimate indivisible fragment it is, by definition, "the gene" that is treated in the abstract discussions of population genetics.

Williams' argument may be elaborated as follows. We have decisive empirical reasons to think that natural selection is a process that typically spans many generations with only modest differences in fitness attaching to the alternative competing forms. Although for expository reasons I have considered a balanced lethal system (Section 7.4), this set of fitness relationships is generally believed to be grossly atypical; Darwin was basically correct in thinking that the adaptations we observe in nature are the result of a slow accumulation of changes. For a selection process to span many generations, the variants being selected must exist over that length of time. After all, to be selected an item has got to exist. But then Williams' remarks about the impermanence of gene complexes and phenotypes on the one hand and the permanence of "meiotically dissociated" genes on the other establishes the single gene's claim to be the unit of selection.

Although this "argument from replication" leads to the same conclusion as the "representation argument" of the previous section, it really is quite different, in that the main premise describes certain well-known empirical facts about the replication process, rather than merely deducing a consequence from the standard definition of evolution as change in gene frequency. However, as before, it is worthwhile to consider two formulations of the argument—one that uses the concept of *selection for properties*, the other using the concept of *selection of objects*.

How does the argument work when it is construed as an argument to the effect that we ought to think of natural selection as selection for and against single genes? We want to identify the cause of the changes that occur during a selection process that spans many generations. This causal factor (or factors) must be present in every, or nearly every, generation. It is the impetus that keeps the system evolving in a definite direction. But the only such causal factor is the single gene. So selection is always selection for or against the possession of single genes.

Next, the formulation in terms of how the effects of evolution by natural selection are to be recorded: Which objects are selected; which items are such that there is selection *of* them? It would be absurd to keep score by recording the array of total genotypes or phenotypes that are present in each generation. An organism's total genotype and total phenotype are here today, gone tomorrow. No salient facts about the trends of evolutionary change will emerge from this sort of bookkeeping.

To evaluate this argument, in either of its formulations, we need to be clear about what we mean by saying that "the same gene" is present through many generations of a selection process. We do not mean that *numerically* the same physical object has existed that long. Rather, we mean that the genes in one generation are qualitatively identical with—i.e., are similar in relevant physical respects to—those found later. To introduce some philosophical jargon—let us say that genes in different generations may be *instances* of the same *gene-kind*. A gene today and a gene several generations ago are not identical gene-instances. So when Williams talks about the same gene persisting through the generations, he is talking about *gene-kinds*, not *gene-instances* (Wimsatt 1980).[27]

Even when we imagine that selection acts on single genes, there is no need for copies of a gene in one generation to be descended from copies of that gene in an earlier generation. The gene-instances do not have to be identical by descent. Mutations may periodically introduce new "lineages" of the gene in question. A model of selection plus mutation will correctly predict that a *gene-kind* will go to fixation, regardless of whether the gene-instances are identical by descent or not. Persistent selection for or against an allele across many generations simply requires that there be copies of the allele in every generation for selection to act upon. Fidelity of replication is just one obvious and important method to ensure that this is true.

Once we see this, doubts arise concerning Williams' remarks about units alternative to the single gene. In the standard model of heterozygote superiority discussed in Section 1.4, heterozygotes were assigned a fitness value, even though the diploid genotype of each individual organism is destroyed at meiosis. This poses no difficulty for the model, however, since Mendelism reunites what Mendelism rends asunder. Homozygote and heterozygote genotypes are *reassembled* during every generation, therefore providing a steady supply of gene *combinations* for natural selection to act upon.

Indeed, the same may be said of phenotypic characteristics. Granted, an individual's resistance to malaria is the result of an interaction between the organism's genetic endowment and the environment. But, if individuals with varying degrees of resistance to malaria crop up in every generation, natural selection *for resistance to malaria* may have a multigenerational time frame in which to work.

In short, remarks about genes' being the unit of *replication* establish very little about what the units of *selection* are. Basically, if something is to be affected by selection in a multigenerational selection process, the requirement is that it cannot be here today and gone tomorrow. But this leaves an immense amount of leeway, for both gene complexes and phenotypic characters. This holds true, whether we are interested in using the concept of *selection for properties* to record the causes of evolution by natural selection, or want to use the concept of *selection of objects* to keep the books concerning the effects of evolution by natural selection.

So if a selection process goes on for a long time, the unit of selection cannot be here today and gone tomorrow. However, this requirement

27. I do not use the customary philosophical terminology of "type" and "token" here, because it is too easy to confuse "gene-types" with "genotypes."

of permanence must be understood in terms of kinds, not instances. Note also that it is a constraint on the existence of (multigenerational) selection. A more substantial requirement applies if we consider not just selection, but *evolution* by natural selection. I noted in Section 5.2 that evolution by natural selection imposes a *stability* requirement, one that goes by the name of heritability. But as noted there, phenotypic characters can exhibit the requisite match between parents and offspring (as can gene combinations); indeed, phenotypes can do so even when there is no "genetic basis" for the trait at all (remember the point about cowboyhood).

As mentioned in Section 7.4, Dawkins (1978, 1982) has complicated his picture of the unit of selection problem, now seeing two quite different issues where *The Selfish Gene* saw only one. Instead of arguing that the gene is *the* unit of selection, Dawkins (1982, p. 82) distinguishes the issue of replicator selection from the issue of vehicle selection. The latter is the locus for the problem of group versus organismic selection and is an empirical matter that Dawkins regards as solved. The former, on the other hand, is a terminological issue, one to be settled by a useful stipulation rather than by an empirical argument. In this vein, Dawkins argues that chunks of DNA are plausibly treated as replicators, whereas organisms and groups are possible vehicles.

Dawkins uses the argument from replication, discussed in this section, to argue that genes, rather than organisms or groups, are replicators, But he provides an additional reason for seeing things this way. A replicator, he says must be capable of passing the changes in state it acquires along to descendants. Genes can do this, of course, but because the Lamarckian inheritance of acquired characteristics is mistaken, organisms cannot. This means that organisms, whether sexual or asexual, are not replicators, not even highly inaccurate ones (Dawkins 1982, pp. 97–99).

Dawkins does not say why he imposes this requirement on replicators. What is it in our models of selection processes that demands this? Wouldn't genes still be replicators if they occasionally indulged in Lamarckian behavior? By stipulation the answer is *no*; what is unclear is the point of this stipulation. Furthermore, it is unclear how Dawkins' argument against the claim that organisms are replicators is to proceed. He describes how a mutilation of an organism fails to be transmitted to its progeny. But there are mutilations and there are mutilations. Small incisions are still incisions; an incision in an organism's kidney is presumably an incision in the organism. Why shouldn't a mutation count as something that happens to an organism as well as to the gamete in which it occurs? If this cannot be prohibited (arbitrary stipulation aside), then organisms *can* count as replicators in Dawkins' sense.

Besides permanence and the capacity to transmit acquired changes to descendants, a replicator (at least an active one) must have a third characteristic, according to Dawkins (1982, p. 91). It must exert phenotypic *power* over the vehicles it inhabits, thereby causing itself to increase or decrease in frequency. Can a single nucleotide at a given locus be said to have a phenotypic effect? Dawkins replies that "while it is undoubtedly meaningful to speak of a single nucleotide as exerting power in this sense, it is much more useful, since the nucleotide exerts a given type of power only when embedded in a larger unit, to treat the larger unit as exerting power and hence altering the frequency of its copies." We will see in the next chapter whether the fact of context dependence can be accommodated so easily by the perspective of genic selection. At any rate, Dawkins goes on to remark that the requirement of phenotypic power should not be interpreted as showing that very large genetic units—e.g., the whole genome—are units of selection. The reason, of course, is the permanence argument discussed earlier.

I suggest that Dawkins has two conflicting considerations at work in his view of what the unit of replication is. The idea that a replicator must be relatively permanent leads us to view smaller genetic units rather than larger ones as the units of replication. The idea of phenotypic power, on the other hand, is manifestly a causal requirement. I will argue in Chapter 8 that certain familiar facts about dominance and epistasis show that smaller genetic units—say, a given nucleotide at a given locus—frequently fail to play a determinate causal role in natural selection.

The term "gene" has had multiple meanings in evolutionary biology. Sometimes it is used to describe larger, sometimes smaller, lengths of DNA. This ambiguity is not, per se, a problem. However, if I am right that Dawkins' two criteria for what a replicator is pull in opposite directions, this does present a problem for his version of genic selectionism. We are owed, at the least, some criterion that describes in a given biological context what weight to place on the conflicting requirements of permanence and power.

Looked at in one way, Dawkins' twofold view is really not substantively different from the one presented earlier in *The Selfish Gene*. Replicators are the units of selection, properly so-called. Vehicles are devices constructed by replicators; vehicles exist for the good of the replicators they house. As noted before, group selection is represented as a *kind* of gene selection, not as an *alternative* to it. Groups are candidate vehicles that genes may construct and exploit for their own advantage.

There are vehicles and replicators, according to Dawkins, but the replicators are where the action is.

I propose that we grant that the single gene is the unit of replication.[28] If we also grant that evolution requires change in gene frequency, we can easily see how evolution by natural selection—even via *group* selection—must involve the selection *of* genes. In this sense, group selection is a sort of gene selection, not an alternative to it. However, these various matters can be settled without considering empirical questions about which kinds of selection processes have mattered most in evolution. It is this latter controversy, still inadequately characterized, that we must try to understand.

7.6 Adaptation and Artifact

My discussion of the units of selection controversy has so far been almost completely negative. After briefly describing the historical origins of the debate, I analyzed two influential lines of argument that have been used to defend the idea that the single gene is the unit of selection. The first involved the ideas of parsimony and representability (Sections 7.3 and 7.4). The second based its case on considerations concerning the properties a unit of replication must have (Section 7.5). Both of these I found wanting.

We now need to reorient ourselves—to remind ourselves what the question of the units of selection was originally about. The historical origin of the units of selection controversy is also its conceptual center. Williams began by reacting against group selection explanations. Group selection hypotheses and organismic selection hypotheses are not interchangeable and equivalent ways of describing nature; the difference between them is objective, not aesthetic. Nor does the question of the units of selection concern the role of the Mendelian gene as a mechanism of transmission. Rather, the recent phase of the units of selection controversy began with the problem of altruism. Organismic selection and group selection make opposite predictions about the existence of individually disadvantageous but socially beneficial traits. Group selection and individual selection are possible selective forces that can push in opposite directions. The units of selection controversy began as a question about causation.

One of the central ideas of Williams' book was that the current utility

---

28. The fact that there can be extrachromosomal hereditary mechanisms should not be forgotten. Genes are not the *only* units of replication. And even if we restrict our attention to genes, the question noted above concerning how much of a nucleotide sequence to count as a single gene still remains.

of a characteristic is no sure guide to the causes of its historical genesis. In analyzing natural phenomena that had been offered as evidence for group selection, Williams frequently remarks that a characteristic may be good for the group without being present *because* it is good for the group. That is, a trait can benefit the group without being a group *adaptation*. The benefit to the group is an *artifact*, not a cause, of its prevalence.

Williams' discussion of dominance-subordination hierarchies (p. 218) is a good example of this strategy of argument. Williams describes Guhl and Allee's (1944) study of hens. When a group of hens is first formed, there is a good deal of fighting and competitive behavior. After a dominance hierarchy forms, however, the violence markedly decreases and the average level of well-being, as measured in food consumption and egg laying, increases. Guhl and Allee took this as evidence for the claim that the dominance hierarchy is a group adaptation; it evolved because it promotes the well-being of the group.

Williams' contrary proposal is that the dominance hierarchy is a consequence of the advantageous compromises "made by each individual in its competition for food, mates, and other resources. *Each compromise is adaptive, but not the statistical summation*" (p. 218, emphasis mine). In other words, individual selection produces a set of individual behaviors that collectively generate a dominance hierarchy. This dominance hierarchy may well confer group benefits. But this is no argument, according to Williams, for group selection.

The discussion of communal responses to the threat of predators among musk oxen (pp. 218–219) follows the same pattern. When a herd is attacked by wolves, it will form a circle. The bulls station themselves at the periphery and confront the attackers, while the females and calves take a safer position at the interior. The adult males are apparently behaving altruistically, sacrificing their own safety for the good of the group. How do we explain this without the hypothesis of group selection?

Williams offers two possible accounts. First, if a high proportion of calves in the herd are offspring of a bull, then the bull will act in his own reproductive interest by defending the herd. Ideally, of course, the bull would defend only his own offspring. But adaptations in nature are not so exquisitely fine-tuned as that. Here Williams implicitly appeals to the idea of informational economy that we examined in Section 6.2. The assumption is that the storage capacity of the genome is limited enough that a more fine-tuned adaptation would be prohibitively expensive; it isn't worth the organism's while to encode the conditional instruction "protect an individual if a predator attacks precisely when that individual is a relative." Far better to use the simpler rule "protect an individual if a predator attacks." We have already noted the highly conjectural character of this use of the principle of economy of genetic information; however, it is the strategy of argument that is of interest here. Williams' first explanation of the musk oxen's behavior, then, is that the defensive formation of the bulls is no more altruistic, evolutionarily speaking, than is parental care.

But what if there are several unrelated bulls in the herd? Why should they defend calves *most* of which are not their offspring? Williams offers no evidence either way on the genetic relationships found in a typical herd. His second explanation is intended to apply when individuals in the herd are not closely related. Imagine that each individual has a stimulus threshold that determines whether it responds to a threat by counterthreat or by flight. Individual selection will ensure that whether an organism responds one way or the other will depend on its own strength and the perceived strength of the attacker. But this means that there is a certain sort of danger that will elicit counterthreat in large bulls and flight in calves and cows. Williams' hypothesis is that wolves are a danger of this kind. The same sort of story, he suggests, may explain why bighorn sheep flee from attack by having the ewes and lambs head directly for high rocky havens while the rams lag somewhat behind.

The reader will notice why some principle like parsimony is required to bring this elaboration of alternative hypotheses to a decisive conclusion. That an individual selection explanation is *possible* doesn't show that it is *true*. But my present point in describing these examples is to show how important it is to Williams' argument that group benefit doesn't automatically imply group adaptation. Williams uses the concept of group adaptation so that it implies group selection (Section 6.2). So, to counter group selection hypotheses, one needs to carefully distinguish a possible *effect* of an evolutionary process (group benefit) from one of its possible causes (group selection).

We can use this idea to analyze, and reject, one conceivable and initially plausible characterization of what group selection is. It is the idea that group selection exists when and only when groups vary in fitness.

Selection, we know, requires variation. So group selection must involve variation among different groups. A species that is organized into a single group cannot, as a matter of definition, sustain a group selection process within itself.[29] Now suppose further that the groups differ in *fitness*. It is a somewhat delicate matter to define what group

29. Although it may, of course, conceivably participate in a selection process with groups from other species.

fitness means, but for the moment, let's understand this idea as follows. Just as an organism that has a higher chance of mortality than another organism is less fit (on that score), so a group that has a higher chance of extinction than another group is less fit (on that score). This is the component of fitness that reflects chances of mortality, applied to groups. Second, just as an organism reproduces when it creates another organism that is numerically distinct from itself, so we can think of group reproduction as occurring when a group sends out migrants that found a new colony. Thus groups differ in fertility when they differ in their expected number of colonies. Mortality and fertility being the two components of fitness I will discuss here, we now know what it means for groups to exhibit variation in fitness.

Now I want to consider a set of six populations, each of which is internally homogeneous for height. The first population contains individuals that are all one foot tall, the second is made of two-footers, and so on. Suppose that the growth dynamics of each of these populations is as follows. Each has the same "critical mass"; when the population reaches a certain census size, it can get no larger. At this point, it sends out its "excess individuals" as migrants. They then found their own colony, whose growth dynamics are the same as the parent group. Each colony is started by individuals from the same parent group. I assume throughout that like produces like, as far as height is concerned.

I want to consider two possible selection processes that this set of populations might experience. The first involves only *individual* (organismic) selection. Suppose that individuals are selected for *being tall*. Ignoring the influence of chance, mutation, migration, and so on, we can make the following prediction: The six-footers will outreproduce the five-footers, who will outreproduce the four-footers, and so on. As a result, populations of six-footers will found more colonies and will go extinct less often than populations of five-footers, and so on. Hence, if we start off with one population in each height category, at a later date we will find that the proportion of six-foot populations will have increased from one-sixth to some greater share.

In this case, it seems clear that there is variation in fitness among the populations. Indeed, there is even *heritable* variation in fitness, since a population in one height category always founds an offspring colony in the same height category. Indeed, *all* the variation is between groups; the groups are each internally homogeneous. Nevertheless, there is no group selection here. *Individual* selection—selection for being tall—was the evolutionary force at work. It follows that (heritable) variation in the fitness of groups is not sufficient for the existence of group selection.

This conclusion reflects the point made in Section 3.2 that the concepts of fitness and selection are not interchangeable. The children's sorting toy, I hope, has made the distinction between the *fitness of objects* and *selection for properties* abundantly clear. In that case, we had fitness differences among balls of different color, but no selection for color. In the example just considered, there are fitness differences among groups, but not selection for group properties. It is true, in this story, that groups are selected; the six-foot populations are singled out for favorable treatment under the selection regimen. But *selection of groups* doesn't imply *selection for group properties*.

I now want to consider a second sort of selection process that the six populations might undergo. Suppose that an organism's chance of surviving and being reproductively successful is determined not by its own height but by the average height of the group it is in. Here organisms do better or worse in virtue of the phenotype of the group they belong to. In this case, I would argue, we have a genuine case of group selection. Groups are selected for their average height.

These two possible scenarios concerning the tall and short populations allow us to discern some characteristics that seem to mark the difference between group and organismic selection. In both stories, it is true that individuals reproduce differentially and so do groups. In both cases, we can assign probabilities of reproductive success to individuals as well as to groups. That is, whether group or individual selection is the driving force, both individuals and groups have fitness values. The difference between the two scenarios arises from the *sources* of those fitness values. To distinguish group from organismic selection, we must consider *why* there are fitness differences of a certain kind. The crucial distinction arises at the level of causality.

There is a special twist to the two scenarios that is worth considering. Notice that the group selection hypothesis (that there is selection for groups that have greater average height) and the individual selection hypothesis (that individuals are selected in virtue of their being tall) are predictively equivalent. Given that the populations are internally homogeneous, and that each obeys the rules of growth dynamics described, the system of populations will change in composition in a certain way, no matter which of the two hypotheses is true. This, it should be remembered, is not a general feature of the difference between group and individual selection hypotheses. Never forget that altruism will be selected against by individual selection but that group selection may provide a force in the opposite direction. Still, it is worth considering how, in the present example, a scientist could figure out which of the two hypotheses is true, if both are consistent with the observed changes in the ensemble of populations.

We confront here a problem that may arise for any theory of forces. If a particle accelerates, this implies that some net force must have impinged; but whether the force in question was electromagnetic, gravitational, or whatever is left open. The trajectory of the particle, like the trajectory of the system of populations described above, is consistent with many causal hypotheses. Two techniques are available for finding out which causal mechanism was actually at work.

The first involves manipulating the system. Suppose we rearrange the populations, creating some that are composed of individuals of different heights. We could then compare what happens to a six-foot-tall individual in a population with one average height with what happens to a six-foot-tall individual in a population with a different average height. If we ran a series of comparisons like this, we could accumulate evidence about whether an individual's fitness is fixed by its own height or by the average height of the group it is in.[30]

A second mode of investigation is available, one that doesn't require intervention into the system. In the physical analogy concerning the accelerated particle, we could find out what forces were at work by looking for *sources* (Section 1.5). Is the particle near a large mass? If so, this will create a gravitational force. Is the particle near a moving magnet? If so, this may suggest that it is affected by an electromagnetic force. We can see what forces a system experiences by examining its environment.

What might this procedure look like in the present biological case? How might we look at the environment to see whether there is selection for groups having a greater average height or for individuals who are tall? Suppose predation were the main source of selection, and that predators take bites out of a group of very small organisms rather than singling out prey organisms one at a time. This might give us evidence for thinking that it is statistical properties of the group that make it more or less vulnerable to predators. A large organism in one group might have a very different vulnerability to predators than a large organism in another group, owing to the fact that the containing groups differ. An analysis of *how the selection process works* can help us discriminate between the two selection hypotheses mentioned.

In this hypothetical example of the tall and the short, there is a perfect correlation between an organism's height and the average height of the group it belongs to. Yet, I claim that there is a substantive difference

30. Of course, to use the data that accrue from the manipulated system as evidence about what was going on in the system prior to manipulation, we need to make a substantive assumption. We need to assume that the manipulation of the system did not itself alter the forces at work. This is by no means inevitable, and requires empirical confirmation of its own.

between the hypothesis that there is selection for being tall and the hypothesis that there is selection for belonging to a tall group. Before concluding that this is a distinction without a real difference, the reader should notice that the idea here is none other than one that is standard in discussions of *free riders* (Section 3.2). Pleiotropically connected characteristics can be such that there is selection for one without there being selection for the other. Genes that are linked together may likewise be such that one confers an advantage, whereas the other does not. The distinction I propose concerning the two scenarios that may hold true for the tall and the short populations is not something that arises only in the rarefied atmosphere of hypotheses about higher-level selection. Causation is not something that suddenly *becomes* important here but has no place in more familiar conceptual settings. The point is that we must not *forget* the importance of causal considerations when we consider the problem of the units of selection.

Group selection is sometimes defined as evolution caused by the differential extinction or proliferation of groups.[31] It should be clear by now how this characterization falls short of the mark. Besides the fact that selection does not imply evolution, there is a more fundamental difficulty. For one thing, groups can go extinct and proliferate *by chance*, thereby producing evolution. Suppose we therefore exclude this possibility by amending the definition so that it requires that the differential extinction and proliferation be nonrandom. Even now, we have failed to distinguish changes in an ensemble of groups produced by group selection from changes in that ensemble caused by individual selection. The proliferation of tall populations causes an increase in average height in the ensemble of populations, regardless of whether group or individual selection drives the system. Williams' (1966, p. 16) example of the fast and slow herds of deer, which I discussed in the Introduction, establishes the same point. If one herd is fast and another slow and if predators eat all the slow deer, then a group has gone extinct. What is more, the extinction was nonrandom and caused a change in the composition of the two-group ensemble. The extinction of the slow herd effected an increase in the average speed of the deer population. It is also true, of course, that the death of the slow deer produced this change. I conclude that group selection isn't just evolution via the nonrandom extinction and proliferation of groups. The idea additionally requires some reference to *why* groups go extinct and why some pro-

31. Wade (1978) attributes this definition to Wright (1945), Wynne-Edwards (1962), Maynard Smith (1964), Williams (1966), and Lewontin (1970). These authors describe examples that would count as genuine cases of group selection, if their empirical assumptions about the populations involved were correct. I suspect that the concept of group selection they were using has more structure than the proposed definition captures.

liferate. This is something that the simple definition under consideration fails to capture. Williams' distinction between adaptation and artifact is as elusive as it is important.

7.7 Group Selection without Altruism

In this section, I want to clarify the relation of the concept of altruism to the phenomenon of group selection. Altruism, in the form of "prudential restraint" in reproduction and in the consumption of resources, was Wynne-Edwards' (1962) reason for postulating group selection. Williams argued against the idea that group selection has played much of a role in evolution by trying to show that group benefits do not in general exist *because* they benefit the group. But what exactly is the connection here? If altruism really were an illusion, would it follow that group selection can be discounted? Or, put differently, does group selection always have to be *for* characteristics that individual selection is *against*?

We must distinguish an evidential and a definitional formulation of this question. It may be true that our powers of natural observation are such that we will never be able to discover cases of group selection, unless they happen to involve selection for altruism.[32] But this would not mean that group selection must always involve selection for altruism. My interest here is mainly in the second, definitional question, although the evidential issue will arise as well.

Rather than making up a hypothetical example to illustrate the point I have in mind, I want to describe two cases that have struck biologists as plausible examples of group selection. One involves natural observation; the other involves experiment. Both help show that group selection need not work in a direction opposite to that of selection at the organismic or genic levels.

The first example is the one that Williams (1966, pp. 117–119) cited as the only authenticated case of group selection of which he was aware. It is the investigation by Lewontin and Dunn (1960) of the segregator-distorter *t* allele in the house mouse *Mus musculus*. Males who are heterozygous for this gene do not produce 50 percent *t*-bearing sperm. The *t* allele distorts the segregation ratio and cheats Mendel's laws (to use Crow's 1979 apt expression). The sperm pool of heterozygous males contains up to 95 percent *t*-bearing sperm. This kind of process is called "meiotic drive" or "segregation distortion." We seem to have here a selfish gene par excellence; the *t* allele subverts the ability of its competitors to find their way into the next generation.

---

32. We will see that this skeptical worry is a bit exaggerated in any case.

This advantage of the *t* allele in gamete formation is opposed by its effects on organisms. Homozygosity is either lethal or results in male sterility. Lewontin and Dunn (1960) combined these two facts about the *t* allele (that it excels in the game of gamete formation but hurts organisms in which it is found in double dose) to compute what the frequency of the gene ought to be in nature. Observation revealed that the predicted value was too high. They concluded that there must be a third force that has the effect of reducing the *t* allele's frequency.

They conjectured that this third force is a form of group selection.[33] House mice live in small, semi-isolated, local populations (demes). When all the males in a deme are homozygous for the *t* allele, the deme goes extinct. The males therefore exert a "poisoning effect" (the expression is Williams') on the females in the group. The females in such groups are likely to contain a higher frequency of the *t* allele than females in other groups, and so their failure to reproduce further reduces the frequency of the gene.

Notice that a female in such a deme has a fitness equal to zero, simply because she belongs to a group in which all the males are homozygotes. Just as in the hypothetical example of selection for groups according to their average height (Section 7.6), we see that group selection in the mouse case involves an important sort of *context dependence*: Otherwise identical individuals may have different fitnesses because they live in different groups, and otherwise dissimilar individuals may have identical fitnesses because they live in the same group. The significance of this idea of context dependence will be discussed in the next section.

So, according to the analysis of the *t* allele proposed by Lewontin and Dunn, three forces affect the frequency of the *t* allele. (1) Meiotic drive serves to increase the trait's frequency, (2) organismic selection against male homozygotes works against the trait, and (3) group selection against groups having 100 percent male homozygotes also works against the trait. Notice that the organismic and group components of this process act in the same direction. The trait is bad for organisms (males) and bad for groups. Although we have yet to develop a precise criterion for saying at precisely what levels of organization the different aspects of this process act, one thing should be clear: There is nothing in the nature of the idea of group selection that requires it to act in a direction opposite to that of organismic (or genic) selection. Group selection does not have to be selection for altruism.

---

33. The conceptual point I want to make here does not depend on this explanation's being correct. The goal is to see why Lewontin and Dunn's (1960) proposed mechanism is a case of group selection.

The same conclusion can be extracted from another example of group selection, this one from Michael Wade's (1976) selection experiments on the flour beetle, *Tribolium castaneum*. Wade's aim was to see what changes could be produced by group selection. His experimental setups are of interest here because they help clarify the conceptual question of what sort of process group selection is supposed to be.

Wade created and monitored four selection processes at once (Figure 12). In each, he started with 48 populations containing 16 beetles each. At the end of 37 days, he did his selecting. In the first treatment, he selected for groups having a high census size. The largest group was chosen and was used to found as many populations of 16 individuals each as it could provide. He then went to the second largest group and drew new populations from it until it was used up. This continued

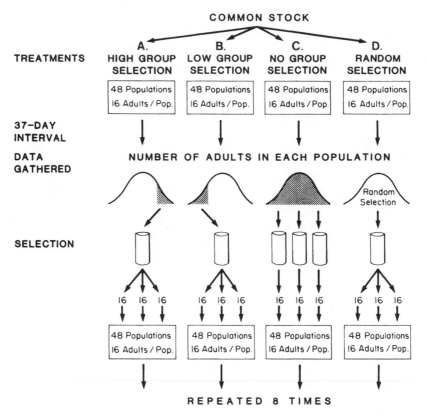

Figure 12. The four selection treatments used in Wade's (1976) *Tribolium* experiment. Treatments (A), (B), and (D) involved both group and individual selection, whereas (C), a "control" treatment, involved no group selection (adapted from Wade 1976).

until a second generation of 48 groups, each containing 16 individuals, was established. Thirty-seven days were allowed to pass, and then the selection was performed again. After eight "generations" of colony formation, the average population size was ascertained.

The second experimental treatment also involved group selection, but in the opposite direction. Here, groups were selected for their *small* census size. The smallest group was used to found colonies, then the next smallest, until 48 new groups were constructed.

In order to control for the activity of individual selection, Wade ran the third experimental treatment to preclude the activity of group selection. He began, as usual, with 48 canisters. At the end of the 37-day interval, he drew a single propagule of 16 individuals from each canister. *Within* each canister there was, of course, a good deal of individual selection. But because each group was assured of contributing precisely one group of 16 individuals to the next generation, the possibility of selection among groups was precluded.

The fourth treatment Wade called "random selection." After 37 days, a group was selected at random and used up to found new colonies of 16 individuals each. Then, a second group was drawn at random and used to found new colonies, and so on, until 48 new groups were created. Again, after 8 generations, Wade observed what changes had occurred.

It might appear that Wade's fourth treatment involved no group selection at all. After all, when organisms survive and reproduce at random, we do not call this "random selection." In a way, "random selection" is a contradiction in terms. How, then, could the fourth treatment be a case of group selection, since groups are "selected at random" to contribute to the next generation? It is true that chance determines whether a group will contribute (reproduce) or not. But if a large group is chosen to reproduce, it will found more colonies than a small one. Hence, under this process of "random selection," the expected number of offspring varies from group to group. The same situation can arise among organisms: suppose that two organisms have identical chances of reproducing, but the first will produce twice as many offspring as the second, if reproduction occurs at all. The two organisms differ in fitness, although they are equally likely to reproduce.[34]

In the first two treatments, both group and individual selection occurred. Within each canister, individuals competed with each other. And every 37 days, some groups had a better chance of proliferating than others, owing to their phenotypes (high or low census sizes). The

34. I owe this point of clarification to David Wilson.

third treatment simply involved individual selection with no group selection. The fourth, like the first, involved individual selection as well as group selection for large census size.

What changes occurred under these four selection regimens? The third treatment (the one in which individual selection acted alone) produced a dramatic reduction in population size, owing to factors like lengthened developmental time, reduced fecundity, and increased cannibalism.[35] The second treatment (in which groups were selected for being small) produced an even greater change in the same direction. The reason presumably was that individual selection depressed the census size (as it had done in the third treatment), and group selection depressed it further. When it comes to creating an evolutionary change, two forces are better than one.

I will postpone a further analysis of Wade's results until Section 9.2; the present point is that group and individual selection may act in the same direction. Hence the issue of group selection is not settled by any argument that restricts its attention to the phenomenon of altruism.

7.8 The Analysis of Variance[36]

In the process of rejecting definitions of group selection in terms of heritable variation in the fitness of groups (Section 7.6) and altruism (Section 7.7), we have come across an idea that seems central to the idea of group selection. In the hypothetical example of the short and tall populations, as well as in the more realistic cases of the dynamics of the *t* allele and Wade's experimental regimens, the idea of *context dependence* loomed large. In each of these cases, we noted that an organism's chances of surviving and reproducing were crucially influenced by the sort of group it was in. This suggests an acid test for group selection: If phenotypically identical individuals have different fitnesses in virtue of differences between the groups they are in, then we have group selection. Or, less elliptically stated, the idea to be considered is this: *group selection exists exactly when groups vary in fitness, and the fitness of an organism depends on the sort of group it is in.*[37]

---

35. This finding is not as paradoxical as it may seem—as if organisms were being selected for their lack of fitness. When selection favors organisms that are bigger or use up more food resources or eat other organisms, this may reduce the census size of the population. The discussion of selection and improvement in Section 6.1 shows how this may result.
36. Much of the material in this section is drawn from Lewontin (1974a), which provides an excellent characterization of the limitations of the analysis of variance as an analysis of causes. For statistical details on how the method works, see Sokal and Rohlf (1969).
37. As before, we will understand group fitness as including both its chances of surviving (avoiding extinction) and reproducing (founding new colonies). We eventually will want to examine this idea more closely, but this interpretation will serve for the present.

As promising as this idea may seem, it still falls short of the mark. How close it comes to being adequate depends on how expressions like "in virtue of" and "depends" are understood. We will see that the idea is most adequate when these ideas are understood in a rather rich, causal sense (although even then the characterization has its defects). But this lesson will only become clear once we see the limitations of a more attenuated reading. This interpretation of the idea of "context dependence" is the one provided by the statistical method known as *the analysis of variance* (ANOVA).

Analysis of variance is a statistical method used to discern the relative contributions that various causal factors make to producing an outcome. A standard example is the administration of fertilizer and water, both in varying degrees, to a field of corn. Suppose the land is divided into small plots and each is given either one ounce or two ounces of fertilizer, and either one gallon or two gallons of water, during the period of study. The average heights of corn plants receiving the four possible corn treatments may then be represented in an ANOVA table:

| | | Amounts of water | |
		W-1	W-2
Amounts of	F-1	1	3
fertilizer	F-2	3	5

This hypothetical result of the four treatments intuitively means that increasing the fertilizer from one ounce to two, and increasing the water from one gallon to two, each were positive causal factors in promoting plant growth. The analysis of variance represents the causal role of these two factors by showing how the average height (H) of plants in a "cell" (i.e., in a given combination of water and fertilizer treatments) departed from the average height (A) of all plants, from the average height of plants that received as much water (W) as those in the cell in question did, and from the average height of plants that received as much fertilizer (F) as the plants in the cell in question did. What this "explanation in terms of deviations from the mean" actually has to do with explanation and causality we will see in a moment. For now, I want to lay out the basics of how the computation proceeds.

If we assume that each of the four cells contains the same number of plants, we may compute the "marginal" averages and the "grand" average as follows:

| | | Amounts of water | | |
		W-1	W-2	Marginal averages
Amounts of	F-1	1	3	2
fertilizer	F-2	3	5	4
Marginal averages		2	4	3 = grand mean

Now suppose you wish to explain why the plants receiving two ounces of fertilizer and two gallons of water averaged 5 feet of growth. The analysis of variance explains this result in terms of its departure from the grand mean and from the two marginal averages:

$$(H - A) = (H - W) + (H - F)$$
$$(5 - 3) = (5 - 4) \ + (5 - 4)$$

The two terms to the right of the equality sign are the so-called main effects of the two treatments. Each is supposed to help explain why the results of the particular treatment departed from the grand mean. Notice that the height in any of the three other cells of the table could be characterized in the same way.

Even if we suppose that amounts of water and fertilizer are the only factors that distinguish one plot of land from another in this case, it still is possible that the two main effects do not explain all of the departure of the cell from the grand mean. For example, suppose the data were as follows:

		Amounts of water		
		W-1	W-2	Marginal averages
Amounts of	F-1	1	3	2
fertilizer	F-2	5	9	7
Marginal averages		3	6	4.5 = grand mean

In this case, looking again at the lower-right cell, we find that

$$(H - A) \neq (H - W) + (H - F)$$
$$(9 - 4.5) \neq (9 - 6) + (9 - 7)$$

A third term needs to be introduced, the "water-fertilizer interaction" (WF), which is given a value (in this case of -0.5) to make the two sides balance:[38]

$$(H - A) = (H - W) + (H - F) + WF$$
$$(9 - 4.5) = (9 - 6) + (9 - 7) - 0.5.$$

---

38. I will ignore how the variance within each cell affects the confidence with which one may hold various hypotheses. My concern here is not with the justification of causal claims but with what those causal claims mean. The reader worried by this issue might want to assume that there is little or no dispersion within the cells.

Although the nomenclature "main effects" and "interactions" may suggest that the former always have more causal importance than the latter, this need not be true. For suppose that the field of corn had produced the following results:

| | | Amounts of water | | |
		W−1	W−2	Marginal averages
Amounts of	F−1	1	3	2
fertilizer	F−2	3	1	2
Marginal averages		2	2	2 = grand mean

Here, there are no main effects of water and fertilizer; the departure of a cell from the grand mean is wholly explained by the water-fertilizer interaction.

We can generalize a little on this example and describe the conditions under which a treatment is said to exert a main effect and the conditions under which there is an interaction between the two treatments. Although the example considered here has only two levels in each treatment (two amounts of water and two amounts of fertilizer), the criterion is easily generalized. Suppose the values in the table were:

| | | Amounts of Water | |
		W-1	W-2
Amounts of	F-1	w	x
fertilizer	F-2	y	z

Fertilizer, say, will exert a main effect precisely when the two marginal averages for the fertilizer treatments—$(w+x)/2$ and $(y+z)/2$—are different. And the two treatments interact precisely when $(x−w) \neq (z−y)$. So to have zero interaction, there must be some constant quantity that may be added to w to obtain x, and to y to obtain z. Hence, when there are no interactions, the relationship of the values in the cells is said to be purely *additive*.

Rearranging the equation that takes account of the two main effects and the interaction term, we obtain

$$H = A + (H − W) + (H − F) + WF.$$

This equation might encourage the belief that the analysis of variance explains the score in each cell (H)—e.g., why the plants in a certain plot averaged one foot in height. But this will not generally be so, since the right-hand side of the equation describes the relation of that result with averages obtained over the entire field of corn. If the growth in the plot considered did not causally depend on the growth patterns in other plots of land, then it will be false to say that the plants in one plot averaged one foot in height *because* the grand mean, the main effects, and the interaction were thus-and-so (Lewontin 1974a). Mathematically, the two sides of the equation must balance, and so in a purely computational sense, H may be said to be a function of the three terms on the right. But if the dependence is merely mathematical (i.e., is not also causal), then the story told by the analysis of variance will not explain why the plants in the plot in question averaged one foot of growth or why they varied by a certain amount from the grand mean.

Earlier in this book, I quoted Fisher's (1930) ironic remarks on how Darwin's Malthusianism (Section 6.1) and the idea of selection working for the good of the species (Section 7.1) reflect philosophies that have outlived their usefulness. I cannot resist a bit of mimicry: The use of the word "explains" in connection with the analysis of variance is a vestige of the natural state model (Section 5.3), according to which variation in a population is explained in terms of interfering forces causing individuals to deviate from type. If this were the underlying causal structure, the analysis of variance might truly be said to explain variation. When it is not, the analysis of variance may have its uses, but explanation of variation is not one of them.[39]

The application of this idea to the idea of group selection is straightforward. The suggestion mentioned at the beginning of this section was that group selection occurs precisely when the fitness of an organism depends on the kind of group it is in. We might represent this as an ANOVA table, in which organismic phenotypes or genotypes constitute one "treatment" and group characteristics constitute the other. The cells of the table record the average fitness of organisms having both properties.

		Group characteristics	
		G-1	G-2
Organismic	O-1	w	x
characteristics	O-2	y	z

Group characteristics exert a *main effect* on organismic fitness when $(w+y)/2 \neq (x+z)/2$. And there will be group-organismic *interactions*

---

39. This is not to say that the analysis of variance may not have other explanatory uses. For example, Fisher's fundamental theorem (Section 6.1) shows how the rate of change in fitness depends on the additive genetic variance; the analysis of variance thus can be used to explain a population's response to selection.

precisely when $(w-x) \neq (y-z)$. Group characteristic might be viewed as causally efficacious when either the group main effect or the interaction term is nonzero. This is how the analysis of variance might be used to characterize the idea that group selection exists precisely when the fitness of an organism "depends" on the kind of group it is in.

Just to nail down the implications of this proposal, let's use it to characterize one of the examples of group selection discussed before. Wade's group selection experiment seems to fall neatly into place when described in the ANOVA format. We might imagine a pair of genotypes O-1 and O-2 that occur in both large and small groups. Suppose that within each group, O-1 is at a competitive advantage when compared with O-2. If groups are selected for their large census size, the ANOVA table might look like this:

		Group characteristics	
		Large group	Small group
Organismic	O-1	5	3
characteristics	O-2	2	1

Group size has a considerable main effect here. And I chose the numbers so that there also is group-organismic interaction.

Although the analysis of variance may yield intuitive results for some cases, its limitations are immediately evident when we look at others. For example, the hypothetical example of the short and tall populations discussed earlier reveals a rather straightforward defect of the ANOVA characterization that I will call the *absent value problem*.

The groups in this little story were internally homogeneous for height. This means that one cannot take the organismic characteristics (being six feet tall, being five feet tall, etc.) and the group properties (average heights) and fill in much of an ANOVA table. Because the actual populations do not exemplify all of the treatment combinations, the table of fitness values will have more gaps than entries:

		Group averages in height					
		6	5	4	3	2	1
	6	1.6	-	-	-	-	-
	5	-	1.5	-	-	-	-
Individual	4	-	-	1.4	-	-	-
heights	3	-	-	-	1.3	-	-
	2	-	-	-	-	1.2	-
	1	-	-	-	-	-	1.1

With data like this, an analysis of variance cannot be carried out.

If one is considering group selection in a case in which the group property is determined by the frequencies of some set of traits within it, where each population has some frequency between 0 and 1 (*non-inclusive*), it may be possible to fill in the ANOVA table and carry out the analysis. But suppose the populations are not so conveniently arranged. Suppose each has gone to fixation with a different trait. For example, one might be considering a group selection hypothesis about the maintenance of sexual reproduction. Here the comparison might be between two populations, the first of which is composed solely of asexual organisms, the second of which is composed only of sexual ones. Analysis of variance cannot be used to characterize this situation, since some of the entries in the ANOVA table will be filled in with "not defined."

It is the ANOVA's obsession with the actual that gets in the way here. In the parable of the tall and the short populations, there is a perfect de facto correlation between an individual's height and the average height of the group the individual belongs to. This holds true whether there is individual selection for being tall or group selection for belonging to a tall group. To discover which of these selection hypotheses is true, we want to ask a *hypothetical* question. What would happen if populations were *not* internally homogeneous? But here we enter *terra incognita* as far as the analysis of variance is concerned. The units of selection problem concerns causes; the analysis of variance is not designed to cut through misleading correlations to see to the causal heart of the matter.

There is a second, perhaps deeper, defect in the ANOVA proposal. Let us suppose that the ANOVA table can be filled out entirely. That is, let us suppose that each combination of treatments—individual characteristic + group context—is exemplified somewhere in the ensemble of populations. Even so, the analysis of variance can mislead. It is possible to have a main effect due to group context or a group-organism interaction without there being any group selection at all. Some examples will show how.

Let us be clear on what the ANOVA criterion demands. Suppose I look at two populations that are each polymorphic for traits A and B but differ with respect to some other characteristic. I might see that A is fitter than B in each population but that the magnitude of the fitness difference varies from population to population. If so, the ANOVA table would imply that there is a main effect due to the group characteristic. Or perhaps I would see that A is fitter than B in one group and that the reverse is true in the other. In this case, the ANOVA table would say that there is a group-organism interaction.

Neither of these fitness relationships by itself implies the existence of group selection. If the two populations are each subject to slightly different individual selection pressures, it may emerge that A is the fitter characteristic, although the magnitude of the fitness difference is different in the two groups. We already have considered the phenomenon of frequency-dependent selection (Section 6.1). Recall the mimicry example. Suppose that we have two populations of Limenitis that contain no mimics. A mimic mutant is introduced into the first, and then several generations later, a mimic mutant is introduced into the second. Although mimics are always fitter than nonmimics in the population, the difference between their fitness values declines as the mimic becomes more common (Figure 9, p. 184). If we examined the two populations several generations after the mutants were introduced into each, we would find a "main effect" of group context. In both populations, mimic would be fitter than nonmimic, but the difference would vary from one group to the other. However, this should not be taken to imply that group selection is at work. One might simply have two cases of frequency-dependent individual selection, with a slight "time lag."

The same argument applies to the group-organism interaction: A is fitter than B in one population, but the reverse is true in the other. This relationship also may occur without any hint of group selection. Consider the classic work of Levene, Pavlovsky, and Dobzhansky (1954). They found that the so-called Arrowhead homozygote on the third chromosome of Drosophila pseudoobscura is fitter than the Chiricahua homozygote, under laboratory conditions in which just these two chromosome types compete. But when there is a three-way competition between these two genotypes and the Standard chromosome type, the fitness relation of Arrowhead and Chiricahua is reversed.

Suppose we were investigating the Arrowhead and Chiricahua homozygotes in two natural populations. Natural observation might generate an ANOVA table in which Arrowhead is fitter than Chiricahua in the first but Chiricahua is fitter than Arrowhead in the second. This may simply be due to the presence of Standard in one population but not in the other. A nonzero interaction term in the ANOVA table is no sure sign of group selection.

The discussion of underdominance in Section 6.1 provides another example of this type. In the constant-viability model we examined there, an unstable equilibrium ($\hat{p}$) obtains when the heterozygote is the least fit of the three genotypes. If we consult the allelic fitness functions displayed in Figure 8 (p. 181), we see that A is fitter than a in one frequency range ($p > \hat{p}$), while the reverse is true in the other ($p < \hat{p}$). Now imagine two populations that differ in their frequency of A. In one, A is fitter than a, while in the other a is fitter than A. If we drew up an ANOVA table in which group gene frequencies served as one treatment and the allelic alternatives A and a were the other, we would find a group-gene interaction. Group selection does not follow, of course. These two populations simply show how frequency-dependent individual selection in different populations can produce context dependence without group selection.

The intuitive reason for this failure of the analysis of variance should be fairly clear. The analysis of variance does not explain *why* the pattern of variation in fitness has the shape it does. Group selection may generate a nonzero main effect of group membership or a nonzero group-organism interaction. Or these very same quantitative relationships may be due entirely to individual-level processes. Group selection is not to be defined as a particular pattern of fitness relationships among organisms; the heart of the idea is to be found in a mechanism that can generate certain fitness relationships. But this question of mechanism is precisely what the analysis of variance is incapable of detecting.[40]

So there can be group-organism interaction and a group main effect without group selection. What about the converse? Can group selection exist in the absence of these ANOVA properties? The absent value problem provides one case in which this is possible. But even when all cells in the ANOVA table are filled in, group selection does not require these ANOVA properties. One can have group selection balanced by other selective forces, with the net result that all individuals in the ensemble of populations have identical fitness values. The entries in the ANOVA table are *overall* fitnesses. *Component* selective forces can exist without producing any *net* differences in fitness. I conclude that the ANOVA criterion is neither necessary nor sufficient for group selection.

The failure of the analysis of variance definition of group selection arises from the fact that it ignores causality. In this, it has a defect in common with the proposal considered in Section 7.6 that defined group selection in terms of heritable variation in the fitnesses of groups. That groups differ in fitness leaves open *why* they do so. That organisms have context-dependent fitness values (either in the sense of group context being a main effect or in the sense of a nonzero group-organism

40. Think of two populations that live on opposite sides of the universe, each driven by its own internal dynamic. Suppose that one group ends up outreproducing the other. If the two populations are not subject to a common causal influence, then it makes no sense to view them as participating in the same selection process. The tabulation of the fitnesses of the various traits represented in the populations is as it may be. But no condition governing just the mathematical relationships of those fitnesses will be sufficient to guarantee the presence of group selection.

interaction term) does the same. The analysis of variance is an imperfect guide to causation.

Biologists who are familiar with the fact that the analysis of variance can mask rather than reveal causal relationships may wonder why I have labored this point at such length. The reason is that it has been absorbed about as well as the motto "correlation and causation are different." Everyone, if asked, would readily assent; but one frequently finds pieces of reasoning that show that such truisms have not really been absorbed at the level of practice. I cannot count the number of times I have been told that if different traits have gone to fixation in different groups (or species), then a selection process that further modifies the representation of those traits in the ensemble of groups cannot occur at the individual level. This, of course, is simply to equate the difference between individual and group selection with the difference between intra- and intergroup variation in fitness. Indeed, this comment about the units of selection problem is offered about as frequently as the idea that group selection means that an organism's fitness depends on the kind of group it is in. I hope by now that it is clear why I think that this condition is also insufficient to establish the existence of group selection.[41]

What role can the analysis of variance play in the idea of group selection? We know that a necessary condition for evolution by natural selection is heritable variation in fitness. Recall that this is a requirement on *evolution* by natural selection, not on natural selection per se (Section 7.4). Just as an immobile billiard ball may be acted upon by a multiplicity

41. In Sober (1981b), I criticized Wimsatt's (1980) definition of a unit of selection essentially on the ground that the analysis of variance is often an insensitive guide to causal structure. Here is Wimsatt's proposal: "A *unit of selection* is any entity for which there is heritable *context-independent* variance in fitness among entities at that level which does not appear as heritable context-independent variance in fitness (and thus, for which the variance fitness is *context-dependent*) at any lower level of organization." Wimsatt (1981, pp. 149–151) argues that my criticisms are not accurate and that his proposal, suitably clarified, is equivalent to one I offered in Sober (1981b). In particular, Wimsatt stresses that his characterization should apply to the conditions that *actually* obtain in the populations considered; one should not imagine some counterfactual situation that the populations might have been in (but were not) and then see what the criterion of context dependence would imply. This stress on attending to the *actual* array of fitness values is precisely what undermines the accuracy of the ANOVA criterion; I still believe that it has the same effect on Wimsatt's criterion. It is not to be doubted, however, that further stipulations may be imposed that ensure that an ANOVA criterion will coincide with a plausible causal analysis. My feeling is that these further requirements will be sufficiently substantive to permit one to conclude that the analysis of variance is not what lies at the heart of the idea of a unit of selection. I believe that precisely the same comments apply to Arnold and Fristrup's (1982) use of Price's (1970) covariance analysis of selection as a key to the concept of a unit of selection.

of forces, so a static population may be prevented from changing by the interposition of mutually nullifying selection pressures. However, for a system in which selection is the only force at work, heritable variation in fitness is both necessary and sufficient for evolution. In the case of evolution by group selection, there must be heritable variation in the fitness of groups. However, the analysis of variance does not provide the tools for expanding this necessary condition for evolution by group selection into a condition that is both necessary and sufficient. Nor does it completely describe what group selection is, independently of the question of whether in a particular context group selection produces evolution. To take these two further steps, we must turn directly to an analysis of the concept of cause.

Chapter 8

Causality

In the preceding chapter, I criticized certain arguments for genic selectionism (Sections 7.3 to 7.5) and definitions of group selection (Sections 7.6 to 7.8). My criticisms were rather nonpartisan, in that they
are apt to alienate just about everyone. Different definitions of group
selection abound. If group selection must involve altruism, and altruism
is defined to mean a characteristic that cannot evolve by natural selection
at all, then the issue of group selection is handily disposed of. If, on
the other hand, group selection simply demands that an organism's
fitness depend on the kind of group it is in, then the problem is also
deftly solved, but this time in the opposite direction. A similar range
of facile solutions is available in the case of genic selection. If the unit
of selection must survive the disassembling of the genome that occurs
in meiosis and recombination, then the single gene must be the unit
of selection. If, on the other hand, the unit of selection must be the
object that "directly" interacts with the environment, then neither genes
nor gene complexes can be units of selection when they lever themselves
into the next generation by influencing the phenotypes of organisms.
Each of these quick arguments exploits a view of what a unit of selection
is. Instead of thinking these lines of reasoning solve the problem of
the units of selection, we should stand the arguments on their heads;
the facility with which they answer the question shows that their conception of the question is faulty.

A common thread running through my criticisms has been that the
arguments or definitions failed to get to the causal heart of the matter.
In this chapter, I will begin to set forth my own ideas concerning how
the idea of a unit of selection ought to be understood. To do this, I
will not simply use the concept of causation as if it were perfectly
transparent and required no explication on its own. I will provide an
account of how causation ought to be understood. First, though, I want
to sketch an intuitive picture of how the idea of a unit of selection
works before anchoring it to a deeper understanding of causality. Then,
in the next chapter, I will apply my account to a number of currently
controversial problems in the theory of natural selection.

8.1 Object and Property

In every case of evolution by natural selection, regardless of the level at which the process occurs, there will be selection *of* genes.[1] What is distinctive about genic selection is that there must be selection *for* a gene. The latter idea, in turn, is to be understood causally; "selection for a gene" means that possessing the gene *causes* superior survival and reproduction. But survival and reproduction *of what*? A gene may facilitate the survival and reproductive success of the organisms in which it is housed. However, examples like meiotic drive show that genes can be beneficial without benefiting organisms. An organism with a single copy of the *t* allele does not, on that score, enjoy any fitness advantage over an organism that has no copies of it. Rather, it is the *chromosome* in which the *t* allele occurs (or the *t* allele itself) that gains an advantage in its contribution to the gamete pool.

So the account of genic selection will have two components. First is the requirement that possessing the gene must be a cause of differential survival and reproduction. Second, the concept must allow some latitude with respect to what it is that enjoys the fitness advantage that results from possessing the gene. This may be an organism, a chromosome, or any of a number of other possibilities that we will explore in due course.

Similar remarks apply to the idea of group selection. For the group to be a unit of selection in a particular selection process, there must be selection for a group property. For there to be selection for a group property, the group property must cause superior survival and reproduction. But the question again arises: survival and reproduction *of what*?

In that part of the dynamics of the *t* allele due to group selection,[2] we saw that belonging to a group whose males are all homozygous for the *t* allele causes an *organism* to fail to reproduce. Females, in particular, gain a fitness advantage by belonging to groups that have lower frequencies of male homozygotes. Here group selection means that group properties play a causal role in determining the fitness values of *organisms*. But the sort of pluralism we noted above with respect to the idea of genic selection can also arise in this instance: Group selection requires that group properties play a positive causal role in survival and reproduction. But it is left open exactly what it is that enjoys this

---

1. Here I continue to play along with the assumption that evolution requires change in gene frequencies, although the limitations of this view of evolution have already been set forth (Section 1.2).

2. Assuming, still, that Lewontin and Dunn's (1960) explanation is correct.

fitness advantage. Organisms may do so, but so may objects at other levels of organization.

The logical structure of these ideas is precisely what the ordinary notion of causation predicts. We talk about the baseball causing the window to break. We also talk about curiosity killing the cat. In the former case, we ascribe causal efficacy to an object; in the latter, we do so to a property. However, it is quite obvious that both claims are elliptical. The baseball broke the window in virtue of its *being thrown*. Curiosity killed the cat in virtue of *the cat's* being curious. It is objects having properties that cause objects to have properties. We may abbreviate our description of causal relations by talking about an object (e.g., the baseball) or a property (e.g., curiosity) as the cause. However, when we avail ourselves of this convenience, we should not lose sight of the underlying nature of the causal relation we are describing.

Applying this lesson to the ideas of group and genic selection is a bit confusing. When we talk of group selection, is the core idea that there is selection for group *properties* or that groups are the *objects* that are selected? When we talk of genic selection, is the core idea that there is selection for genic *properties* or that genes are the *objects* that are selected? This ambiguity—between selection for properties and selection of objects—has done much to confuse discussion of the units of selection problem.

The first step toward clarification is to bring the idea of *selection for properties* to center stage. Group selection means that group properties cause differential survival and reproductive success. Genic selection means that genic properties are causally efficacious in this way. But this first step is not enough. It remains to be said what the *objects* are that these causally efficacious properties attach to. Organisms have genic properties, but so do chromosomes. Organisms belong to groups having certain group properties, but groups themselves also possess group properties. Group and genic selection *require* that certain sorts of properties be causally efficacious; they are a bit more flexible when it comes to saying what objects those properties attach to.

As a preliminary characterization of ideas that eventually will be described more fully, let's assume that *organisms* are the objects that matter. They are the *benchmarks* in terms of which group and genic selection are calibrated.[3] That is, group and genic selection conflict over which sorts of properties are causally efficacious but agree that it is organisms that benefit or suffer in virtue of their having some properties

---

3. *The Oxford Universal Dictionary*, third edition, defines *benchmark* as follows: "A surveyor's mark, cut in rock, or other durable material, to indicate the starting or other point in a line of levels for the determination of altitudes over the face of a country . . .".

rather than others.[4] Later on, we will think more carefully about which objects other than organisms may serve as benchmarks for selection processes.

So the idea is that there is group selection for groups that have some property P if (and only if)

1. Groups vary with respect to whether they have P, and

2. There is some common causal influence on those groups that makes it the case that

3. Being in a group that has P is a positive causal factor in the survival and reproduction of organisms.

The analysis of genic selection goes in parallel. There is selection for possessing the gene P if (and only if)

1. Organisms vary with respect to whether they have P, and

2. There is some common causal influence on those organisms that makes it the case that

3. Possessing the gene P is a positive causal factor in the survival and reproduction of organisms.

The first clause in each definition should be fairly transparent. There can't be selection at a given level unless there is variation at that level. The second requirement—the idea of there being an ensemble of objects, each subject to a certain causal influence—is designed to avoid a problem described in the previous section. Objects that are disconnected from each other causally (say, because they are at opposite ends of the universe) are not subsumed by a common selection process, no matter what may be said of how their fitnesses compare with each other. What is required is that there be some common factor—perhaps due to the objects' competing with each other, perhaps due to their being affected by a common cause—that endows the trait in question with the selective significance it has.

As mentioned above, clause 3 discerns what sort of a selection process is at work by using *organisms* as a benchmark—both group and genic selection are defined in terms of how belonging to a given kind of group or possessing a particular gene functions as a causal factor in the life prospects of organisms. Although this restriction will be lifted in the more general treatment to be presented later, there is an important

4. A remark for philosophers: Note that theses concerning the importance of group or individual selection involve ontological commitment to the existence of properties. Predicates occur in opaque contexts in causal locutions. Claims about the importance of group selection quantify into those contexts. The items in the domain of quantification, therefore, cannot be sets of objects. See Sober (1981a) for further discussion.

reason to at least begin with this idea. I noted in Section 7.6 that differential proliferation of groups is no sure sign that group selection was at work. Similarly, one gene may make more copies of itself than its alternative alleles do, but this is no sure sign that selection for that gene caused the change in gene frequency (Section 7.4). One way to say whether changes in groups or genes are due to one form of selection rather than another is to see how the forces impinge on organisms.

The main task at present is to explicate what "positive causal factor" means in clause 3. The simplest way to do this is by way of an example entirely removed from evolutionary theory. Once the basic outline of the idea is delineated, we will return to the units of selection problem, plug the idea into the characterizations of group and genic selection just offered, and see what happens.

8.2. Coronaries and Correlations

What does it mean to say that smoking is a positive causal factor in the production of heart attacks among adults in the United States? To begin with, it does *not* mean that everyone who smokes must have a heart attack. Although the tobacco industry may well have wished to construe the causal claim in this extremely demanding way, the Surgeon General did not; and he was right. Smoking may be a positive causal factor, even though not every smoker is stricken. There are two reasons why. The obvious one is that people may differ in their physical constitutions, so that smoking may have different effects on different people. Second, people may be indeterministic systems. If this were true, two individuals alike in physical constitution might both smoke (to exactly the same degree), with only one of them having a heart attack. If quantum-mechanical uncertainty at the micro-level can "percolate up" to affect events at the macro-level (and presumably nothing could absolutely ensure that this will not occur), this would be a real possibility (Section 4.2.). But we needn't assume that determinism is false in order to think about what it means for smoking to cause coronaries; people are different, and that is reason enough to expect that not every smoker will be stricken.

We also don't want to define the causal relation as meaning that there is a correlation between smoking and heart attacks in the population. One often hears the homily "causation and correlation are different," but it will be useful to spell out why this is so. First of all, what does it mean to say that smoking and heart attacks are correlated? To simplify matters, let's think of smoking as a yes/no property (you either are a smoker or you are not) and of having a heart attack in the same way. Then there is a positive correlation between smoking (S)

and coronaries (C) in the population of U.S. adults if and only if the proportion of coronaries among smokers exceeds the proportion of coronaries among nonsmokers. I will use "%(C/S)" to represent the percentage of coronaries that occur among smokers. So the claim of positive correlation simply amounts to this: %(C/S) > %(C/not-S). Equal percentages would mean that there is no correlation between smoking and heart attacks. If smokers had heart attacks less frequently than nonsmokers, then smoking and coronaries would be negatively correlated.

Some other properties of the concept are worth noting. Notice that if everyone (both smokers and nonsmokers alike) has a coronary, there can be no correlation between smoking and heart attacks. The same holds true if everyone smokes. Furthermore, even if only 5 percent of the smokers get heart attacks, there still may be a positive correlation between smoking and coronaries, provided that a smaller percentage of nonsmokers are stricken. Of course, if no smoker gets a heart attack, there can be no positive correlation.

Now we may spell out the truism that correlation and causality are different in a concrete way. To begin with, correlation is a symmetrical relation, whereas causation is not. That is, it is a consequence of the definition of correlation that if smoking is positively correlated with coronaries, then coronaries are positively correlated with smoking. This tells us that the correlation of smoking and coronaries cannot be *sufficient* for the former to cause the latter. For if correlation were sufficient, we would conclude from the correlation not just that smoking causes heart attacks but that heart attacks cause smoking.

Thus correlation isn't sufficient for causation. Is it necessary—must there be a correlation between smoking and heart attacks if the former is to cause the latter? Again, the answer is *no*. It can be true that smoking causes heart attacks even when smokers get heart attacks no more frequently than nonsmokers. How is this possible? Suppose that smokers exercise more than nonsmokers. If exercising were a *negative* causal factor with respect to coronaries (i.e., if it tended to prevent heart attacks), then the frequencies of coronaries in the two subpopulations (smokers and nonsmokers) might be the same. However, this would not lead us to deny that smoking is a positive causal factor in producing coronaries. Smoking *is* a contributor, but its effects are counteracted by a second factor (exercise) acting in the opposite direction. So correlation isn't necessary for causation.

In this situation the causal connection of smoking and heart attacks is "masked" or "confounded." One would not realize that the causal connection obtained just by answering the question of correlation. Because smoking is correlated with exercising, smoking fails to be correlated with heart attacks. One correlation prevents another.

This difference between correlation and causation holds true even if we imagine that the connection of smoking and exercise is not at all accidental. Suppose that there were some biological characteristic (a gene, perhaps) that causes individuals to both smoke and exercise. It would then be no accident that smoking and exercise are correlated or that smoking and coronaries are not. This possible causal setup is shown below. An arrow with a plus sign means a positive causal relation, with a minus sign it means a negative causal connection, and a broken line indicates a correlation.

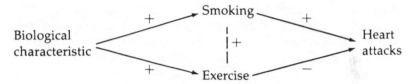

Sometimes so-called spurious correlations are simply due to sampling error. If one looked at a small sample of smokers and nonsmokers and found that the sampled smokers exercised more frequently than the sampled nonsmokers, this sample *might* fail to be representative of the population from which it is drawn. But if the causal arrangement is as shown above, the correlation is not due to sampling error but reflects the causal facts. Even if you examined *all* the individuals in the population, you still might find no correlation between smoking and heart attacks. And even if the population were indefinitely large, the absence of correlation between smoking and coronaries would still persist. It isn't only sampling error that can make correlations fail to reflect causation.

Since correlation is not an adequate explication of the concept of cause, it is not an adequate explication of the concept of *selection for properties*, which is our causal concept par excellence. In the children's sorting toy (Section 3.2), there was a correlation between a ball's being green and its finding its way to the bottom of the toy. However, there was no selection for being green; this was not what caused the green balls to descend. Yet it is entirely correct to say that the green balls *were selected*. There was *selection of* those objects. Although correlation is not an explication of *selection for properties*, it is precisely what is involved in the idea of *selection of objects*.

Evolution by natural selection requires change in gene frequencies.[5]

---

5. Recall that this is a simplifying assumption, objections having been registered in Section 1.2.

For these frequencies to change in virtue of selection, there must be a *correlation* between the possession of genes and the differential survival and reproduction of certain objects. In the case of individual, Darwinian selection, organisms differ in fitness in virtue of their different phenotypes. As long as those phenotypic differences correlate with genetic differences, evolution by natural selection is possible. By the same token, for group selection to produce evolution, there must be a correlation between genetic characteristics and the phenotypes of groups. In the dynamics of the *t* allele (Section 7.7), the differences between groups with respect to their chances of extinction reflected differences in frequencies of the *t* allele. This is how group selection produces evolution. The correlation of genetic characteristics with organismic or group phenotypes in these processes is what is represented in the different fitness values assigned to single genes. The existence of such a correlation is an absolute criterion for evolution by natural selection, no matter what the unit of selection is. But correlation is no sure sign of causation, any more than *selection of* is a sure sign of *selection for*. We must look elsewhere for a positive characterization of what this causal notion amounts to.

Given that causation and correlation are different, the reader may wish to consider the following passage in which Dawkins (1982, p. 12) describes what he takes causal claims about genes to mean:

> If, then, it were true that the possession of a Y chromosome had a causal influence on, say, musical ability or fondness for knitting, what would this mean? It would mean that, in some specified population and in some specified environment, an observer in possession of information about an individual's sex would be able to make a statistically more accurate prediction as to the person's musical ability than an observer ignorant of the person's sex. The emphasis is on the word 'statistically', and let us throw in an 'other things being equal' for good measure. The observer might be provided with some additional information, say on the person's education or upbringing, which would lead him to revise, or even reverse, his prediction based on sex. If females are statistically more likely than males to enjoy knitting, this does not mean that all females enjoy knitting, nor even that a majority do.
>
> It is also fully compatible with the view that the reason females enjoy knitting is that society brings them up to enjoy knitting. If society systematically trains children without penises to knit and play with dolls, and trains children with penises to play with guns and toy soldiers, any resulting differences in male and female preferences are strictly speaking genetically determined differences!

They are determined, through the medium of societal custom, by the fact of possession or non-possession of a penis, and that is determined (in a normal environment and in the absence of ingenious plastic surgery or hormone therapy) by sex chromosomes.

In assessing this analysis of causality in terms of "statistically more accurate prediction," the reader should bear in mind that sex and fondness for knitting are correlated precisely when the frequency of fondness among males differs from the frequency of fondness in the population at large. The assumptions about transitivity and context dependence (Section 7.2) are also worth noting.

Shifting back to our example of smoking and coronaries, we now must ask what account can be given of the idea that "smoking is a positive causal factor in the production of heart attacks among adults in the United States." Rather than comparing the overall frequencies of coronaries among smokers and nonsmokers, we must undertake a more fine-grained analysis. We must attend to how smoking affects *each individual*'s chance of getting a heart attack. Would each individual's chance of having a heart attack be greater if he smoked or if he didn't? We ask this of the individuals who exercise as well as of those who do not. If each individual's risk of having a coronary were increased by smoking, we would conclude that smoking is a positive causal factor in the production of heart attacks.

Notice that according to this proposal, smoking might be a positive causal factor in the production of coronaries even though smoking and coronaries are not correlated. Indeed, it might be true that smoking causes heart attacks even when smokers are stricken *less* frequently than nonsmokers. This could happen if exercise were a powerful inhibitor of coronaries and smoking were a mild contributor. If smoking and exercise were themselves joint effects of a common cause, as shown in the previous diagram, the overall frequency of heart attacks might be lower among the smokers.

Correlation looks at *overall* probabilities. It compares the probability of heart attacks among smokers with the probability of heart attacks among nonsmokers. The causal concept we are considering looks at probabilities on a case-by-case basis; we see whether each individual would run a higher risk of a coronary if he smoked. The overall probability need not reflect the individual probabilities; this fact is sometimes called Simpson's paradox.[6] Even though the overall frequency of coronaries among smokers may be *lower* than the frequency among non-

---

6. Named for E. H. Simpson (1951). According to Skyrms (1980, p. 107), the idea was discussed by F. Y. Edgeworth, Pearson, Bravais, and Yule.

smokers, it may nevertheless be true that each individual runs a *greater* risk of a coronary, if he or she smokes.

Since Simpson's paradox will be important to some of the ideas presented in what follows, it is well to burn it into consciousness by briefly describing another example. Cartwright (1979) describes how the University of California at Berkeley was suspected of discriminating against women in admission to graduate school. Women were rejected much more frequently than men, and the difference was large enough that the discrepancy could not be dismissed as due to chance. The frequencies thereby pointed to a difference in probability:

Pr(rejection / female) > Pr(rejection / male).

But when the investigating committee examined the academic departments, it discovered that within each department, women were rejected no more frequently than men. That is, for each department D,

Pr(rejection / female & applied to D) = Pr(rejection / male & applied to D).

How to reconcile these two statistics? The reason was that women tended to apply to departments that had high rejection rates for both sexes. Because there was a correlation between being female and applying to a department to which it was hard to gain admission, there was an overall correlation between being female and being rejected, even though being female did not increase any individual's chance of being rejected. The overall frequencies gave a misleading picture of the underlying causal facts.

In assessing the impact of smoking on heart attacks, we must see how the probabilities are related both among those who exercise and among those who do not. Presumably other background conditions besides exercising must be taken into account. For example, diet may be a contributor to heart attacks; so may stress. We want to know how smoking affects the probability of heart attacks within each assortment of yes/no background properties: if an individual exercises *and* diets *and* experiences stress, if the individual exercises *and* diets *and* does not experience stress, if the individual does not exercise *and* does not diet *and* does not experience stress, etc. Each possible constellation may be represented as a maximal conjunction (B_1, B_2, etc.) of causally relevant background factors. The requirement may therefore be formulated as follows: Smoking (S) is a positive causal factor in the production of coronaries (C) if and only if

Pr(C/S and B_i) > Pr(C/not-S and B_i), for each i.

That is, for smoking to be a positive causal factor in the production of coronaries is for smoking to raise the probability of coronaries in each causally relevant background context.[7]

It is of some interest to contrast this interpretation of causality with one that implicitly figures in some experimental investigations. A standard way to investigate whether smoking causes coronaries would be to establish a control and test group. The individuals in the two groups would be physically similar, except that individuals in the test group would smoke, while those in the control group would not. A significantly higher incidence of heart attacks in the test group would be evidence in favor of the causal hypothesis. This familiar experimental procedure suggests the following interpretation of the notion of cause (defended by Giere 1979, 1980):

> Smoking causes coronaries in the population of U.S. adults if and only if there would be more coronaries if everyone smoked than there would be if no one did.

According to this interpretation, the causal structure of the actual population (in which some individuals smoke and some do not) is determined by what would happen in two hypothetical populations, one composed of 100 percent smokers, the other of 0 percent smokers.

This characterization of causality may be adequate in some contexts, but it is not universally correct (Sober 1982a). If you are at a baseball game at which everyone is seated, it may be true that each individual would have a better view if he or she stood up. Standing up is a positive causal factor, in the circumstances, for improving one's view of the game. But if everyone stood up, the average quality of view would be no better than if no one did. The causal efficacy of standing up is frequency-dependent. The view of standers is best when standers are rare.[8]

I discussed frequency-dependent natural selection in Section 6.1. Cases of this sort show the difference between the causal concept proposed here and the one that thinks the question "what would happen if everyone did it?" is the acid test. Consider the hypothetical fitness function shown in Figure 10 (p. 185) in which the average fitness in the population ($\overline{w}$) is the same when mimicry is at fixation as it was

7. This proposal has been extensively discussed in the philosophical literature. For defense and elaboration, see Good (1961–1962), Suppes (1970), Rosen (1978), Cartwright (1979), Skyrms (1980), and Eells and Sober (1983). For criticisms, some of which are addressed in the last-named paper, see Hesslow (1976), Salmon (1980), and Otte (1981). The idea is also familiar in the social sciences—see, for example, Simon (1957) and Blalock (1961) for discussion.
8. I have been unable to track down where I read this example. My apologies (and thanks) to its inventor.

before mimicry is introduced. When mimics are rare, the trait is selectively advantageous. It is then a positive causal factor in survival and reproduction. When mimicry is rare, each individual would be better off as a mimic. But it is not true that individuals would on average be better off if every individual were a mimic than they would be if no one were. The example of the "spoiler" trait (Figure 11, p. 186) makes the same point. At every possible frequency, trait S is advantageous; no matter what the population frequency is, an individual is better off having trait S than it would be with trait A.[9] Intuitively, S is a positive causal factor in surviving and reproducing, regardless of the population frequency. Yet, if everyone had S, the individuals would be *worse* off than they would be if everyone had A. By focusing solely on the 100 percent and 0 percent cases, one misses an important element of causal structure. Although this may not be a problem in deciding if smoking causes coronaries, the situation is quite different when the frequency of a trait modulates its causal efficacy.

8.3 Fine-Tuning

The probabilistic characterization—that causes augment the chances of their effects in all causally relevant background contexts—requires further scrutiny in two ways. First, the idea of a "causally relevant background condition" needs to be clarified. And second, we need to examine more closely whether the probabilistic idea really does provide a plausible necessary and sufficient condition.

Unless the idea of a causally relevant background condition is suitably restricted, the probabilistic characterization is very implausible. Let's switch examples for a moment to see why. I call you on the telephone (C). This makes your phone ring (R). The ringing then causes you to answer your phone (A). Here we have a causal chain from C to R to A:

$$C \longrightarrow R \longrightarrow A$$

Each cause raises the probability of the effect immediately following

---

9. Strictly speaking, fitness functions do not imply that *each individual* would be better off with one trait rather than another; they only describe the fitness values of traits (i.e., the average fitness of individuals with the trait in question) at the frequencies represented. However, it is often natural to think of fitness functions as representing the former kind of fact as well. For example, it is not implausible, in the mimicry case, to say that each individual would be better off as a mimic than it would be as a nonmimic. In using fitness functions to discuss causality, I will assume that this sort of interpretation is appropriate.

it. What are we to say of the relation between my calling you (C) and your answering (A)? Intuitively, it seems that we would want to say that my calling caused you to answer your phone. But if your phone is ringing, the chance of your answering is not affected by whether or not it is I who called. That is,

$$\Pr(A \;/\; C \text{ and } R) = \Pr(A \;/\; \text{not-}C \text{ and } R).$$

So, if the phone's ringing is counted as a causally relevant background condition in this process, we can't conclude that my calling caused you to answer the phone. The obvious solution is to say that the phone's ringing is *not* a causally relevant background condition.

The solution I propose is as follows. One must time-index the factors whose causal relation is at issue. If we want to know whether C at time t_1 is a positive causal factor in producing A at t_3 (where t_1 is before t_3), we may hold fixed only those factors that are present at t_1. *Intermediate* events (at t_2) do not count as relevant causal background conditions; for my call to cause you to answer, it isn't required that my phoning have an impact on your answering above and beyond the impact that the ringing of the phone has (Eells and Sober 1983).

Of course, it is certainly possible that my phoning *does* pack this extra punch. Perhaps when it is I who phones you, a little red light starts flashing on your telephone. Your eagerness to answer (naturally) is augmented by this signal beyond what it would have been if the phone had rung without the light's flashing. The constraint on causally relevant background conditions doesn't prohibit this from being true; the restriction merely allows my phoning to cause you to answer without requiring this extra Rube Goldberg connection to be in play.[10]

The idea of a "causally relevant background condition" presents another problem.[11] I just considered what might count as a background condition with respect to the impact of my phoning (C) on your answering your telephone (A). Now let's refocus. What counts as a relevant background condition when we want to know whether your phone's ringing (R) causes you to answer (A)? Must the phone's ringing raise the probability of your answering, given, for example, the fact that I dialed your number? It is easy to imagine cases where this will not be possible. Suppose my dialing *guarantees* that your phone will ring (i.e., confers on that event a probability of 1). Then there will be no way

10. Another, less homespun, way for my phoning you to have an impact on your answering beyond the impact that the ringing of the phone has is for the causal chain to violate the Markov condition; perhaps the past can have an impact on the future above and beyond the influence that the present has. We may leave this possibility to quantum mechanics to sort out.
11. Discussed in Salmon (1980), Otte (1981), and Eells and Sober (1983).

for the ringing to *raise* the probability of your answering above what it was, given that I dialed your number. In fact, the required inequality, namely, $Pr(A / R$ and $C) > Pr(A / $ not-R and $C)$, will not be well-defined, since $Pr($not-R and $C) = 0$, and so the right-hand expression in the inequality is not defined.[12] So if we let my calling be part of the background context, then, in this case, it can't be true that the phone's ringing causes you to answer.

To make matters worse, let's suppose that your phone won't ring unless I call you. That is, imagine that my calling is a necessary *and* a deterministic cause of your phone's ringing. Then, in addition to the above inequality not being well-defined, neither will the following: $Pr(A / R$ and not-$C) > Pr(A / $ not-R and not-$C)$. Once again, if my calling is part of the background context, then it won't be true that the phone's ringing causes you to answer.

This problem may be solved in the way suggested before. We already have said that in assessing the impact of an event X on a later event Y, events intermediate between X and Y do not enter into the appropriate background context. Let us say the same for events that precede X. So the causally relevant background conditions that must be taken into account in assessing the effect of X on Y are just the conditions that obtain simultaneously with X. In the smoking/coronary example, we want to know whether people's smoking at time t causally contributes to their having heart attacks later. To find this out, we see whether their physical conditions *at time t* are such that their chances of coronaries are greater if they smoke than they would be if they didn't.[13]

There is another way to formulate this constraint that does not involve the concept of simultaneity. In assessing the influence of C on E, one holds fixed any factor that is not itself a cause or effect of C.[14] That is, if the causal relation is as follows (where the arrows indicate causal relevance, whether positive or negative), one evaluates the impact of

12. The conditional probability $Pr(X / Y)$ is standardly defined as $Pr(X$ and $Y) / Pr(Y)$, which means that $Pr(Y)$ must be nonzero, if the conditional probability is to fall within 0 and 1.

13. It is quite true that this characterization of what is included in the "causally relevant background conditions" presupposes that an event at one time does not have the entire future history of the world packed into its description. I do not count this as an *objection* to the proposal, however, but rather as an element of it that deserves further philosophical investigation. Determinism and indeterminism are theses that have the same presupposition.

14. In holding fixed all such factors, one will, of course, be holding fixed many that are irrelevant to the effect. This can do no harm in the assignment of causal roles, however. What is essential is that one *not* hold fixed factors that are effects or causes of C and that one hold fixed all other factors that are relevant to E (Eells and Sober 1983).

C on *E* relative to maximal conjunctions of the background factors, *R*, *S*, and *T*:

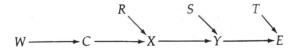

Again, the point is to not hold *W*, *X*, or *Y* fixed.

What one takes the ensemble of background conditions to be in a given causal process depends on what factor is being evaluated. If I want to know whether smoking causes heart attacks, I will not consider *smoking* to be a background factor; there is no way that smoking—the putative causal factor—could increase the chance of a coronary in both the presence and the absence of *smoking*. Yet, if I wanted to check what causal role *exercise* plays in the production of heart attacks, I would consider whether exercise raises the chance of a coronary in both the presence and absence of *smoking*. Smoking is properly taken as part of the background in this case, but not in the previous one. This merely reflects the fact that one cannot compute the probability of a heart attack, conditional on a contradictory set (smoking and not-smoking). If one wants to gauge the causal role of a factor *F*, the conjunctions of background factors must each be consistent with the presence and absence of *F*.

These two formulations of what a causally relevant background condition amounts to are worth comparing. The former uses the concept of simultaneity; the latter explicitly uses the concept of cause. The former may seem to be preferable in that it avoids introducing a circularity into the picture of causality here presented. However, my hunch is that its plausibility depends on the further assumption that cause and effect cannot be simultaneous. As we will soon see, there are other assumptions about causality that must be true if this theory is to provide both necessary and sufficient conditions. Cartwright's (1979) formulation of the probabilistic theory is, as she observes, a circular one. She notes, and I agree, that this circularity does not deprive the account of its interest and utility.

These clarifications go part of the distance toward meeting the objections to the probabilistic idea defended here (Salmon 1980, Otte 1981). However, two residual problems remain. Rather than using them as a tool for repairing the definition, I will just point out that the idea has two factual presuppositions.

The first problem shows that the proposal assumes that when two events have a common cause, the probability relation of the cause to one of the effects differs from the probability relation that obtains between the two effects. Biologists will probably not find this limitation at all troubling, since the presupposition is deeply engrained in common sense and in most scientific work (except, of course, quantum mechanics). Philosophers may be more interested in this issue.

Suppose that two pistols rarely fire but that if the first one fires at a given time, the second always fires a moment later. What is more, the second one never goes off unless the first one did a moment before:

$$Pr(F_2 / F_1) = 1 > Pr(F_2 / not\text{-}F_1) = 0.$$

Obviously, the information that the first has fired raises the probability we assign to the firing of the second. But suppose, as a matter of fact, that neither event causes the other. The two pistols are linked to a common cause. As usual, an arrow indicates causation and a broken line denotes correlation:

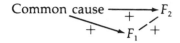

What is there in the probabilistic criterion to prevent us from concluding that the firing of the first pistol causes the second to fire?

According to the probabilistic account, if the first firing is not a cause of the second, there must be some specification of a causal background condition in which the first firing does not augment the probability of the second. Indeed, there *may* be such a circumstance; suppose that the two guns probably fire when a button is pushed (*B*) that is linked to both. The pushing of the button *screens off* the second firing from the first:

$$Pr(F_2 / F_1 \text{ and } B) = Pr(F_2 / not\text{-}F_1 \text{ and } B).$$

There *may* indeed be such a common cause.[15] But there is no guarantee in advance that the correlation of the two firings isn't just a "coincidence." The probabilistic condition asserts, in effect, that the events at a given time that give the most information about a future event are its causes. This may or may not be so.

Common sense may seem to demand that the perfect correlation of the two pistols cannot just be a "coincidence." One must cause the other, or there must be a deterministic common cause that coordinates their behavior. This is not the place to consider whether common sense is committed in this way. However, it is pretty clear that quantum mechanics displaces this principle. The theory predicts that pairs of particles produced in Einstein-Podolski-Rosen experimental setups will

---

15. Or one might say: there *may* be an event on the causal chain from the common cause to F_2 that screens off F_1 from F_2.

have a perfect negative correlation with respect to their spin. If there were a screening-off common cause, it would have to confer on each particle's state a probability of 1. But there need not be any such (deterministic) common cause, and the various no-hidden-variable proofs make it plausible to think that there is no such common cause.[16] Since the biological applications I will make of the analysis of causation will not be affected by this sort of consideration, I will not revise the proposed criterion to take account of it.

The second limitation arises when causes are necessary and deterministic relative to their effects. Suppose that C deterministically causes both R and F and that R and F cannot occur unless C is present. Assume that R and F are simultaneous. Presumably, this setup should not rule out the possibility that R and F each causally contribute to the occurrence of an event E:

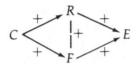

However, notice R cannot raise the probability of E in both the presence and the absence of F. The reason is that R and F imply each other, and so the probability of "R and not-F" and of "not-R and F" are not defined. As Salmon (1980) and Otte (1981) have remarked, perhaps the probabilistic theory of causation should be thought of as an account of *probabilistic* causation—i.e., as applicable only when causal relations are not deterministic. This limitation will not affect applying the theory to the units of selection problem, since selection in finite populations should always be thought of as conferring probabilities of survival and reproduction between 0 and 1 (*non*inclusive).

I now want to discuss a more down-to-earth objection to requiring that the causal factor should raise the probability of the effect in *every* causally relevant background context. Suppose some other physical condition, apart from smoking, *guarantees* the occurrence of a coronary. If an individual has that physical condition, smoking cannot boost the probability of a heart attack any higher than it already is.[17] Yet it would

16. Reichenbach (1956) and Salmon (1975) formulated the so-called principle of the common cause. Suppes and Zinotti (1976) and Van Fraassen (1980) have laid out the quantum-mechanical argument against it. See Sober (1984a) for further discussion, which defends a reformulated principle of the common cause and connects it with the use of a parsimony criterion in phylogenetic inference.
17. The example of *overdetermination* is a special case of this sort. If Holmes and Watson each, simultaneously, shoot Moriarty through the heart, can we say that Holmes' shooting caused the death and that Watson's shooting did too? Neither shot raised the probability of death.

be overly restrictive to conclude that smoking is not a positive causal factor for heart attacks in a population that happens to include some individuals with the condition. To take account of this sort of case, we should relax the requirement in the following way: The causal factor must raise the probability of the effect in at least one background context and must not lower it in any. I will follow Skyrms (1980) and call this less demanding version of the probabilistic characterization the "Pareto-style" formulation.

So if smoking increased some individuals' chances of coronaries while leaving others unaffected, smoking might nevertheless be a positive causal factor for coronaries. What is still precluded, however, is that smoking increase some people's chances and reduce the risks that others run. One can't conclude, in this case, that smoking is either a positive or a negative causal factor; there is no such thing as *the* causal role it plays vis-à-vis heart attacks. One may, of course, consider the two relevant subpopulations separately and characterize the causal role that smoking plays in each. But there is no way to describe the causal role of smoking in the population as a whole if the direction of the probabilistic inequality is reversed by switching from context to context.

An example of this sort is provided by certain drugs that are called "stimulants." These are sometimes given to children diagnosed as "hyperactive"; although the drugs are stimulants *for adults*, they are depressants when given to children. What is the causal role of such drugs in the population as a whole? The analysis of causation asserts—plausibly, I think—that there is no such thing.

It might be suggested that the Pareto-style requirement could be weakened still further. Smoking could still count as a positive causal factor for heart attacks in a population if it raised the chance in the (statistically) "normal" (N) cases and lowered it, if at all, only rarely (i.e., in the not-N cases).[18] The problem with this idea is that it runs the risk of equating causation with correlation. The suggestion could be satisfied simply because on average smoking increases the probability of attacks. This implies that smoking will be a positive causal factor if the following inequality holds:

$$\text{Pr}(C \text{ / } S \text{ \& } N)\text{Pr}(N) + \text{Pr}(C \text{ / } S \text{ \& not-}N)\text{Pr}(\text{not-}N) >$$
$$\text{Pr}(C \text{ / not-}S \text{ \& } N)\text{Pr}(N) + \text{Pr}(C \text{ / not-}S \text{ \& not-}N)\text{Pr}(\text{not-}N).$$

This requirement is equivalent to demanding that

$$\text{Pr}(C \text{ / } S) > \text{Pr}(C \text{ / not-}S),$$

which is precisely what it means for C and S to be correlated. Again,

18. Assume for convenience that normality and smoking are probabilistically independent.

notice that to say that two factors are correlated is just to say that, *on average*, one raises the probability of the other.

In order to evaluate the adequacy of the proposed characterization of causality, we need to recognize that it is intended to characterize *population-level* causal claims like "smoking is a positive causal factor in the production of coronaries in the population of U.S. adults," not *individual-level* causal claims like "Harry's smoking caused him to have a coronary." These two causal concepts differ in several ways. First notice that for Harry's smoking to cause him to have a coronary, it must be true both that he smoked and that he had a coronary. But for smoking to be a positive causal factor for heart attacks in a population, it need not be true that any smoker actually has a heart attack. Indeed, it doesn't even have to be true that anyone actually smokes. It suffices that smoking *would* place each individual at greater risk.[19]

Second, although I think that a positive causal factor must raise the probability of its effect (or must raise it in at least one context without lowering it in any) in the case of population-level causation, this need not be true in the case of individual-level causation. I modify an example due to Deborah Rosen (cited in Suppes 1970, p. 41) that illustrates this difference. Suppose a golf ball is rolling toward the cup, when a squirrel comes along and kicks it away. The ball then ricochets against several objects and, improbably enough, drops into the cup. The hole-in-one was obtained "the hard way." Let us suppose that the probability of the ball's going into the hole was *reduced* by the squirrel's kick.[20] Yet it is hard to deny that the squirrel's kick caused the ball to go into the hole.

So much for the individual-level causal claim. What of the population-level claim? That is, do kicks of the sort the squirrel administered constitute positive causal factors in the production of holes-in-one? No, they do not. If we replicated this setup several thousand times, we would see why. In the control group, the ball is rolled toward the hole and no squirrel kick occurs. In the test group, the ball rolls toward the cup and then receives just the sort of kick that the squirrel provided. If the kick really did *reduce* the probability of a hole-in-one, we would expect there to be *fewer* holes-in-one in the test group than in the control group. Kicking, we conclude, is a *negative* causal factor.[21] So

19. Comments from Part I concerning the inadequacy of identifying probability with actual relative frequency continue to apply. A coin can have a certain probability of landing heads, even if it is never tossed and even if it is tossed but never lands heads.
20. Don't assume that if we only had a more detailed specification of exactly what the squirrel did, we would see that the ball had to go into the cup. This is just to assume that determinism is true and obscures the point of the example.
21. This difference between individual and population-level causal claims is described in Good (1961–1962), Skyrms (1980), and Eells and Sober (1983).

individual-level causes don't have to raise the probability of their effects, but population-level causes do.[22]

A purely evolutionary example makes the same point. A characteristic is introduced as a mutation and strong selection increases its frequency. Just as the trait reaches a frequency of, say, 0.4, the environment changes, and the trait is rendered neutral. Subsequently, the trait, by chance, goes to fixation. The change in the environment reduced the probability that the characteristic would reach fixation. It had a better chance of doing this when favored by selection than it subsequently had when drift alone was at work. But when we ask what caused the characteristic to reach 100 percent, it seems perfectly correct to answer: random genetic drift. Sampling error can be causally efficacious, even though it sometimes will interpose itself into a historical process so as to reduce the probability of the outcome it in fact produces.

The distinction between population-level and individual-level causal claims has an important bearing on the interpretation I am giving of the units of selection controversy. When we think about genic selection, for example, we are asking about the population-level causal role of a gene of a given kind. We are not considering the individual-level question of whether the copy of the gene that I have in a certain cell in my left elbow happens to have a particular effect on me. The question is about gene-*kinds*, not gene-*instances* (Section 7.5). Recall that population genetics assigns a fitness value to a gene by averaging over the various contexts in which the gene may appear (Section 7.4). Therefore, the things to which these values are assigned cannot be physically unique objects. What is being assigned a value is, as it were, a thing whose parts are spread through the population.[23] Genic selectionism

22. As noted in Eells and Sober (1983), the meaning of the individual-level causal claim is hard to nail down, when the squirrel example is compared with the following one (due to Cartwright 1979), which has the same probabilistic structure. A plant is sprayed with a defoliant that has a 0.9 chance of killing it. Yet the plant lives. We do not say that the plant's being sprayed caused it to remain alive; we say that it lived *in spite of* its being sprayed. How is the ball's being kicked by the squirrel different from the plant's being sprayed by the defoliant? An appallingly vague hunch occurs to me: The squirrel's kick initiated a *new* causal process that led to the ball's dropping in the hole. But the spraying of the plant did not initiate a new process that led to the plant's remaining alive. Rather, the spraying was an event that failed to interfere with the processes that were already under way; it was *these* that kept the plant alive. Perhaps individual causal claims need to be understood by taking the idea of a causal process as fundamental (a suggestion made for slightly different reasons by Salmon 1980). Similar ideas about causal processes were deployed in the discussion of disjunctive causes in Section 3.1.

23. Dawkins' analogy of the rowing team (Section 7.2) obscures this fact, since in that case it is physically unique oarsmen that compete with each other, rather than *kinds* of oarsmen whose causal efficacy is being gauged. We will return to this point in Section 9.1.

assigns causal efficacy to genes in the way the Surgeon General assigned causal efficacy to smoking.[24]

The distinction between individual-level and population-level causality makes all the difference in the world when we consider how "context dependence" affects the truth and falsity of causal claims. It is a truism that causes bring about their effects only in virtue of certain background conditions being satisfied. Striking a match may have caused it to light, but the striking was not in itself sufficient for the lighting: Oxygen had to be present and the match had to be dry. It is no criticism of an individual-level claim that the cause's bringing about its effect depended on the context's being right (Section 7.2).

Population-level causal claims are a different matter, however. As noted above, if a factor augments some individuals' chances of an effect but diminishes the chances of others, the factor will not play a determinate causal role in the population as a whole. Here context dependence of a certain sort is enough to defeat the causal claim.

As noted in Section 7.2, a standard criticism of genic selectionism, made by Mayr (1963) and Gould (1980a), among others, is that a gene contributes to an organism's phenotype, and thereby to its fitness, only in virtue of the presence of other genes. A gene may be advantageous when set against one genetic background, deleterious when set against another. If genic selectionism were a claim about individual-level causation, this would not count as a criticism. But, since genic selectionism concerns the causal role of a *kind* of gene in a population undergoing natural selection, the fact of context dependence is of the first importance. As we will see in the next section, it is enough to refute the general claim that all selection is selection for or against single genes.

The distinction between individual-level and population-level causal claims also is important to the question of the transitivity of causal chains. We may use the terminology of Section 7.5 and distinguish event-*instances*, which are the subject matter of individual-level causal claims, and event-*kinds*, which are the subject matter of population-level causal claims. If three event-instances a, b, and c are such that a caused b and b caused c, does it follow that a caused c? I can think of no compelling counterexample to this assertion, and no plausible theory about individual-level causal claims that implies that transitivity fails.

However, the analysis of population-level causal claims described in this section implies that *that* sort of causal relation is not in general

24. Of course, there is always the gloss of the idea of genic selectionism that converts that doctrine into a truism. Here causation is replaced with correlation, and the representation argument (Section 7.4) shows that all cases of selection, including group selection, are "really" cases of genic selection. This tactic has, I hope, already been disposed of.

transitive. There is an intuitive reason why this should be true. Consider the biological condition that causes both smoking and exercising. Smoking, we assume, causes coronaries, whereas exercise tends to prevent them. Now if exercise were a very strong preventer of coronaries and smoking were only a mild cause of them, we might have a failure of transitivity. The biological condition causes smoking. Smoking causes coronaries. But it isn't true that the biological condition is a positive causal factor in the production of coronaries. The condition lowers the risk of heart attack by producing two effects that themselves exert opposite influences of unequal magnitude on the occurrence of heart attacks.[25]

Although several colleagues have offered me examples that purport to show that individual-level causal chains can fail to be transitive, only one has struck me as in any way plausible. Byerly (1979, p. 68) says that Montaigne puzzled over the following line of reasoning:

> Eating ham makes a person drink; drinking quenches thirst; hence eating ham quenches thirst. Eating ham causes a person to drink and drinking causes the thirst to be quenched. But eating ham, even if it be a *ceteris paribus* sufficient condition for the thirst quenching, seems inadmissible as the cause of the thirst quenching.

I suggest that the probabilistic theory delivers an entirely convincing analysis of this example, if it is thought of as reporting a population-level relationship. Construe the problem as a chain with four links: Eating at t_1 causes thirst at t_2 which causes drinking at t_3 which causes the absence of thirst at t_4. It is plausible that eating ham at t_1 might fail to be a positive causal factor for the absence of thirst at t_4. Merely suppose that individuals would have the same chance of being thirsty at t_4 whether or not they ate ham at t_1. This could be true, even if eating ham at t_1 is a positive causal factor for being thirsty at t_2. However, none of this suggests that transitivity can fail for individual-level causal claims. If Michel's eating ham at t_1 set in motion a process of the sort just described, then it seems to me that his lack of thirst at t_4 *was* produced by his eating ham at t_1. His lack of thirst traces back, we

---

25. Although population-level causal chains need not be transitive, two simple sufficient conditions have been identified in the literature. Suppes (1970) notes that when all causal connections are deterministic, causal chains must be transitive. Eells and Sober (1983) provide a more general sufficient condition, which demands, roughly, the following: C at t_1 causes E at t_3 if each effect at t_2 of C that bears on E at all is positively causally relevant to E. The proof involves assumptions besides the "unanimity" of the intermediate causal factors; the chain must be Markovian and certain independence assumptions must be satisfied.

might say, to his eating.[26] The difference between population-level and individual-level causality makes all the difference.

Because causality at the population level need not be transitive, we must not blithely assume that the following defense of genic selectionism is available:

> *Of course*, each gene (unless it is selectively neutral) causally contributes to (i.e., is a positive causal factor in the production of) the organism's phenotype, and, *of course*, an organism's phenotypic properties causally contribute to its survival and reproductive success, so, *of course*, it must follow that each gene causally contributes to survival and reproductive success. Hence, it will always be correct to think of genic selection as the cause of evolution by natural selection.

The three *of course*'s do not follow as a matter of course. The two premises of this little argument, as well as the conclusion alleged to follow from them, may look plausible. But there is a lot less to them than meets the eye, as we will now see.

26. I believe the causal relationship between event-instances is the same as that between a parent organism or species and its offspring. The characteristics of the parent may be positive or negative causal factors for the traits that the offspring exhibits. Either way, the parent *produces* the offspring (Sober 1984a).

Chapter 9

Consequences

The flurry of heart attacks, telephone calls, squirrel kicks, and ham sandwiches in the previous section should not obscure the essential point. The claim that smoking is a positive causal factor in the occurrence of heart attacks in a particular population was given the following gloss: smoking must raise at least one individual's chance of a coronary, and must not reduce anyone's risk. This less demanding "Pareto formulation," I argued, is preferable to the somewhat simpler requirement that smoking increase everyone's chance of a heart attack. But on either formulation, it is neither necessary nor sufficient for the causal claim to be true that smokers are stricken more frequently than nonsmokers. Correlation, we must bear in mind, is different from causation.

In Sections 9.1 and 9.2 I apply the probabilistic characterization of causality to a number of examples, many of them ones I have discussed before. In Chapter 7, I suggested what I hope are plausible interpretations of a variety of selection processes. The time has come to go from plausibility to precision. The probabilistic analysis will be justified to the degree that it helps clarify our interpretation of various selection processes.

Sections 9.3 and 9.4 take up examples of selection processes not already discussed. In 9.3, the concepts of altruism, kin selection, and inclusive fitness are scrutinized and some recent developments in sex-ratio theory are discussed. In 9.4, the controversial ideas of species selection and punctuated equilibrium are connected with the understanding of natural selection developed so far. The net effect, I hope, will be to show how a multiplicity of models and conjectures currently discussed in evolutionary theory have a common underlying structure; each can be viewed as employing the idea of selection presented here, fleshed out with different empirical side assumptions.

At the end of the nineteenth century, mathematicians working in geometry looked at their subject and saw before them a multiplicity of theorems. They noted that some of these theorems remain true when the objects described (e.g., a circle) are deformed quite radically; others remain true only when the transformations are more restricted. Such observations gave rise to the so-called Erlanger program—the attempt to consolidate the different branches of geometry by seeing how their theorems are affected by different sets of transformations. The goal was not to discover which geometry correctly describes the structure of physical space but to see how the various theories and theorems fit together in a systematic way. I have pursued a similar problem with respect to the units of selection problem. Evolutionary theory is alive with a variety of models and conjectures. Besides the problem of deciding what the units of selection are, there is another, more preliminary, question that I have thought worthy of attention. How are the different ideas about the units of selection related to each other? Is there some root idea of what a unit of selection is that underlies the various proposals? Is there some normal form that all the theories have in common, differing from each other in terms of how this or that parameter is set? Rather than trying to say here what the units in fact are, I have tried to extract the base conception of what it would take to be a unit of selection. The question of how selection in fact proceeds, like the question of the geometry of physical space, philosophy must leave to empirical science.

9.1 The Selfish Gene or the Artifactual Allele?

How to apply the probabilistic idea of causation (Chapter 8) to the thesis of genic selection? To say that there is selection for a given gene at a particular locus is to say that possessing that gene is a positive causal factor in survival and reproduction. This, in turn, requires that the allele must not decrease fitness in any context, and must raise it in at least one.

What is a genetic context? An allele's various genetic contexts include the genes that may be found at other loci. In addition, a specification of the genetic context of a single gene must also take into account the other genes that occur at the same locus. An a allele, for example, may be next to another copy of itself, and therefore be part of the homozygous aa genotype; alternatively, it may be next to a copy of A, in which case it occurs in heterozygous form. For the moment, we will set aside the influence of the genes that happen to exist at other loci and concentrate only on the single locus at which these two alleles may occur.

We now can see in a more precise way why the example of heterozygote superiority failed to count as a case of genic selection (Section 7.4). In this case, recall, the Aa heterozygote is fitter than both the AA and the aa homozygotes. This means that one dose of the a allele is better than two, and is also better than zero; the same holds for the A

allele. In this case, the *a* allele does not have a unique causal role. Whether the gene *a* will be a positive or a negative causal factor in the survival and reproductive success of an organism depends on the genetic context. If it is placed next to a copy of *A*, *a* will mean an increase in fitness. If it is placed next to a copy of itself, the gene will mean a decrement in fitness. Here, the gene is like the drug mentioned in Chapter 8 that is a stimulant for adults but a depressant for children. Although the gene does play a determinate causal role in different subgroups, it has no univocal causal role in the whole population.

In the simple one-locus/two-allele model we are considering, what would be required for genic selection to occur? Bear in mind that we are still assuming that an organism's fitness is determined only by its genotype at this single locus. For genic selectionism to be correct in this model, all that is required is that the heterozygote be intermediate in fitness. The ordering in terms of doses of the gene—zero, one, or two—must reflect the impact of substitutions on individual fitness. More precisely, if selection for gene *A* is to occur, each individual *i* must be such that:

$$\Pr(i \text{ survives } / i \text{ has } AA) \geq$$
$$\Pr(i \text{ survives } / i \text{ has } Aa) \geq$$
$$\Pr(i \text{ survives } / i \text{ has } aa)$$

with strict inequality, for at least one *i*.[1]

It is important to be clear how this causal analysis differs from the one provided by the analysis of variance, which I criticized in Section 7.8. The reason genic selection is ruled out in the case of heterozygote superiority is *not* that heterozygotes are, on average, fitter than homozygotes. This may be true but need not be. Imagine that there is a correlation between that genotype and a deleterious genetic condition (*D*) at another locus. The causal structure would then be as follows (the plus and minus arrows as usual represent causal influence; the plus broken line represents positive correlation):

1. It is convenient to think of each *organism* as constituting a causally relevant background context here. Identical twins will thereby be considered twice, but this redundancy is harmless, of course. A less innocuous consequence is that by focusing only on organisms that actually exist in the population, we ignore causally relevant background contexts that *can* exist but do not. So, for example, we might wish to consider all the possible genotyes at other loci, even though a population that contains alleles *B* and *b* at a locus might at a given time contain no heterozygotes. One of the virtues of the probabilistic analysis exploited here is that it permits a narrower or a broader construal of background contexts. My critical points against reductionism in the units of selection controversy apply to either sort of account. For example, heterozygote superiority is a counterexample to genic selectionism whether one considers only genetic backgrounds that actually exist in the population or possible backgrounds as well.

Aa genotype $\diagdown$ $+$

$+$ $\Big|$ $\longrightarrow$ Reproductive success

Genotype D $\diagup$ $-$

In this case, heterozygotes might occupy a position of intermediate fitness. The causal role of a characteristic does not have to be reflected in the relation of the average fitness of individuals having it to the average fitness of individuals having other properties. Again, we must not confuse causation with correlation. Rather, the causal role of a property is to be identified by asking what its effect is or would be on each individual in the population. Smoking causes coronaries, not because smokers are more frequently stricken but because smoking places each individual at greater risk.

It may turn out, of course, that the analysis of variance generates the same answer to the causal question as the characterization suggested here. Indeed, the same may be said of a characterization of causation in terms of correlation. The point is just that it need not, and by examining examples in which it need not (e.g., the internally homogeneous populations of Section 7.6), we can see why a pattern of variance is no guarantee of a pattern of causation.

Although the analysis of variance does not yield a decisive criterion for what the unit of selection is, it does provide a heuristic guide. Finding that Aa heterozygotes on average are fitter than AA and aa homozygotes raises the suspicion that there is selection for Aa. But this need not be true, since the causal action might be elsewhere in the genome. The heterozygote genotype Aa might be selected without there being selection for it (Section 3.2). Still, we may look at the interactions described in ANOVA tables as useful indicators.

When we do this in the context of models of selection that take more than one locus into account, the possibility of *epistasis* provides another way in which the perspective of genic selectionism may fail to be true. Suppose that the fitness ordering of the genotypes at one locus is influenced by the genotypes present at another. If each locus has two alleles, there are nine possible two-locus configurations. Lewontin and White (1960) (described in Lewontin 1974b, pp. 279–281) investigated just this sort of relationship between two chromosome inversions found in the grasshopper *Moraba scura*. On each of the chromosomes of the *EF* pair, the Standard (*ST*) and Tidbinbilla (*TD*) inversions may be found. On the *CD* pair, Standard (*ST*) and Blundell (*BL*) are the two alternatives. The fitness values of the nine possible configurations were estimated from nature as follows:

	Chromosome CD		
	ST/ST	ST/BL	BL/BL
Chromosome EF			
ST/ST	0.791	1.000	0.834
ST/TD	0.670	1.006	0.901
TD/TD	0.657	0.657	1.067

Notice that there is heterozygote superiority on the CD chromosome if the EF chromosome is either ST/ST or ST/TD, but that BL/BL dominance occurs if the EF chromosome is homozygous for TD. Similarly, the TD homozygote is superior when in the context BL/BL but is inferior otherwise.

These fitness values represent average differences in viability. What would it take to show that the unit of selection is at a level of organization higher than that of the single-locus genotype? Simply imagine that the inequalities in the above table reflect the impact of substitutions on the probability of mortality of each organism. For example, suppose that each individual that is ST/BL on the CD chromosome would be fitter if it had ST/TD on the EF chromosome than it would be if it had TD/TD. If the inequalities in the above table reflected these sorts of individual facts, it would follow that no genotype at a single locus is a positive causal factor in survival and reproduction. It is not implausible to suspect that the average viabilities measured in nature reflect this sort of relationship of an organism's genotype at many loci to its viability.

There is no reason to think that epistatic interactions among loci are rare in nature. Although the more complex the genetic control of a phenotypic characteristic is, the less likely it is to have been studied and understood, biologists commonly model polygenic effects in a way that undermines the plausibility of genic selectionism. Consider a phenotype that has an intermediate optimum and is controlled by the alleles at many loci. Birth weight in human beings appears to be a case in point. The usual model posits a large number of loci at which "plus" alleles may be found; the more of these an organism has, the higher its phenotypic score will be.[2] An intermediate phenotypic optimum means that an intermediate number of plus alleles in the genome is best. However, this implies that whether a plus allele at one locus is favored or disfavored depends on how many plus alleles there are at other loci. Some individuals experience that allele as a positive causal factor in survival, others as a negative one. The allele resembles the

2. That is, holding constant environmental factors like nutrition, for example.

drug that is a stimulant for adults and a depressant for children. There will be no such thing as *the* causal role it plays in the population as a whole. Hence the unit of selection will not be the single gene (Sober and Lewontin 1982).

Having noted some cases in which the thesis of genic selectionism fails, it is well to note that the criterion suggested here for genic selection does not swiftly dispose of that idea as a conceptual impossibility. There is nothing incoherent about the idea that there are genes that always have the same directional effect on fitness, regardless of the contexts in which they occur. A dominant lethal (one that kills any organism in which it occurs, regardless of its genetic context) would be selected against. And as mentioned above, when fitness is settled simply by which of three genotypes is present at a single locus, genic selection occurs when the heterozygote is intermediate in fitness.

We now are in a position to sort out Dawkins' (1976) rowing analogy (Section 7.2). A crew coach, recall, is trying to choose the best team and has the rowers assemble themselves into different teams and then compete. Using the distinction between *selection of objects* and *selection for properties*, it is clear that the coach selects oarsmen *and* teams (Section 3.2). *Selection of objects* is a liberal concept; here, as in biology, there are several correct answers concerning which objects get selected. However, this fails to settle the question of which properties were selected for. Was the coach selecting for combinations of rowers? Was he selecting for particular rowers?

We need not psychoanalyze the coach to find out. Let us suppose that he is sensitive simply to the composition of crew teams and whether they win. To simplify, suppose there are only two positions in a boat and that each position has only two individuals competing for it. Individuals A and a compete for the first slot while B and b compete for the second. The coach makes the teams compete against each other a large number of times (to reduce chance to a level of insignificance) and records the frequency with which the different teams win. Let us record the results in a table:

		Second position	
		B	b
First position	A	w	x
	a	y	z

There is a very important feature of this competition that we must note. A, a, B, and b are individual oarsmen. There is no way that the AB team can race against Ab. There are only two possible competitions: AB versus ab and Ab versus aB. So the table which represents the frequency with which the teams won must be such that $w + z = x + y$. If this were mistakenly construed as an ANOVA table, we would conclude that there is no interaction between the two positions!

Dawkins' analogy is a wonderful one. What could be more similar to genes on a chromosome than rowers in a racing shell? A rowing shell with two candidates for each of two positions is simply a haploid organism with two alleles at each of two loci. But here is the important difference between chromosomes and racing shells that makes Dawkins' analogy deeply misleading: Genes are capable of epistatic interactions. When we assess the survivorship of individuals with the four possible genotypes, there is no a priori guarantee that $w + z = x + y$. If this requirement is obeyed, there is no objection to thinking of selection acting at each locus. However, it is not impossible that we should obtain fitness values estimated from survival rates such as these:

		Second position	
		B	b
First position	A	0.3	0.4
	a	0.2	0.1

Here it is reasonable to talk about genic selection for A and against a.[3] But genic selection will not be correct as a description of what occurs at the "second position." B is superior to b when a is at the first position, but the reverse is true when A is there.

An ANOVA table for the selection process can exhibit interaction effects and therefore show that the unit of selection is not the single gene. When the racing teams compete, however, it is a given that $w + z = x + y$. This difference has a very simple basis. In Dawkins' story, it is *individual* rowers, not *kinds* of rowers, that combine to form teams that then compete. There is no way the AB team can race against the Ab team, since one rower can be on only one team in any given race. However, in natural selection, we consider gene-*kinds*, not gene-*instances*. In the example population, all four genotypes are exemplified at the same time and compete against each other. If we modify Dawkins' example to remove this misleading feature, we can see how the coach may well be selecting for combinations of rowers, not for single rowers. Let the coach be a mad scientist who clones his favorite rowers and makes them race against each other in all combinations.

---

3. That is, if the inequalities in the average viabilities reflect what would be good or bad for each individual in the population.

As mentioned in the previous section, the concept of cause has two components. It is *objects* having *properties* that are causally efficacious. True, we talk of baseballs breaking windows and of curiosity killing cats, but these remarks are elliptical: It was the baseball's striking the window and the cat's being curious that did the causal work. The causal analysis of genic selection reflects these two components. The first and invariable requirement is that genic *properties* cause enhanced survival and reproductive success. However, to say this leaves open what the *objects* are to which these genic properties attach. As a provisional statement of the idea, I assumed that organisms are the *benchmarks* in terms of which genic selection is calibrated. This construal means that genic selection occurs when genes are positive causal factors in the survival and reproduction of *organisms*. But other benchmarks are possible. One case in which genic properties are causal factors for objects other than organisms is meiotic drive. A driving gene in a heterozygote secures for itself an enhanced representation in the gamete pool. This need not be good or bad for the organism.[4] Rather, it is the chromosome in which the driving gene occurs (and, of course, the driving gene itself) that benefits. In meiotic drive, there is selection for a particular gene, but possessing that gene need not be a causal factor in the survival and reproduction of organisms. Here is genic selection without an organismic benchmark.

A slight complication must be noted. If meiotic drive is a genuine case of genic selectionism, it can't be true that a given allele is advantageous in one context but deleterious in another. Does this sort of context invariance hold in the case of segregation distorters? In *Drosophila*, suppressor genes at other loci are known to exist that diminish or eliminate the biasing effect; whether there are suppressors of this sort in the house mouse is not known (Crow 1979). Such suppressors help restore Mendelian equity by bringing the segregation ratio back toward 50/50. If this is all the modifiers do, then meiotic drive still counts as a case of genic selection. But if a modifier could turn a driving gene into a *yielding* gene—making it obtain only *minority* representation in the heterozygote gamete pool—then genic selection would be refuted. For if modifiers worked in this way, the so-called driving gene would be like the so-called stimulant drug mentioned earlier. The gene would drive in one context, while being driven upon in the other. The gene would then have no univocal causal role. Were such modifiers to exist, we would have selection for gene combinations (not single genes),

---

4. In the case of the *t* allele in the house mouse discussed in Section 7.7, there does happen to be an effect on the organism—homozygotes are sterile. But this consequence is not entailed by the idea of meiotic drive.

calibrated in terms of a genic benchmark. In the absence of such interactions, however, we may conclude that a segregation distorter allele
provides a neat case in which a genic property (being a segregator
distorter) is a positive causal factor in the success of a gene.

Another example of genic selection in which the causal factors are
not calibrated organismically is the hypothesis of "selfish DNA," put
forward by Doolittle and Sapienza (1980) and by Orgel and Crick
(1980). They introduced this idea to explain why so much DNA has
no known organismic function. The guess is that large amounts of DNA
make no difference to an organism's phenotype. Yet even when strands
of DNA are not causal factors in the survival and reproduction of
organisms, they may nevertheless experience selection of another sort.
Perhaps strands of DNA differ in their abilities to jump from locus to
locus. A "jumping gene" that has superior stability and colonizing
ability may expand its representation within the lifetime of a single
organism. After reproduction levers the strand into the next generation,
it can colonize some more. A process of this sort would explain why
we observe so many repetitions in the genetic code. It isn't that repetition
makes any difference to the organism; rather, multiple copies exist in
virtue of an autonomous selection process at the molecular level.

Although Crick, Doolittle, Orgel, and Sapienza acknowledge their
debt to Dawkins (1976), naming their idea in homage to Dawkins'
influential book, it is important to see the difference between their idea
and the concept of genic selection discussed so far. Dawkins thinks of
genic selection as the correct way to describe *all* selection processes.
For him, all genes are selfish genes. Doolittle et al., on the other hand,
think of selfish DNA as a very special sort of phenomenon, one whose
interest is that it bypasses the organism altogether.

One novelty in the idea of selfish DNA is the way it computes success
and failure. Ordinarily, we think of an allele as having at most two
copies of itself in a diploid organism. According to the usual system
of head counting, its frequency in the population is calibrated so as to
reach its maximum (100 percent) when each individual is homozygous
for the allele *at the locus*. But the idea of selfish DNA opens up much
vaster vistas into which an allele might expand. The upper bound for
a diploid population of size N is not $2N$ copies but $2NM$, where M is
the number of loci per organism.[5]

Selfish DNA may have no effect on an organism's phenotype as it
begins expanding into portions of the genome that are entirely redundant. However, it cannot colonize the genome indefinitely without

5. And even this upper bound is not absolute, since there is no reason in principle why
the amount of DNA cannot be expanded.

sooner or later coming into conflict with organismic fitness. A few extra copies of a given allele may do an organism no harm, but if the entire genome were filled with copies of that allele, there almost certainly would not be much of an organism around to look at. Here, selection for good replicators at the level of selfish DNA and selection for genes that enhance the fitness of organisms may clash. The result is not the triumph of all-powerful selfish DNA, but a compromise: Some of the genome (as far as we now know, at least) has no organismic function and may therefore be an arena dedicated to competition among strands of selfish DNA, whereas other parts of the genome are in the business of building an organism and are thereby subject to potentially conflicting pressures that arise from two levels of organization.

Selfish DNA, then, counts as a case of genic selection. As with the dominant lethal and the segregation distorter *t* allele, we have here a genic property that causes enhanced survival and reproductive success. But what are the objects that possess said causally efficacious property? Not an organism, as in the case of the dominant lethal. Not a chromosome, as in the case of meiotic drive. Selfish DNA obliges us to think of genes themselves as the objects that benefit or suffer from the properties they have.[6]

We have found several kinds of cases in which genic selectionism is true. Let us review what they have in common and how they differ. In *all* cases of genic selection, a genic *property* is a positive causal factor in survival and reproduction. Cases of genic selection may differ, however, with respect to what the *objects* are that these genic properties attach to. Genic selection may impinge upon organisms, upon chromosomes, upon gametes, and upon genes themselves. In the case of the dominant lethal, selection against the gene means that organisms die and that certain chromosomes and genes thereby fail to make copies of themselves. In the case of segregation distortion, however, the number of offspring produced by a heterozygote need not be affected by the presence of the driving gene; however, a chromosome's chance of finding its way into the gamete pool is affected by whether it contains a driving gene. So distorting the segregation process means that chromosomes and genes are selected, but organisms need not be. Finally, selfish DNA is a case in which genes are selected but organisms and chromosomes need not be.

We have here, yet again, a consequence of the distinction between

---

6. I should note that the same complication that arises by imagining modifiers that turn driving genes into yielding genes in the case of segregation distortion applies here. If a gene's relative jumping ability is context-dependent in the right way, then selfish DNA will not be a case in which single gene *properties* have determinate causal roles.

selection of and *selection for* (Section 3.2). In genic selection, there is selection for genic properties. It is a further question, however, as to what objects are selected. But this difference is old hat to us by now. In the case of the sorting toy, we get a single answer to the question of what properties are selected for. The answer is *smallness*. But when we ask what objects are selected, we are happy to tolerate a multiplicity of answers. The green objects are selected. The small objects are selected. And so on.

As useful as the sorting toy is for grasping certain biological distinctions, it does not prepare us for the idea that genic selection may involve benchmarks at different levels of organization. In thinking about the toy, there was never any doubt that it is *balls* that descend. They are the objects; the only question is what properties are selected for. But evolutionary theory has opened up a much broader range of conceptual possibilities. Darwin began by thinking of organisms as the objects that differentially survive and reproduce in selection processes, but the concept of selection has subsequently been much enriched.[7]

In this account, I have allowed some scope for the claim of genic selectionism. It is a kind of causal process that is not unknown. But genic selection is far from being the universal form that all selection processes must take. Heterozygote superiority at a single locus and epistatic interactions among loci constitute counterexamples to the general thesis that all selection is selection for and against single genes.

As we have seen, it is always open to the advocate of genic selection to reconstrue such processes in terms of selection coefficients that attach to single genes. The strategy of averaging over contexts is the magic wand of genic selectionism. It is a universal tool, allowing *all* selection processes, regardless of their causal structure, to be represented at the level of the single gene. In the model of heterozygote superiority, the constant viabilities assigned to the three genotypes are combined to yield a fitness value for each allele that is frequency-dependent. The magnitude of the allelic fitness is simply a weighted average of the fitnesses of the two genotypes in which it may occur; as the frequencies of the three genotypes change under the impact of selection, so does the fitness of the alleles. The allelic fitnesses vanish as soon as they appear, since they are temporary reflections of the changing frequencies of the genotypes in which the alleles occur.

When a car drives down a road, it may cast a shadow on the shoulder. The shape of the shadow is caused by the shape of the car. Yet, if the

---

7. We can see now in retrospect one limitation of the definition of fitness discussed in Section 1.4. Only if organisms are the benchmarks of the selection process can one define the fitness of a trait in terms of the average fitness of the *organisms* possessing it.

conditions are right, one can predict the shadow's shape at one time from its shape at a previous time. The change in shape of the shadow may be described as if it were an autonomous process. One can define mathematical coefficients that allow the computation of the shadow's shape at one time from its shape the moment before. It will turn out, almost certainly, that the coefficients that link moments t_1 and t_2 of the shadow's history differ markedly from the coefficients that link moments t_2 and t_3. But life is complicated, one might say; the way a shadow at one moment influences the state it assumes at the next is subtle and sensitive to context. Since one can describe the shadow's changes as determined by its own previous state, isn't it more parsimonious to use this one-level description rather than bring in some altogether novel variable, namely, the car on the road?

There is, of course, another way of describing the process. Granted, mathematical coefficients may be defined that connect the shadow at t_1 to the shadow at t_2, and the shadow at t_2 to the shadow at t_3, and so on. But states of the shadow are artifacts, not causes. The form of the shadow at one time does not *cause* its form at the next moment. Rather, it is the car (and the sun) that determines the shadow's metamorphosis.[8]

Genic selection coefficients are frequently no more causally efficacious than the shape of the shadow. Both are artifacts, not causes. One does not need to introduce the idea of group selection to see that changes in gene frequencies may be produced by processes that take place at higher levels of organization. The chief challenge to genic selectionism is not the idea of adaptations existing for the good of the species but the failure of beanbag genetics. Even if all selection occurred within the context of a single group (and thus the possibility of group selection were ruled out), the relationships considered so far between genes, phenotypes, and fitness would be quite sufficient to undermine genic selectionism.

If beanbag genetics were correct, each gene would produce a single phenotypic characteristic of selective importance. Possessing a gene would then be a positive causal factor for possessing the phenotype in question. Selection for or against the phenotype would then imply selection for or against the gene. No matter that selection acts "directly" on phenotypes and only "indirectly" on genes. This would be no more consequential than the remark that it was the bullet's penetrating the victim's heart that was the immediate cause of his death, whereas the assassin's pulling the trigger was a more distal cause. Indeed, the as-

8. This example is due to Salmon (1975). He calls the shadow's change in shape a "pseudoprocess" and tries to develop a general characterization of this idea.

sassin's action counts as a "deeper," "more fundamental" cause, since the bullet's trajectory was merely a consequence of what the assassin did. In just the same way, if single genes caused the different phenotypes that selection acts upon, it is hard to avoid the impression that they are the more fundamental causal agents.

Beanbag genetics turned out to be mistaken for two reasons. First, there is a one-many relation of gene to phenotype; this is the phenomenon of *pleiotropy*. Second, the relationship is also many-one; this is the phenomenon of *polygenic effects*, in which a given phenotype is the result of an interaction among an ensemble of genes. As we have noted, many biologists have felt that the failure of beanbag genetics refutes the idea that the gene is the unit of selection. Our discussion of causation allows this hunch to be given a precise formulation.[9]

The problem is polygenic effects. If a gene raises the probability of a given phenotype in one context and lowers it in another, there is no such thing as the causal role that the gene has in general. Selection for or against the phenotype may cause the frequency of the gene to change, but this will be due to the *correlation* between gene and phenotype. There will be selection *of* the gene, but no selection *for* it.

The other half of the failure of beanbag genetics—pleiotropy—is not a problem for genic selectionism. If a single gene *causes* a number of phenotypic characteristics, there will be selection for or against the gene depending on whether the phenotypic traits are, on balance, advantageous or deleterious (weighted, of course, by the strength of the gene's positive causal connection with each). So, for example, if a gene were to deterministically cause two phenotypic characteristics, one mildly advantageous, the other seriously deleterious, we would want to say that there was selection against that gene, and the theory of causation advocated here would sanction that judgment.[10]

The account of causation I have defended explains how, in a certain precise sense, polygenic effects provide a counterexample to the reductionist adage that "the whole is nothing more than the sum of its parts." Natural selection furnishes cases in which *ensembles* of interacting

9. "Beanbag genetics" was probably not a position that any population geneticist ever believed. Haldane, for example, realized that the approach was oversimplified, but he took the attitude that the subject had to learn to crawl before it could walk or run. Moreover, from very early on, Fisher and Wright took account of the impact of dominance and epistasis. At the same time, I should emphasize that geneticists often thought, and continue to think, of beanbag genetics as a *useful approximation*.

10. This reflects the argument of Eells and Sober (1983) concerning the transitivity of causal chains. If the intermediate factors caused by C are themselves all positively relevant to the occurrence of E, then, given a few other simple assumptions, transitivity follows. However, the unanimity condition is not necessary for the chain to be transitive.

genes are causally efficacious even though single genes are not. An ensemble of genes is, to be sure, made only of genes; there is no mysterious extra added ingredient—an *élan vital*—that infuses the ensemble with powers that the individuals lack. But this fact about *composition*—that an ensemble of genes is made of genes and of genes alone—cuts no ice when the question is one of *causal efficacy*. The point is that ensembles may have determinate causal roles in selection processes, even when single genes do not. This fact makes all the difference between the "selfish gene," which manipulates organismic phenotypes for its own advantage, and the "artifactual allele," which changes frequency in consequence of the effects of ensembles of genes on the fitness values of organisms.

9.2 Group Selection in Focus

In this section, I want to examine some of the consequences of the definition of group selection suggested in Chapter 8. The formulation offered there makes use of what I called an organismic *benchmark* and is thereby subject to the same limitation as my initial characterization of genic selection. Just as in the previous section, I lifted the organismic restriction to accommodate examples like meiotic drive and selfish DNA under the rubric of genic selection, so, in Section 9.4, a similar amendment will be needed to make sense of the idea of species selection. For now, though, let us see how group selection can occur in virtue of group properties being causally efficacious via their impact on *organismic* fitness.

My suggestion was that group selection occurs in a set of populations exactly when there exists some property P such that

1. Groups vary with respect to whether they have P, and
2. There is a common causal influence on those groups that makes it the case that
3. Being in a group that has P is a positive causal factor in the survival and reproduction of organisms.

Clause 1 is needed, since there cannot be group selection without variation among groups. Clause 2 is required because populations that are causally disconnected from each other are not subsumed under a common selection process, no matter what the relation may be between their fitnesses or the fitnesses of their member organisms. Finally, the meaning of "positive causal factor" in clause 3 was discussed in Sections 8.2 and 8.3.

An example introduced in Section 7.6 concerned a set of internally homogeneous populations that had different average heights. The example was constructed so that group selection—for groups of greater average height—would yield the same changes in the ensemble of populations as individual selection for being tall. How does the definition suggested here discriminate between these two possibilities?

If there is individual selection for being tall, then being tall must be a positive causal factor for survival and reproduction in the ensemble of populations. This will be true if each individual's chance of surviving and reproducing is greater if he or she is tall than it would be if he or she were not. But the Pareto formulation of the analysis of causation doesn't require quite this much. The necessary and sufficient condition is simply that a positive causal factor must raise the probability of the effect in at least one causal background context and must not lower it in any. Tallness must do for survival and reproductive success what smoking does for heart attacks.

If being tall is the only causally efficacious property here, then we can see how it is possible to have individual selection for being tall without having group selection for groups of greater than average height. If your height settles how fit you are, then the average height of the group you are in matters not. Your membership in a group of a given average height is not, in this case, a positive causal factor; the group phenotype does not affect the fitness of any organism.

The probabilistic structure of this case of individual selection is as follows. For each individual x, and for each average group height i, it must be the case that

> Pr(x survives and reproduces / x is tall & x is in group of average height i) $\geq$ Pr (x survives and reproduces / x is not tall & x is in group of average height i),

with strict inequality for at least one i. Here, the different average heights of the groups are causal background contexts against which we assess the causal role of the individual property of being tall.

The probabilisitic comparison required here tells us something about how the probability concept must be construed. In the ensemble of populations considered, there are no tall individuals in short populations and no short individuals in tall populations. Yet, to carry through the proposed analysis, one must be able to say how a short individual's being in a tall population *would* affect its life prospects. Again, we see that probability cannot be equated with actual relative frequency (Section 1.4).

We can look at the example of the tall and the short from the other direction. What would be required for group selection, rather than individual selection, to be at work? If the group phenotype conferred a fitness advantge on organisms, then an individual of a given height would have a higher fitness if she were in a taller group than she would if she were in a shorter one. In this case, the individual property of being tall matters not. If you are in a group with a given average height, you will have the same fitness regardless of whether you are short or tall.

The probabilistic relationships under group selection are quite different from those shown above that obtain for individual selection. When groups are selected for their greater average height, we have the following situation. For each individual x and for each individual height h,

Pr(x survives & reproduces / x is in a tall group & x has height h) $\geq$

Pr(x survives & reproduces / x is not in a tall group & x has height h),

with strict inequality for at least one h.

It is certainly possible for one of these probabilistic relationships to obtain without the other. This is the basis for the difference between the two kinds of causal process. It also is possible for *both* probabilistic relationships to obtain, in which case one would have group and individual selection acting in the same direction.

This example was presented initially to show how two quite distinct sorts of selection processes can, in a special circumstance, have identical consequences. If a group sends out colonizing propagules when it reaches a certain critical mass, and if new groups are always founded by organisms from the same parent population, and if like produces like as far as height is concerned, the two hypotheses predict the same trajectory of change. That there are cases like this should come as no surprise, once the difference between group and individual selection is seen to be a difference in causal structure. After all, the balls that fall to the bottom in the selection toy (Section 3.2) are both green and small. Because the properties are correlated, the hypothesis that there is selection for being green and the hypothesis that there is selection for being small make identical predictions about the changes that will occur once a good shaking is administered. In the height case, there is a correlation between being tall and being in a population of tall individuals. No wonder the two hypotheses can generate identical predictions. But there is no mystery here; there is no fact about these two cases that is inaccessible to investigation. We know from the way the toy is built that there is selection for size, not color. A parallel investigation into the ecological factors impinging on the ensemble of populations may reveal whether there is selection for being tall or for being in a tall population.

Organisms constitute the benchmark here; they help determine the difference between group and individual selection. Let us be clear on how they do this. If we listed the fitness of each individual, we would find no difference among the fitness values of individuals in the same group; all the variance is between-group variance. *But this would be true regardless of whether individual or group selection is at work!* We need to consider not simply the fitnesses that organisms *actually* have but the fitnesses they *would* have if they were in different groups, or if they had different heights. We need to consider counterfactual, not just actual, circumstances. It is in this sense that we use organisms to mark the difference between different sorts of selection processes.[11]

Lewontin and Dunn's (1960) group selection account of the dynamics of the *t* allele in the house mouse (Section 7.7) receives an intuitive analysis from the probabilistic construal. Recall that males homozygous for the *t* allele are sterile. This means that homozygous males exert a "poisoning effect" on the females in the group; if all the males are homozygous, the females fail to reproduce. In this case, belonging to a group in which all males are homozygous for the *t* allele is a negative causal factor in reproduction. So this is a case of group selection, in the sense of our definition.

It is clear that each *female* experiences this sort of group context as a negative causal factor. But what about the males? Homozygote males are unaffected by the sort of group they are in; they have a fitness equal to zero in any case. But this is not a problem for the definition of group selection, since the Pareto version doesn't require that *every* individual suffer. However, what of the other males?

To ask this question is to neglect one of the crucial constraints on what counts as a causally relevant background condition. If one wants to assess the bearing of smoking on heart attacks, one does not see how smoking affects the probability of coronaries in both the presence and absence of *smoking* (Section 8.2). The causally relevant background conditions must be logically independent of the causal factor considered. In the *t* allele example, the causal factor considered is membership in a group all of whose males are homozygous for *t*. This implies that

---

11. The analysis of variance, as I construed it in Section 7.8, is a way of partitioning measurements on *actual* organisms only; it is not sensitive to the sort of counterfactual considerations to which an analysis of causation must attend. However, if it were reformulated so that the ANOVA table represented the fitness values that *would* obtain in certain counterfactual circumstances, it would no longer be subject to the absent-value problem. Nevertheless, it would still be neither necessary nor sufficient as an analysis of causation.

the genotypes of *males* in the group cannot form part of the background. It suffices simply that each female has her chance of reproducing diminished by belonging to such a group.[12]

Another example that falls neatly into place is Wade's experimental work on *Tribolium* (Section 7.7). In each of the first two group selection treatments, two sorts of selection processes occurred simultaneously. Within each population, competition among beetles led to changes in population size; individual selection for increased cannibalism, lengthened developmental time, and reduced fecundity had the effect of lowering group numbers. The selection of groups for their large or small census size created an additional selective force that either acted in the same direction as individual selection or counteracted it. In the large group selection treatment, membership in a big group was a positive causal factor in reproduction; in the small group selection treatment, membership in a small group was a positive causal factor. As far as the group selection component of the process went, individuals in the same group had equal chances of finding their way into the next generation.

Wade's experimental treatments involved the extinction of groups and the founding of new groups. One striking fact about the experiment is that Wade manipulated groups in much the same way that standard selection experiments manipulate organisms. But it is worth noticing that the definition of group selection used here does not require that groups go extinct or found new colonies. Let's rewrite Wade's experimental protocol to see how this is possible. Suppose that individuals in groups with small census sizes were subject to random mortality; the experimenter simply randomly kills some fixed percentage of the individuals in such groups. Individuals in large groups, let us imagine, will be exempt from this procedure. This would be group selection in our sense—membership in a large group would be a positive causal factor in an organism's survival and reproduction—though it might turn out that no group founds a new colony and no group goes extinct.[13] Group selection here manifests itself simply in differences in the census

12. I am grateful to Gregg Mitman for raising this point.
13. It might be supposed that smaller groups run greater risks of going extinct and that the treatment just described counts as group selection only because this is true. However, there is no reason to expect, in general, that groups of size N are more prone to extinction than groups of size $N+1$. It is by no means impossible that in some circumstances large group size should cause population crashes and thus imply a greater risk of extinction. Indeed, this is precisely the situation that Wynne-Edwards (1962) conjectured was a standard setting for the operation of group selection. In any event, suppose, just to get the proposed example in focus, that the difference in group size discussed here does not lead to differential risks of group extinction.

sizes of groups. This conceptual possibility is left open by the fact that it is *organisms* that serve as the benchmark of group selection in the definition. We will explore the consequences of revising this organismic orientation in Section 9.4.

The definition of group selection also explains why simple context dependence of fitness is not thereby a case of group selection. In Section 7.8, I considered several examples in which frequency-dependent selection acts within each of two populations. Let us consider a case of this sort in which selection favors whichever of traits A and B is more rare. In the first population, A is favored; in the second, B is at an advantage. This two-population system does not count as an example of group selection, even though an organism's fitness crucially depends on the composition of the population it is in.

There are two reasons for this. First, for all I've said so far, the two populations may be causally unconnected with each other (the distant ends of the universe objection). Clause 2 of the definition of group selection will then not be satisfied. But let us get this problem out of the way. Suppose the two populations are subject to a common predator and that the predator's search image is designed to pick out prevalent prey organisms. In consequence, rarity within a prey population represents a selective advantage. This places the two populations under a common causal influence. Even so, the two groups do not participate in a group selection process, for a reason having to do with clause 3.

Although the group composition has an impact on an individual's fitness, it is not true that individuals in the same group experience that group phenotype in the same way. Membership in a group in which A is common *hurts* A individuals and *helps* individuals that are B. Membership in a group having a given composition is *not* a positive (or a negative) causal factor, because the group phenotype raises some individuals' chances of survival and reproduction and lowers those of others. The situation here is precisely analogous to the example of the drug that is a stimulant for adults and a depressant for children (Section 8.3). There is no such thing as *the* causal role that the group property plays with respect to survival and reproduction. So dependence of fitness on group context is no sure sign of group selection.

The definition of group selection also may be applied to Sewall Wright's (1931, 1945, 1978) influential, and much misunderstood, models of how evolution might proceed in a system of small semi-isolated demes. Wright's investigations stand in contrast to the perspective developed by R. A. Fisher according to which evolution by natural selection mainly occurs in a single large breeding population. As mentioned in Section 4.2, Wright's models provide a greater opportunity for chance to play a role, in virtue of the fact that he considers not a large population of size N but a set of small populations that add up to N.

As noted in Section 6.1, frequency-*in*dependent selection can be expected to raise the average fitness of organisms in a population. So, when a population begins a selection process at a given gene frequency, selection will drive the population up an adaptive gradient until it reaches a gene frequency at which there is nowhere to go but down. Selection will then serve to hold the population at this "adaptive peak."

This way of describing selection does not require the topography of gene frequencies to have a single adaptive peak. Indeed, in discussing the simple one-locus two-allele model of heterozygote inferiority (Section 6.1), we noted that selection will lead to fixation of one *or* the other of the two alleles. Figure 8 (p. 181) indicates *two* adaptive peaks in this one-locus model. More generally, if we think of selection acting simultaneously at many different loci, there may be numerous adaptive peaks; the adaptive topography may be a terrain of peaks and valleys.

Since this situation is often misleadingly described, let us take care to see how multilocus theory can imply numerous adaptive peaks. Let us imagine two loci, each with two alleles, that are on different chromosomes. If each exhibits underdominance and selection proceeds independently at each, there will be four adaptive peaks. If we represent possible gene frequencies at one locus along the x axis and possible gene frequencies at the other along the y axis, the z axis may represent values of $\bar{w}$. The adaptive surface would look like a sheet that is lifted at its four corners as in Figure 13.

Notice that there are *at most* four adaptive peaks in the two-locus two-allele system described here. With underdominance at only one locus, there are two peaks; with no underdominance, there is only a single peak. Of course, increasing the number of loci increases the dimensionality of the adaptive "surface"; it should be clear how underdominance at other loci can generate multiple equilibria.

If we were to drop the requirement that selection is frequency-independent, the conclusion stated above about the number of adaptive peaks might no longer hold. However, the idea of an adaptive surface is that selection drives a population to the adaptive peak that is in its "zone of attraction." Selection always pushes *up*. If the zone of attraction is not the globally optimal one, then selection will *prevent* the population from achieving what is (globally) best. This dynamical assumption— that selection increases $\bar{w}$—holds if selection is frequency-independent. But under frequency-dependent selection it may or may not (Section 6.1). One cannot use the adaptive surface idea with frequency-dependent

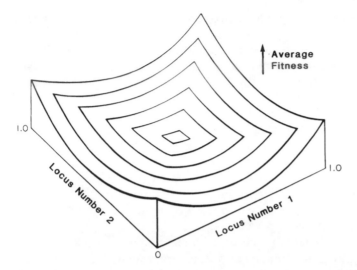

Figure 13. The adaptive surface for two independent loci, each exhibiting heterozygote inferiority. A population's initial gene frequencies at the two loci determine which of the four adaptive peaks it ascends. The elevation of a point represents the value of $\overline{w}$ exhibited by a population with the corresponding array of gene frequencies.

selection unless one makes further assumptions about the shapes of the fitness functions.

If a deme is small, there is a real possibility that chance will displace it from one side of a valley to another. A large population will find it much more difficult to escape from its zone of attraction than a small one. What is more, jumps across valleys are harder for drift to achieve if the population is at the top of a tall peak than if it is at the top of a shorter one. So drift is a mechanism that provides more scope for attaining a globally optimal, and not just a locally optimal, array of gene frequencies. Random genetic drift provides an opportunity for the population to "explore" the adaptive topography.

Another factor in the dynamics of an ensemble of demes is *migration*. If a population attains a globally adaptive peak, it may sent out migrants to less favored local demes. These migrants may bear advantageous gene combinations and therefore may help make over the other demes. Again, a force distinct from individual (within-deme) selection can have effects contrary to what one would predict from the action of individual selection alone.

Wright systematically investigated the properties of a system in which small semi-isolated demes interact. The adaptive peak idea is only one of the tools he developed in this project. He considered, for example, the effects of frequency-dependent selection, and other factors that cannot be summarized in terms of the adaptive topography idea. One such exploration was his brief discussion in Wright (1945) of how an altruistic characteristic might evolve. Chance might allow the altruistic trait to achieve fixation within a group, and migration might then repeatedly inject the altruistic character into other groups, until altruism became common in the ensemble of populations. Note that in this situation it is *chance* that takes the altruistic character to fixation within a deme; by definition, individual selection acts in the opposite direction. What is more, we will see in the next section that altruism involves frequency-dependent fitness functions in which $\bar{w}$ does *not* increase under selection within a group. Here we have an application of the shifting-balance theory in which the idea of demes ascending adaptive peaks plays no role.

When local populations evolve to their adaptive peaks and then invade each other differentially as a function of their respective values of $\bar{w}$, should this be understood as a case of group or individual selection? The problem is that the process of jumping between zones of attraction, ascending adaptive peaks, and sending out migrants appears to be completely describable in terms of drift, individual selection, and migration. The idea of group selection seems altogether superfluous.

Yet one *could* redescribe this process in terms of group selection. Some groups are fitter than others, in that they have reached higher adaptive peaks. These fitter groups then send out more migrants, which then "take over" the groups they colonize. In view of these two apparently interchangeable descriptions, how are we to decide whether the shifting balance process involves group or only individual selection?

Wright's models describe how population structure can make a difference in evolution. The processes involved make outcomes probable that would not have been expected if one were considering a single large panmictic (randomly mating) population. But the efficacy of population structure is, as we have seen, not sufficient for group selection. After all, we know that population size is itself an important factor in determining the role of random genetic drift in evolution, but this does not lead us to classify drift as a special form of selection. And in the example just discussed of the two populations undergoing frequency-dependent *individual* selection, we saw a case in which population structure has an impact on fitness without there being group selection.

When a local deme reaches its equilibrium, the different alleles in the population become equal in fitness. That is to say, as the populations go to their local optima, additive genetic variance in fitness within each deme is destroyed. The variation in fitness that remains will then be *between* the different demes that occupy different adaptive peaks.

This situation, I hope, has a familiar ring to it. It is precisely the situation of the populations that are internally homogenous for height. In that case, the individuals in a population were identical in fitness *and* phenotypically identical (in height). In analyzing that example, I argued that one could not tell whether there is individual or group selection simply from the fact that all the variation in fitness is between groups. This fact about fitness is compatible with there being individual selection for being tall and with there being group selection for groups of greater average height. Precisely the same diagnosis applies to the Wright model. Individuals in populations occupying lower adaptive peaks will produce fewer offspring on average than individuals in populations at higher adaptive peaks. To see whether this is due to individual or to group selection, one must ask *why* these differences obtain.

The example of heterozygote inferiority, which is the simplest case of multiple (i.e., two) adaptive peaks, makes this clear. According to the fitness function shown in Figure 8 (p. 181), demes are fixed at either 100 percent A or 100 percent a. We may imagine an ensemble of small local demes, each going to fixation at one of these two values. But, since A has a higher average fitness, we may suppose that populations in which A is fixed send out migrants more frequently than populations in which a is 100 percent. If we chose the parameter values carefully, we might expect that eventually all the populations will end up with A at fixation, a result that might not have occurred in a single large population with the same fitness function.

Although this scenario can occur in the system of populations that Wright considered, it does not involve a process of group selection. The definition of group selection suggested here explains why. Membership in a group in which A predominates is *not* a causal factor. Indeed, the *constant*-viability model we are dealing with implies that an individual's fitness is not affected by the kind of group it belongs to. If you are an AA homozygote, you have a certain fitness value regardless of what your group is like; the same holds for the other two genotypes. This is not to deny that AA individuals are fitter than aa individuals, and so a population in which A is 100 percent will have a higher average fitness than one in which a has gone to fixation. With the right patterns of colonization, this may mean that AA individuals leave their home groups and enter new ones more often than aa individuals do. But again, whether or not they do this is not influenced by the kind of group they are in.

So the shifting balance process can occur without group selection. What is required for group selection to occur in the context of Wright's models? To be sure, a system of small semi-isolated local demes provides a favorable setting in which group selection may operate. And Wright's

(1945) scenario for the evolution of altruism clearly involves group selection. However, I would suggest that the idea of populations ascending adaptive peaks, far from implying the occurrence of group selection, is a conceptual obstacle that the idea of group selection must circumvent. Roughly, the problem is that group selection, considered genetically, demands that fitness values be frequency-dependent. The idea one wants to capture is that gene frequencies in a group themselves affect the fitness values of organisms. But this blocks the automatic assumption that selection within a group must increase $\bar{w}$. It may do so even under frequency-dependent selection, of course. But details must be provided to show how.

To ask whether ensembles of populations of the sort handled by Wright's theory must undergo group selection is like asking whether physical objects of the sort handled by Newtonian mechanics must experience gravitational forces. The models have great generality and cover a variety of cases. Some, like the evolution of altruism, involve group selection, but many do not. The mere fact that Wright always attended to the way population structure can modulate the course of evolution hardly shows that his models always involve group selection.

Wright's theories, in virtue of their great generality, can be applied to individual and group selection processes alike. As is standard in population genetics models, they describe the *consequences*, not the *sources*, of various facts about fitness, population size, and migration (Section 1.2). Insofar as group and individual selection differ in virtue of their causal structure, it is unrealistic to think that a population genetical model will *define* what group selection is. Source laws and consequence laws play different roles in a theory of forces.

One final implication of the definition of group selection offered here is worth noting. Fisher (1930, p. 50) argued that group selection must be a weaker force than individual selection, because there are fewer groups in an ensemble of groups than there are organisms in that ensemble (Section 7.1). Williams (1966, pp. 115–116) repeats the argument. Although there is a certain intuitive plausibility to this suggestion, it simply fails to apply to the idea of group selection I have described.

We know how population size affects the confidence we may have that fitter organisms will outsurvive and reproduce less fit ones (Section 4.2). The larger the effective population size, the more confident we may be. The smaller the population, the better the chance that evolution will depart from the trajectory that natural selection predicts. The intuitive transfer of this idea to group selection is as follows: The more groups there are, the smaller the role chance plays. If there are only a few groups in the process, then it is quite possible that fitter groups will not outsurvive and reproduce less fit ones. Fisher offers a comparison of individual and group selection. Since there must be far more organisms than there are groups of organisms, sampling error has a better chance of confounding group selection than it does of confounding individual selection.[14]

This reasoning might be unobjectionable if group selection were not standardly construed by using an organismic benchmark. However, standard thinking about group selection has considered how group selection can modify gene frequencies, where these are computed per capita. The question has always been how group selection can affect the proportion of *organisms* having this or that trait. It is this traditional conception, implicit in discussion of the *t* allele, myxoma (Section 9.3), and altruism itself, that I think is captured by the causal idea I have presented so far. If we look carefully at this characterization, I think we will see that Fisher's "weak force argument" does not bear on the question.

Group selection exists, according to my construal, when group phenotypes have certain sorts of effects on the fitness values of *organisms*. When an ensemble of populations experiences group and individual selection simultaneously, there are just as many entities affected by group as by individual selection. Both causal processes impinge on the same objects; both typically are understood in terms of an organismic benchmark. Although there may well be general empirical reasons why group selection will be a relatively weak force in evolution, the relative number of groups as compared with organisms is not one of them.[15]

Fisher gives a second reason for thinking that group selection matters less. It is *time*. Groups usually go extinct or found colonies at a slower rate than organisms die and reproduce. As we will see in the next section, this can be a very important consideration if group and individual selection oppose each other.

9.3 Altruism and Averaging

Although I argued in Section 7.7 that group selection need not oppose individual selection, the fact remains that one of the main interests of group selection is that it has been invoked to account for the prevalence of altruistic characteristics. In this section, I want to clarify the idea of altruism and describe its place in the causal analysis of group selection.

14. I am guessing a bit about how this argument runs, since it is not spelled out explicitly.
15. Fisher's argument does bear on another formulation of the group selection idea—one in which the organism is replaced by some higher-level entity as the benchmark of selection. Species selection is the example I discuss in Section 9.4.

First, I discuss the connection between Simpson's paradox (Section 8.2) and the evolution of altruism. Then, I examine the concepts of kin selection and inclusive fitness and try to pinpoint their relationship to group and individual selection on the one hand and to altruism on the other. After that, I examine some post-Fisherian ideas on the evolution of female-biased sex ratio. The section ends with a discussion of some of D. S. Wilson's (1980) ideas on the possibility of group selection without altruism.

There is an extremely crude argument against the possibility of altruism that nevertheless exerts considerable influence. It is very simple. A trait will evolve if and only if bearers of the trait are more successful, on average, in transmitting it than are bearers of the alternative. But an altruistic trait cannot evolve in this way, since it is, *by definition*, inferior in fitness. So natural selection cannot take a novel mutant that is altruistic and bring the trait to predominance. And even if a population somehow managed to contain only altruists (e.g., by random drift), it would nevertheless be subject to "subversion from within." A mutant selfish individual would be able to "invade" the population and selfishness, because of its superior fitness, would eventually go to fixation. A population composed entirely of altruists does not exemplify an "evolutionary stable strategy," in the sense of Maynard Smith (1974).

There must be something wrong with this argument, however. As mentioned in Section 7.1, a number of biologists have investigated the circumstances under which altruism could evolve by group selection. They developed models that suggested that group selection could have this effect only for highly constrained sets of paramenter values. Not that the models are beyond criticism.[16] But it is a bit hard to believe that these theorists exerted themselves solving equations and doing computer simulations to answer a question that can be settled so easily by attending to a simple definition.

What the above argument shows is not that altruism is a priori incapable of evolving but that the concept of altruism is a subtle one. Somehow, the following two propositions must both be true:

1. Within each group of individuals, altruists are at a reproductive disadvantage compared with nonaltruists.

2. In the ensemble of groups (within which the group selection process takes place), whether altruists have a higher average fitness

---

16. One of Wade's (1978) points concerning how the models incorporate biologically unrealistic assumptions that a priori bias the case against group selection will be discussed later in this section.

than nonaltruists is a contingent empirical matter, not settled by proposition 1.

The issue of group selection and its connection with altruism cannot be understood until these seemingly contradictory claims are reconciled.[17]

This is the paradox of altruism. The attentive reader may perhaps remember a similar puzzle in the discussion of causality: it is Simpson's paradox (Section 8.2). How can it turn out that smoking *increases* the probability of a heart attack both among those who exercise and among those who do not, and yet smokers have heart attacks *less* frequently than nonsmokers? The problem is to reconcile the following two facts:

1. Within each group of individuals who have the same causally relevant background properties,[18] smoking raises the probability of a heart attack.

2. In the ensemble of groups, whether smokers have heart attacks more frequently than nonsmokers is a contingent empirical matter, not settled by proposition 1.

The key to reconciling 1 and 2—whether they are formulated in terms of altruism and selection or in terms of smoking and heart attacks—is *correlation*.

Altruists are less fit than the selfish individuals with whom they share a group. This fact about their *relative* fitnesses obtains in each group. But suppose that the *absolute* fitness values of altruists (A) and selfish individuals (S) depends on their frequencies within the group as shown in Figure 11 (p. 186). The higher the frequency of altruists, the greater the absolute fitness of altruists and selfish individuals alike. This correlation may ensure that altruists have a higher overall average fitness than selfish individuals, even though in each group, altruists are less fit than their selfish colleagues.

A numerical example may make this fact a bit less mysterious. Think of two populations of equal size, the first at 90 percent A, the second at 10 percent A. Suppose that the fitness values of A and S in the first population are, respectively, 10 and 12 offspring per capita, whereas the fitnesses of the two traits in the second population are, respectively, 2 and 4. Then the average fitness of A across the two populations will be $(0.9)(10) + (0.1)(2) = 9.2$, whereas the average fitness of S will be $(0.1)(12) + (0.9)(4) = 4.8$. So A is less fit than S in each population, although A has a higher average fitness in the ensemble. In consequence,

---

17. I am indebted to Wilson's (1980) insightful analysis of this concept.
18. In this case, among those who exercise and among those who do not.

if we assume that selection is the only force at work (and that the traits are heritable), the frequency of A will increase in the ensemble, although it will decline in each subpopulation.

Simpson's paradox is at the heart of the idea that altruism may evolve by group selection. The higher frequency of heart attacks among non-smokers in the example was due to a correlation of smoking with exercise. The greater reproductive success of altruism in the ensemble of groups was due to a correlation of higher concentrations of the altruistic trait with groups of higher productivity. Without some such correlation, the inferiority of altruism within each group cannot be reversed in the ensemble of groups.

In the above example, altruism can evolve because altruists benefit from belonging to groups in which they are common.[19] Contrast the fitness functions in Figure 11 with those shown in Figure 14, each of which implies that altruism is at a disadvantage both within each group and in the global population. Altruistic traits of this sort simply cannot evolve by natural selection at any level of organization.

Because a frequency-dependent fitness function is essential if altruism

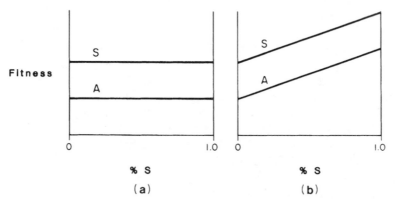

Figure 14. Two possible fitness functions relating selfishness (S) and altruism (A). As in the function shown in Figure 11 (p. 186), both (a) and (b) imply that an altruist in a group must be less fit than a selfish individual in the same group. The difference is that (a) and (b) guarantee in addition that altruism must be overall less fit in the ensemble of groups.

19. In this section, I interpret the example fitness functions as indicating not just the average fitness of organisms with this or that trait but the impact of the trait on each organism's fitness. Where the graph shows one trait to be fitter than another at a given frequency, this I will take to imply that at that frequency, at least one individual has his fitness increased by that trait and no individual has his fitness diminished. In this way, the fitness functions can be taken to have a direct bearing on the causal idea of selection.

is to evolve, it is a mistake to think of the evolution of altruism in terms of populations ascending to the tops of their adaptive peaks (Section 9.2). The adaptive topography, recall, represents values of $\overline{w}$. When selection is frequency-independent, populations are guaranteed to increase their values of $\overline{w}$ under natural selection (Section 6.1). But frequency dependence removes this blanket assurance. Indeed, the fitness function so crucial for the evolution of altruism (Figure 11) shows $\overline{w}$ *declining within each group.*

Even a favorable fitness function—one that allows for the positive correlation of group productivity and the altruistic characteristic—will not be enough for the altruistic characteristic to become stably present in appreciable numbers. In the two-group example I computed above, it may be true in the short run that altruism increases in the global population, although it declines in each subgroup. But this cannot go on forever. Each subgroup is on its way toward the elimination of altruism, which means that the selection process will eventually destroy the correlation that made it possible for altruism to increase in the first place.

In order to avoid this ultimate result, there must be colonization and extinction. It is here that we find the contingent empirical facts that determine whether group selection can introduce and sustain an altruistic characteristic in an ensemble of populations. In this process, time is of the essence: The groups must form new colonies rapidly enough to offset the within-group process that serves to displace the altruistic trait. Given favorable parameter values, altruism may come to exist at some stable intermediate frequency. Even though within-group individual selection will tend to eliminate altruism, group selection may prevent this from happening. Whether it will do so depends on the relative advantages of the two traits within each group, on the slopes of the fitness functions, and on the rate at which new groups are founded.

Also crucial is the question of *how* new colonies are established. If groups are founded by individuals who are alike, this will enhance and preserve intergroup variation and allow group selection to exert a more powerful influence on the advancement of altruism. If, on the other hand, migrants from different groups mix together and then found new groups, between-group variance will be diminished and the evolution of altruism will be more difficult. One of the important points that Wade (1978) makes is that most quantitative models of group selection have assumed that the latter pattern obtains; this is partly responsible for the uniformly negative verdicts that such models returned about the importance of group selection. Both patterns of colonizing are biologically possible, of course. Wade's point is not that these models have made assumptions that never hold true but that they have failed to explore certain biologically realistic situations that may make the evolution of altruism less difficult to achieve. Note that Wade's (1975) *Tribolium* experiment (Section 7.7) does not obey the homogenizing assumption; colonies are always founded by individuals that come from the same parent population.[20]

Although the founding of new groups and the extinction of old ones is a prerequisite for altruism to evolve by group selection, it is not required by the definition of group selection given earlier. There are two reasons for this. The first is that I aimed at defining *group selection*, not *evolution by group selection*. The second is that I defined *group selection*, not *group selection for altruism*. I hope that by this point, the distinctness of these three concepts is clear enough that no argument is needed for treating them separately.

So, understood in a certain way, "altruism" really is impossible. If a trait is to increase in frequency under selection, it must on average do better than the competition. This is true no matter what the unit of selection is. However, to define "altruism" to mean a trait of inferior fitness is to make it uninteresting as well as impossible. In addition, this way of defining the concept is untrue to its actual history, which allows no such easy resolution of the question of whether altruism can evolve. An altruistic trait is simply one that confers a fitness benefit on individuals *in the group* who happen to lack the trait (i.e., on selfish individuals). The interaction of altruistic and selfish individuals within a group must be reflected in their *relative* fitnesses; what happens to their *absolute* fitnesses is a further matter not settled by the definition. This way of understanding the concept reflects the fact that the evolution of altruism is a contingent empirical question.

The fact that altruism must somehow benefit itself if it is to evolve has an interesting bearing on the characterization of group selection I have offered. Group selection, on the construal defended here, requires that a group property be a positive causal factor in the survival and reproductive success of organisms.[21] The Pareto interpretation of positive causal factor requires, in turn, that the group property help at least one individual in the group and not harm any. This may appear to clash with the idea that altruism can evolve by group selection. After all, isn't it part of the idea of altruism that altruists are *harmed* by their actions, whereas selfish individuals are *benefited*? If so, how can the

20. Wimsatt (1980) has observed that mixing migrants diminishes the effectiveness of group selection in the same way that blending inheritance undermines the power of individual selection.
21. In the next section, the definition of group selection will be modified so as to no longer exploit an organismic benchmark. The resulting definition, unlike the present one, accords a fundamental role to colonization and extinction.

property of belonging to a group that contains altruists be said to be a positive causal factor in survival and reproduction, since altruism hurts some and helps others? There appears to be no unique causal role played by altruism, paralleling the example of the drug that is a stimulant for adults and a depressant for children (Chapter 8).

The distinction just drawn between the relative fitness of altruistic and selfish individuals within a group and the overall difference between them averaged over all groups disposes of this objection. It is not to be doubted that an altruist will be at a disadvantage relative to a selfish individual *in the same group*. Yet both may benefit from belonging to a group in which altruists are common. Remember the fitness function of Figure 11 (p. 186). If you are an altruist, you are better off in the group containing 90 percent altruists than you would be in the 10 percent group, *and the same holds true if you are a selfish individual*. In the relevant sense, altruists and selfish individuals are in the same boat when groups are selected for their high frequency of altruists. The group property is a positive causal factor in survival and reproduction.[22]

We have already seen how the strategy of averaging is the *deus ex machina* of genic selectionism (Sections 7.4 and 9.1). Regardless of the level (or levels) at which selection really acts, we may obtain selection coefficients attaching to single genes by averaging over the various contexts in which the genes of interest appear. The same mathematical manipulation may be pressed into service to "reinterpret" apparent cases of group selection and show that they "really" are cases of selection acting at a lower level or organization. An altruistic trait may be present in different contexts. One context for the trait might be a group in which it is common; another context might be a group in which it is rare. Group selection for an apparently altruistic trait may thus be "represented" as individual selection for a trait that is, on average, beneficial for its possessor. This way of arguing against group selection hypotheses is universally available—just as the representability argument (Section 7.4) in favor of genic selection is. However, just as in the case of genic selectionism, the argument's universality is a telltale sign of its vacuity.

Biological discussion of the coevolution of the myxoma virus and the Australian rabbit *Oryctolagus cuniculus* betrays just this sort of ambivalence over the concept of altruism. The Australian government introduced the virus to reduce the rabbit population. It was subsequently

22. The Pareto formulation of the idea of a causal factor does not require that both altruists and selfish individuals benefit from membership in a group containing a high frequency of altruists. It would suffice, for example, if altruists were unaffected and selfish individuals were helped. This circumstance will be relevant to the discussion of biased sex ratios later in this section.

observed that the rabbits gradually evolved an increased immunity to the virus and that the virus itself also declined in virulence. The change in the rabbit population is no surprise from the point of view of individual selection. But why should the virus have declined in virulence? Virulence is a manifestation of the rate at which a virus reproduces within a host's body. A reduction in virulence thus appears to be selectively disadvantageous.

Lewontin (1970, pp. 14–15) suggested this answer:

> The key is that the myxoma virus is spread by mosquitoes, which mechanically transfer a few virus particles to the rabbits they bite. . . . Each rabbit is a deme from the standpoint of the virus. When a rabbit dies, the deme becomes extinct since the virus cannot survive in a dead rabbit. Moreover, the virus cannot be spread from that deme because mosquitoes do not bite dead rabbits. Thus there is a tremendously high rate of deme extinction, with the result that those demes are left extant that are least virulent. This causes a general trend toward avirulence of the pathogen despite the complete lack of selective advantage of avirulence within demes.

Biologists are not in total agreement over whether this is the correct account of the reduction in myxoma's virulence.[23] However, the point I want to make here is a conceptual one: According to this story, viral strains of lower virulence are *altruists* in the crucial sense described above. They are less reproductively successful within each group. But since a group has a better chance of colonizing and of avoiding extinction if its average virulence is low, relatively avirulent strains may be overall more fit than strains of higher virulence. The correlation of group fitness with concentrations of lower virulence provides the opportunity for Simpson's paradox to intrude.

This scenario has not always been interpreted as an example of group selection for an altruistic trait. Futuyama (1979) does not explicitly dissent from any of the empirical assumptions that Lewontin (1970) makes but interprets them in terms of *individual* selection:

> If the fitness of an *individual* parasite or its offspring is lowered by the death of its host, avirulence is advantageous. . . . Because

---

23. For example, Alexander and Borgia (1978, pp. 452–453) wonder whether virulence is correlated with high reproductive rates and whether a rabbit is infected by viruses of different degrees of virulence. They hold that if rabbits were bitten only once and inoculated with a pure strain, then the group selection hypothesis would not be tenable. All the variance would then be between groups. I examine this line of reasoning later in this section.

the virus is transmitted by mosquitoes that feed only on living rabbits, virulent virus genotypes are less likely to spread than benign genotypes. Avirulence evolves not to ensure a stable future supply of hosts, but to benefit *individual* parasites (Futuyama 1979, p. 455, emphasis mine; quoted in Wilson 1983, p. 172).

There is one point here that I don't want to contest. Certainly, the virus doesn't keep its host alive purely because this would be good for the rabbit. Avirulence must somehow help the virus. However, notice how the ground has shifted from Lewontin's group selection explanation to an individual level account: Avirulence is not an altruistic characteristic, according to this line of reasoning, because viruses of lower virulence are on average fitter. This would be unobjectionable, if altruism were simply defined to mean a trait that cannot evolve by selection at any level. But if we wish to retain a distinction between what happens within a deme and what happens in the ensemble of demes, this averaging argument should be rejected along with the loaded notion of altruism that it presupposes.

The definition of group selection developed in Chapter 8 delivers the correct conclusion that the reduction in virulence is a case of group selection (provided that Lewontin's facts are right, of course). Suppose that each rabbit is inoculated with different strains of the virus, some more virulent than others, and that the life expectancy of the host rabbit is determined by the average virulence of the viruses living in it. It follows that being in a group of lower average virulence is a positive causal factor in the reproduction of each myxoma organism. The life expectancy of a host rabbit is influenced by this property of the group, and all the individuals in the group experience that group property in the same way. Regardless of whether you are a high- or a low-virulence virus, you are better off in a group of lower average virulence than you would be in a group of higher average virulence. Within each rabbit, of course, high-virulence viruses will reproduce faster than low-virulence strains. But the trait favored within each group is not necessarily the trait that becomes predominant in the ensemble of groups.

I said before (note 23) that biologists sometimes hold that the group selection account depends on the assumption that a rabbit is inoculated with viruses of different degrees of virulence. Clearly, within-rabbit variation is necessary for there to be any competition *within* a group. However, the claim here is that within-group variation is necessary for the existence of *between*-group selection. This is the very antithesis of the analysis of variance criterion of group selection, which would conclude in this case that there must be group selection and cannot possibly be individual selection (Section 7.8). I have already noted that I dissent from the ANOVA characterization. It is curious that this mode of analysis and one that demands that members of the group not be clones of each other are in such blatant contradiction.

Alexander and Borgia (1978, p. 453) mention that the viruses are asexual and that those within a given rabbit may be clones of each other.[24] They say that

> A clone of identical viruses would be no more appropriately regarded as a group or deme than would be the cells of a metazoan organism, since each member of a clone should evolve to sacrifice as much for a clone member as for itself.

I fail to see why the coefficient of relationship implies that the clones are not separate organisms that together comprise a deme. Human identical twins count as *two* individuals because they are physiologically independent of each other; asexual clones of a virus have the same status.[25]

Whether natural selection treats two organisms differently depends only on their fitness values. It does not matter how they are related to each other. In fact, it doesn't matter whether they are phenotypically the same or different. Recall that a polymorphic population at selective equilibrium may contain different traits with the same fitness value. A population of clones, on this score, has the same status as any population containing organisms of identical fitness.

The question of whether populations of asexual clones experience group or individual selection depends on *why* individuals in the same group have the same fitness value. This was the point of the parable of the tall and the short (Section 7.6). Do individuals alike in height have the same fitness value because of their height or because they belong to a group with a given average height? Do viral clones have the same fitness value because they have the same level of virulence, or because they belong to a group with a given average virulence? The question about clones can have multiple answers, just as it can if the individuals are phenotypically similar though unrelated. The difference between group and individual selection has to do with *why* individuals

24. I am not aware of any evidence for thinking that rabbits are *never* inoculated twice. And even this would not guarantee that myxoma populations are internally homogeneous. The point here, however, is to consider the consequences of the assumption that groups are internally homogeneous.

25. There are some interesting questions here concerning how organisms are to be individuated. Harper (1977) and Janzen (1977), for example, think of a field of dandelions or may apples as the scattered parts of a single organism. Dawkins (1982, pp. 253–264) provides an interesting discussion of this issue.

have the fitness values they do, not with whether they are related to each other.

Before taking up a more intricate case that illustrates the relationship of altruism to group selection—the evolution of female-biased sex ratio—it is well to note how the definition of altruism used here contrasts with another that is sometimes used in discussions of W. D. Hamilton's (1964) idea of *kin selection* and the related idea of *inclusive fitness*. This will throw further light on the relevance of kinship relationships in an ensemble of groups to determining the unit of selection.

The idea of kin selection begins with the thought that an organism may lever its genes into the next generation either by making offspring itself or by helping genetically similar offspring to do so. If I have the gene *A*, the representation of *A* in the next generation may be indifferent as to whether I reproduce or help some other organism with that gene to reproduce.

For a gene to follow this second causal avenue, it must provide the organism in which it occurs with some method of detecting the presence of *A* genes in other organisms. The organism must be "told whom to help," as it were. One obvious trick is to exploit the fact of kinship; you and your kin have certain predictable probabilities of being genetically similar and different.

The idea that a gene may augment its chance of spreading through a population by causing its bearer to help kin has an immediate bearing on the problem of altruism. When you observe helping behavior in the wild, you may be tempted to invoke group selection to explain it. You may think that the group in which the helping behavior occurs outcompeted groups in which it did not. But this intergroup perspective is not required; it is possible for a selection process occurring *within* the group you have observed to have produced the behavior. A gene promoting helping behavior can spread through the population if the recipients of aid are usually relatives. The helping gene can be favored by selection within a single breeding population in much the same way that a gene for fleeing predators or any other behavioral trait might be. The wrinkle is that the helping gene secures its advantage in a way that is just a little bit devious and circuitous.

Hamilton (1964) is widely thought to have replaced the narrow notion of "Darwinian fitness" with the idea of "inclusive fitness." According to the former notion, a sterile organism will have 0 fitness, since its chances of reproducing are nil. But according to the latter idea, its "reproductive chances" may be somewhat higher if it helps its siblings make babies. Since inclusive fitness is a more adequate measure of an *organism*'s fitness, it has seemed natural to conclude that kin selection must simply be a more adequate framework for the workings of *individual* selection.

In a certain sense, Hamilton consolidated an insight that had been available within population genetics for some time. By this, I don't merely mean that Haldane (1932) and others had made quips to the effect that it would be (evolutionarily) rational to sacrifice one's life for two siblings or four cousins. Rather, I have in mind the fact that the fitness of a trait has standardly been defined in population genetics as the average fitness of the organisms possessing it (Section 1.4). If some organisms have a gene that causes them to aid others, one must know who is on the receiving end of this behavior before the fitness of the gene can be computed. The net effect depends on whether bearers of the gene help only each other, donate aid at random, or aid only those individuals who lack the trait. It is perhaps not too much of a simplification to say that Hamilton's insight was that there are many ways for a gene to be advantageous *on average*. It may cause each of the organisms that house it to work atomistically in their own narrow self-interest. Or it may cause a more complicated set of interactions among the bearers of the gene and achieve the same result.

If this is correct, then Hamilton's (1964, p. 1) claim about the relationship of his idea to earlier efforts is seriously misleading, both historically and conceptually:

> With very few exceptions, the only parts of the theory of natural selection which have been supported by mathematical models admit no possibility of the evolution of any characters which are on average to the disadvantage of the individuals possessing them. If natural selection followed the classical models exclusively, species would not show any behavior more positively social than the coming together of the sexes and parental care.

To assess these remarks, we must distinguish the evolution of a phenotypic character from the evolution of a single gene. Classical models of selection show how a *phenotype* can be selected against without being eliminated. In the model of the sickle-cell system discussed in Section 1.4, anemia is on average to the disadvantage of the individuals possessing it; yet, it evolves and is maintained under selection. On the other hand, this and other standard models of selection do imply that a *gene* that is on average deleterious must be eliminated, if selection is the only force at work. Kin selection does not alter this fundamental fact. Indeed, the greater average fitness of a gene is an absolute criterion for increase under selection, regardless of the level at which selection occurs (Wilson 1980).[26]

---

26. Trivers' (1971) idea of *reciprocal altruism* is simply another device for satisfying the same population genetical criterion. A gene may do well by causing its bearers to donate aid to kin. However, it also may be favored if organisms interact with each other repeatedly and have the opportunity to remember who helped them and who did not. A gene that

Hamilton's ideas are of the first importance. However, I would suggest that their true novelty is often misunderstood. A standard interpretation is that biologists had focused exclusively on organisms and Hamilton got them to think about genes (Dawkins 1976). As a historical comment concerning the impact of the inclusive fitness concept on ecology and ethology, this may be correct. But as a conceptual remark about the relationship of Hamilton's ideas to earlier *quantitative* models, it is not. Population genetics had always thought of selection from the genetic point of view. What Hamilton did was to bring the organism back into the mainstream of quantitative modeling. His achievement was to show how the greater average fitness of a gene (or trait) can be underwritten by interesting sorts of interactions among *organisms*.

The concept of inclusive fitness recognizes that a gene may use kinship as a way of "guessing" whether a copy of itself is present in another organism; if two organisms are full sibs or cousins, for example, this implies a certain probability that one will have a certain gene if the other does. A gene can cause the organism in which it is housed to help other organisms. Suppose the gene somehow distributes benefits differentially, in a way that coincides with its degree of relatedness to prospective recipients. We calculate the inclusive fitness of the organism as a function of the cost of donation, the benefits that others receive, and the coefficients of relationship between the donor and each recipient. If the gene is a prudent gambler, it will cause the organism to aid others in a way sensitive to the odds that copies of the gene are present in recipients. Kinship is a clue that a "selfish" gene might use to advance its own self-interest.

Dawkins (1978, 1982) has noted that the use of kinship is just one technique among many conceivable ones. Kin selection is a special case of his *green beard effect*. It is easy to see how a gene might evolve if it caused its bearers to have green beards and to help those who also have green beards. Kinship is just a handy sort of green beard—a detectable sign of the presence of the gene in question.

Kinship is a clue to common descent, not just to genetic similarity. But, as Dawkins (1978, p. 191) has emphasized, what is crucial for the evolution of a gene is not that a gene should cause its bearer to recognize other individuals who have a gene that is identical by descent (i.e., a descendant of the same mutation event). What is criterial is simply the recognition of copies of the same gene, wherever they came from. This

---

is able to cause its bearer to give away benefits when there is a way of ensuring reciprocation might be quite advantageous. This is an application of the commonsense idea that helping others can, if the circumstances are right, be a sophisticated form of rational self-interest. Reciprocal altruism is thus a kind of helping behavior; this behavior need not count as altruistic in the sense of being disadvantageous within a group.

is easiest to see by considering a simple selection model in which mutation is allowed to introduce repeated copies of a green beard gene into the population. Sorting out copies of the gene that are identical by descent from those that are simply identical would be a waste of time from the point of view of getting the gene to evolve.

Inclusive fitness is really not a substitute definition of fitness but a handy rule of thumb that sometimes produces insights. There are two reasons why. First, since the concept is defined in terms of costs and benefits, it already presupposes some antecedently understood notion of fitness. Second, it is at best a good gambler's guide. Relatedness only bears on the *probability* that a gene in one organism is also present in another. Suppose that one gene caused its bearers to behave in accordance with the calculus of inclusive fitness, while another ignored kinship and exploited a more accurate guide to the presence of the gene in others. Imagine two alleles competing at a locus in a haploid organism. Let the K allele make its bearers help kin to degrees sanctioned by relatedness. Let the G allele make its bearers help only individuals with green beards, where an individual has the G allele if and only if it has a green beard. Notice that K individuals will sometimes help G individuals, but G individuals will never help K individuals. A population in which G is at fixation cannot be invaded by a few mutant Ks; upon introduction they are at a selective disadvantage and are eliminated.[27] However, a population in which K is at 100 percent is subvertible from within; introduce a few G mutants and the G trait will sweep to fixation.

This little thought experiment shows why inclusive fitness is not an adequate definition of fitness. K has a higher inclusive fitness than G, yet when selection is the only evolutionary force at work, G goes to fixation and K is eliminated. A correct definition of fitness should not have this consequence. Inclusive fitness does not replace the more traditional idea that a trait's fitness is the average fitness of the organisms possessing it, where those organismic fitness values take account of survival and reproduction (Section 1.4). Rather, inclusive fitness is a concept that in the context of certain models can do the work that the traditional concept could perform, while also perhaps being of heuristic use in making certain aspects of the dynamics salient. Again, this is not to detract from the importance of inclusive fitness but to make its real significance clear.[28]

---

27. That is, the population configuration of 100 percent G individuals is an evolutionarily stable strategy in the sense of Maynard Smith (1974).

28. See Michod (1982) for discussion of the conditions under which the predictions of an inclusive fitness analysis coincide with the predictions of an overtly genetic model.

What of kin selection? The basic idea behind this kind of selection process is that individuals interact differentially with kin and thereby differentially affect their fitness values.[29] Kin selection would be impossible if individuals interacted with each other at random (i.e., without regard to kinship). Note also that kin selection, thus defined, does not require that individuals use kinship as a clue. If an altruist confers benefits on whatever conspecifics happen to be in the neighborhood, this will count as kin selection if relatives live together. What is essential is that kin are *affected* differentially; they need not be differentially affected *because* they are kin.

As described earlier, the evolution of altruism by group selection requires that altruism be correlated with membership in a group having high productivity. If altruists tend to live mainly with each other, then most of the beneficiaries of altruistic donation will themselves be altruists. One simple way for like to live with like is for kin to live with kin. A kin selection explanation of altruism is sometimes believed to be at odds with a group selection account (Hamilton 1964; Maynard Smith 1964, 1974; Dawkins 1978, 1982). I prefer to think of kin selection for altruism as a special case of group selection. Notice that the fitness function of Figure 11 (p. 186) applies to kin selection just as it does to group selection in which the groups are made up of nonrelatives. As Williams and Williams (1957) pointed out, altruism declines in frequency within each kin group. Even among brothers and sisters, altruists do less well than selfish individuals. As is essential in the group selection story, this within-group advantage of selfishness must be offset within the ensemble of groups. If the groups are made up of kin, this can provide the toehold that Simpson's paradox requires.

So if kin selection promotes an altruistic characteristic, it is a kind of group selection. But the definition of kin selection given above does not require this sort of fitness function. The frequency-dependent function shown in Figure 14 (p. 328) may obtain when kin live together. In this case, A-type individuals do worse when they live with each other than they do when they live with Ss. Selection within each kin group works against A; and the more common type A is within a group, the worse those individuals do. The A type will be eliminated, with kin selection helping the process along. Here group and individual selection work in the same direction. The average fitness of A is lower across the ensemble of groups than it would be if the individuals randomly associated with each other.

29. Thus Michod (1982, pp. 40–41) gives the following as a necessary and sufficient condition for kin selection: "To evolve by kin selection a genetic trait expressed by one individual . . . must affect the genotypic fitness of one or more other individuals who are genetically related to the actor in a nonrandom way at the loci determining the trait."

Just as kin selection can be a form of group selection, so it can also be a form of individual selection. Suppose we begin with a number of kin groups each containing the two traits C and D. Imagine that individuals compete with sibs who are unlike them and cooperate with ones who are like them. Let a consequence of this be that majority traits in a sibship are selectively advantageous. Suppose a consequence of this process is that each sibship becomes monomorphic and that the percentage of the two types in the ensemble of sibships differs from what would have obtained if all the individuals had interacted with each other. Still, this cannot be classified as group selection. There is no group property, such as belonging to a kin group in which trait C predominates, that is a positive causal factor throughout the ensemble.[30] Here we have kin selection without group selection. In fact, this instance of kin selection is nothing more than a kind of frequency-dependent individual selection.

Kin selection, therefore, is not a *kind* of selection intermediate between individual selection on the one hand and group selection on the other. The reason is that kin selection is not defined in terms of which sorts of characteristics are selected for. Kin selection is *not* defined so as to require that kinship is itself a causal factor in an organism's survival and reproductive success.[31] Rather, "kin selection" merely means that selectively significant interactions are nonrandom with respect to kinship. This structural property of the global population is important in determining what kinds of characteristics can evolve; for example, it allows altruistic characteristics to evolve that would not be able to if individuals randomly associated. But here kinship plays the role of effecting a certain correlation between interactors, not in specifying which characteristics are beneficial and which are deleterious.

We must not lose sight of the distinction between natural selection and evolution by natural selection. There may be selection for a trait without that trait's increasing in frequency. This will happen if the trait has zero heritability (Section 5.2). Heritability considerations do not affect whether there is selection for a trait of a given sort but whether such selection can produce evolution. Similarly, the fact that individuals interact with kin does not directly bear on the question of which prop-

30. This example has the same structure as the model of underdominance used in Section 7.8 to show that group selection need not occur just because an individual's fitness depends on the sort of group it is in.

31. There is a special sort of kin selection in which certain kinship relations are selected against. Michod (1982, p. 43) describes research showing that human beings are less likely to find mates if they are kin to individuals with severe genetic disorders such as Huntington's disease.

erties are being selected for; rather, it is relevant to the question of whether such selection can bring about evolutionary change.

The evolution of the sex ratio provides a subtle example of how the method of averaging has allowed an apparent case of group selection to be interpreted as individual selection.[32] Fisher's argument (Section 1.5) predicts that a 1:1 sex ratio should be generated under certain conditions by individual selection, in spite of the fact that the resulting population size might not be optimal for the *population*. Williams (1966, pp. 146–156) devotes a good deal of attention to this issue. He argues that if groups benefit by being large, then group selection will tend to produce a female-biased sex ratio. If, on the other hand, an intermediate size optimizes group benefit, then the sex ratio will either be male-biased or female-biased, depending on whether the group's size is above or below the optimal value. This is what one would expect if a population were to use its sex ratio as a device for size regulation.

Williams (pp. 151–152) then turns to natural observation to see whether the group or the individual selection theory of the sex ratio is better supported:

> Despite the difficulty of obtaining precise and reliable data, the general answer should be abundantly clear. In well-studied animals of obligate sexuality such as man, the fruit fly, and farm animals, a sex ratio close to one is apparent at most stages of development in most populations. Close conformity with the theory [of Fisher] is certainly the rule, and there is no convincing evidence that sex ratios ever behave as a biotic [i.e., group] adaptation.

Williams then mentions that in some insects exhibiting a female-biased sex ratio (e.g., the honeybee), the ratio is determined "by the option of the female parent, in a way readily interpretable as an adaptation for her reproductive success."

This discussion anticipated by a single year Hamilton's (1967) explanation of female-biased sex ratios by the hypothesis of *local mate competition*. Hamilton observed that one of the empirical assumptions needed by Fisher's argument is that there must be population-wide competition for mates. But suppose that the population is subdivided into isolated groups, each founded by a single fertilized female. If the sons of a fertilized female can mate only with their sisters, then for a female to maximize her reproductive success, she should produce only as many sons as it takes to fertilize all her daughters. As a consequence,

---

32. I am indebted to D. S. Wilson (1983), Colwell (1981), and Wilson and Colwell (1981) for the ideas discussed here.

the sex ratio in the population as a whole becomes female-biased, not because this is good for the whole population but because it is good for each fertilized female. Individual selection *redux*.[33]

It is easy to see how a female-biased ratio will be generated if each group is founded by a single fertilized female ($N = 1$) and each group lasts only long enough for the daughters to mate with the sons ($G = 1$). But what happens when groups are founded by more than one fertilized female and when groups exist for several generations, sustained by matings between sibs and then cousins? When the number of found-resses and the number of generations is very large, we have the situation described by Fisher, in which a 1:1 ratio should obtain. Intuitively, we might expect that the extreme female bias that obtains for $N = 1$ and $G = 1$ gradually tapers off to equal representation as N and G increase.

The explanation of female-biased sex ratios is quite important to the debate about group and individual selection. On the face of it, the trait is just what one would expect if group selection were in control. The availability of an individual selection alternative thus becomes pressing. Also, it now appears that a female-biased sex ratio is quite common "among small arthropods in subdivided habitats—a description that includes more species than all the vertebrates combined" (Wilson 1983). So the trait in question is anything but rare.

The observation of female-biased sex ratios might have been interpreted as evidence for group selection. But since the data could be accommodated by the local mate competition model, they have not generally been seen as a threat to the individual selection paradigm. Yet there are a few dissenters. Colwell (1981), Wilson and Colwell (1981), and Wilson (1983), for example, argue that the local mate competition model obscures the difference between group and individual selection and embodies the strategy of averaging that I criticized earlier in this section and in Section 7.4. According to Wilson (1983), "biased sex ratios are the best evidence yet in hand for the operation of group selection in nature." The problem with the idea of local mate competition is not that it computes mistaken predictions but that it distorts the causal facts.

The problem here is quite similar to one mentioned earlier in connection with Darwin's discussion of sterile workers in the social insects (Section 7.1). Darwin, like numerous present-day biologists, argued that a female may enhance her reproductive success by producing offspring some of whom are sterile. The fact that some individuals cannot reproduce may appear to call for a group selection explanation;

---

33. Hamilton called the model "local mate competition" to reflect the idea that the foundress produces few sons so as to prevent them from competing with each other.

groups that include some sterile workers do better than groups containing only fertile individuals. However, this group selection explanation appears gratuitous when we consider that the members of the group are siblings. Talk of group benefit may be replaced, it would seem, with talk of benefit to the organism that produces the group.

To decide how to interpret a female-biased sex ratio, we need to attend more closely to the difference between Fisher's argument for a 1:1 ratio and Hamilton's argument for a female-biased one. Fisher's argument makes no assumption about how the sex ratio in a group affects its absolute size; all that matters to his argument is the way the sex ratio of a female's offspring affects her *relative* reproductive success in the group. That is, Fisher's argument does not distinguish between the two fitness functions shown in Figure 15. Each fitness function implies that a sex is at a reproductive advantage when it is in the minority, and thus predicts 1:1 as the stable equilibrium.

The difference between these fitness functions becomes relevant when we examine how sex ratio affects the *absolute productivity* of a group.[34] Function (a) implies that group productivity is maximized at 1:1. Function (b), however, says that a female-biased sex ratio maximizes group productivity. It is function (b) that is presupposed by the hypothesis of local mate competition. According to this idea, a female increases the productivity of her offspring by biasing the sex ratio. Each daughter

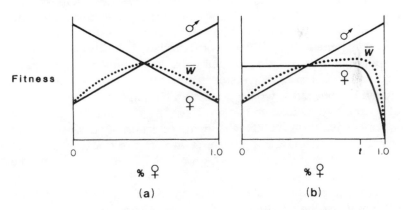

Figure 15. Two fitness functions consistent with Fisher's (1930) sex ratio argument. In each, the minority sex is at a reproductive advantage. In (a), the equilibrium point maximizes group productivity (an increasing function of $\overline{w}$). In (b), however, group productivity is enhanced by a female-biased sex ratio.

34. The dotted line in Figure 15 represents the average fitness of the organisms in the group ($\overline{w}$); just multiply this by the census size to compute the group's productivity.

will be fertilized, regardless of how prevalent daughters are, and thus has a constant fitness value. To be a little more realistic, I have drawn this fitness function to allow for a threshold frequency (t) for females beyond which their expected number of offspring declines because males are scarce.

In both these fitness functions, females are "altruists" when females predominate; the preponderance of females confers on males a reproductive advantage. In the second fitness function, but not in the first, the altruistic trait correlates with high group productivity. What this means is that if we imagine not one but a number of groups, all with the same fitness function, the sex ratio may become stably female-biased, even though males are fitter than females within each group.

Thus the logical structure of this argument is precisely the same as other group selection arguments for the evolution of altruism (e.g., the one considered before concerning myxoma). Within each group, selection pushes toward a 1:1 sex ratio. But the correlation of female bias with group productivity allows Simpson's paradox to gain a foothold.

Let us shift our attention now from the fitness of daughters and sons to the fitness of two sorts of fertilized females. A "Fisherian" female produces a balanced sibship while her "Hamiltonian" counterpart produces a female-biased set of offspring. How do the fitness values of these two reproductive strategies compare? No matter what the size of a deme and no matter how many fertilized females establish it, a Fisherian female will be at a selective advantage over a Hamiltonian female. Just as daughters are altruists when daughters are common, so Hamiltonian females are altruists compared with Fisherians. If there were a population of Hamiltonian females, they would, like any group of altruists, be subject to subversion from within. A mutant Fisherian introduced in their midst would be at an advantage, and the Fisherian strategy would go to fixation (Colwell 1981).

The way to avoid this outcome is group selection. One considers not a single deme that may contain some mix of the two traits but an ensemble of different demes. Although a Fisherian does better than a Hamiltonian within each deme, a population of Hamiltonians does better than a population of Fisherians, in that the former group will have the higher productivity.

How does the definition of group selection introduced earlier apply to this group selectionist account of female-biased sex ratio? We may begin with the sons and daughters in a group. If you are male, it is better to belong to a female-biased group than to one that is male-biased or even. But what if you are a female? Granted, females do worse than males in female-biased groups. However, this is not at issue. The question is whether females are better off in female-biased groups or in male-biased ones.

It is here that the difference between the two fitness functions in Figure 15 becomes crucial. In function (a), being in a female-biased group counts as a *negative* causal factor in each female's reproduction. In function (b), this is not so. The reproductive output of a female, according to this fitness function, is not affected by the sex ratio, as long as males are not too rare (i.e., their frequency is above the threshold $1 - t$). Which of the two fitness functions are we to consider? For the whole idea of local mate competition to get off the ground, the fitness function must be that shown in (b). Function (a) does not allow group productivity to be enhanced by a female-biased sex ratio; function (b) does. This licenses the conclusion that a female-biased sex ratio (in which there are enough males to fertilize all the females) is a positive causal factor in reproduction. It enhances the fitness of some individuals (the males) but leaves the fitnesses of others (the females) unaltered. The Pareto formulation of the concept of cause requires no more than this.[35]

The same conclusion follows if we compare Fisherian and Hamiltonian females rather than daughters and sons. A Hamiltonian female is less fit than a Fisherian female in each group. But each type does better in a group in which Hamiltonians predominate than in one in which Fisherians do. The fitness function (Figure 11) is the familiar one for group selection and altruism. Hamiltonian females are *altruists*. By persistently underproducing sons, they permit Fisherian females to gain the upper hand. The more prevalent Hamiltonians are, the better it is for the Fisherians. But it is also true that a higher frequency of Hamiltonians helps Hamiltonian females as well. So membership in a group having a high concentration of Hamiltonian females is a positive causal factor in an organism's reproductive success. This conforms with the causal characterization of group selection.

Having argued that the group selection interpretation of female-biased sex ratios can be easily accommodated by my proposed characterization of group selection, I now must come to terms with the opposite interpretation. Biologists who think about female-biased sex ratios from the point of view of local mate competition often believe

35. If the fitness function of females in Figure 15(b) had a slight negative slope, group productivity would still correlate with female bias. However, this would block my conclusion that a female-biased sex ratio is a positive causal factor in reproduction, since it would be positive for males but negative for females. Since this function will have the same shape as that of Figure 15(a), I can only say: Close, but no cigar. Concepts of group selection in which organisms no longer serve as benchmarks allow more scope for this setup to count as group selection. One such will be discussed in the next section.

that the relatedness of individuals within the same group makes a crucial difference. Hamilton (1967) thinks of female-biased sex ratios as an adaptive response on the part of a female to diminish competition among her sons. Maynard Smith (1978, pp. 160–161) describes female bias as a response to inbreeding.

The local mate competition analysis does not distinguish the within-group and the between-group components of the causal process. We consider an ensemble of populations, each founded by the same number of fertilized females. We then ask: What would the sex ratio in the ensemble as a whole have to be for no female to be able to improve her reproductive output by producing a mix of sons and daughters different from this grand ratio? Hamilton (1967) calculated the optimal proportion of males to be $(n - 1)/2n$, where n is the number of founding females in each group.

This expression substantiates the speculation mentioned earlier concerning the relation of Fisher's argument to Hamilton's. When n is large, the predicted frequency of males approaches 0.5, just as Fisher predicted. But as n declines, the optimal sex ratio becomes increasingly female-biased. When the group is founded by a single fertilized female ($n = 1$), the algebra dictates that 0 males should be produced. This, of course, is literally false—a slightly inaccurate reflection of the idea that in this circumstance "a female's advantage depends wholly on the number of fertile emigrant females she can produce" (Hamilton 1967, p. 481).

We must look more carefully at this equilibrium ratio to see what sort of selection processes sustain it. The mere fact that no mutant female can improve her chances by departing from the population pattern does not, in itself, indicate *why* the configuration is evolutionarily stable. Recall the billiard ball with zero acceleration; the fact that it is not changing velocity leaves open a multitude of possible explanations of *why*.

Hamilton's equation describes an "evolutionarily stable strategy" (Maynard Smith 1974). It is an equilibrium, not for each subgroup but for the ensemble of groups. A local subgroup obeying Hamilton's equation *can* be invaded. Consider a group founded by six females that produces sons in the ratio $(6 - 1) / 2(6) = 5/12$. If a mutant Fisherian is introduced, it will enjoy a reproductive advantage within the group. Do not lose sight of Fisher's insight: *Regardless of the size of a group, if it has a female-biased sex ratio, an individual producing an equal mix of sons and daughters will be favored by individual selection.*

The equilibrium does not describe evolution within any particular group but within the ensemble of groups. Although within the invaded group, a Fisherian is fitter than Hamiltonian female, Fisherianism will not be fitter than Hamiltonianism when the fitnesses are computed by averaging over all the relevant individuals in all the groups. The sex ratio *in the global population* is not subvertible.

It should now be clear why the local mate competition analysis in no way shows that individual selection alone can sustain a female-biased sex ratio. To say that a Hamiltonian female is acting selfishly is like saying that a low-virulence myxoma virus is acting selfishly. True, before the system reaches equilibrium, the Hamiltonian characteristic may increase its representation under natural selection. So if "altruism" is defined to mean a trait that cannot evolve, one cannot think of Hamiltonianism or low virulence as "altruistic." On the other hand, if "altruism" is given a more reasonable characterization, it is perfectly clear that Hamiltonianism, like low virulence in myxoma, is altruistic. Within each group, the trait is less fit, but because it correlates with group productivity, the trait may have a higher overall fitness. This is how altruism can evolve by group selection.

As mentioned earlier, Hamilton (1967) used the rubric "local mate competition" because the analysis suggests that females enhance their reproductive chances by reducing competition among their sons for mates. They do this by making sons rare. But it is misleading to think of this property of Hamiltonianism as the cause of its predominance. Wilson and Colwell (1981, pp. 890–891), repeating a point established in Colwell (1981), put the point succinctly:

> There is no doubt that, for a given number of founders, the sons of a founding female who produces a female biased progeny will compete less among themselves than will the sons of a founding female in the same . . . group who produces half sons and half daughters. *However, the latter female will invariably leave more of her genes among the grandprogeny than the former, due to Fisher's principle* (emphasis mine).

Similar remarks apply to Maynard Smith's formulation of the local mate competition idea in terms of the frequency of sibmatings. Just as there is no selection within a group to diminish competition between sons, so there is no selection within a group to diminish the amount of sibmatings (Colwell 1981).

Whenever groups are founded by more than a single fertilized female, selection for female-biased sex ratios should be understood in terms of group selection. The same holds true if a group, whatever its initial number of foundresses, exists for more than a single generation. In the former case, a Hamiltonian and a Fisherian foundress may be present in the same group. In the latter case, although there is only one foundress, her descendants may differ in the sex-ratio characteristic and may thereby compete with each other.

What of the case in which groups are founded by single fertilized females and disperse after her offspring mate with each other? Here, it is equally correct to say that there is selection for females who bias their sex ratio toward daughters and that there is selection for groups that are female-biased. We have here a causal chain from foundress to kin group to group productivity. The chain is transitive. Foundresses maximize their number of grandoffspring by producing more daughters than sons. There is no more need to choose between the group and the individual descriptions here than there is to say whether it was Holmes' pulling the trigger or Moriarty's being hit by the bullet that caused Moriarty to die.

We therefore have an answer to the question raised in Section 7.1 about Darwin's analysis of sterile castes in the social insects: If colonies are founded by a single female and disperse after her offspring mate with each other, then it will be equally true that there is selection for females who produce sterile castes and that there is selection for colonies that contain sterile castes. The conflicting images of group and individual selection fuse into one in this special case.

Finally, I want to emphasize that the mere fact that groups have single foundresses does *not* suffice to show that a female-biased sex ratio has an individual selection explanation. True, the individuals in the group will be related to each other. But do not lose sight of the fact that related individuals may differ in their sex-ratio traits. If the group lasts long enough, they may have the opportunity to produce progeny and thereby compete. It is irrelevant that the offspring of a single foundress are siblings. What matters is their *traits*. If one is a Hamiltonian and the other is a Fisherian, they will compete; if they are alike in their sex-ratio characteristic, no selection occurs, whether or not they are related. Kinship is an indicator of whether individuals are similar or different. It does not establish what characteristics will or will not be selected for. The unit of selection question concerns the latter issue, not the former.

It is especially ironic that a female-biased sex ratio should be attributed to individual rather than group selection on the grounds that individuals in the same group will be related to each other. One of the seminal models of *group selection* is Maynard Smith's (1964) haystack model, in which groups are founded by single fertilized females. Maynard Smith introduced this model in order to argue that group selection, thus construed, would have a significant impact on evolution only for a severely constrained range of parameter values. Not that the model is a fully general way of addressing the question of impact (Wade 1978,

Wilson 1983). The point is that it is not disqualified from being a case of group selection simply because of the way it assumes groups are founded. Virtually every model of group selection relies on small group starting size to generate differences among groups by sampling error. The inevitable consequence of this idea is that there will be considerable inbreeding within groups. A group begun by a single foundress is at the extreme end of this continuum, but the structure still conforms to the guiding idea of how group selection might do some evolutionary work.

Examples in which internally homogenous groups differ from each other are often the ones that are most controversial when it comes to deciding how they should be interpreted. This situation may obtain in the myxoma case if we imagine that a rabbit is infected by a single population of clones. It also may arise in the local mate competition account of female-biased sex ratio, if the number of foundresses and the generation time of groups is kept low. I have already mentioned that I reject two standard analyses of this sort of case. Those who define group selection in terms of the existence of within- and among-group variance conclude that group, not individual, selection occurs if groups are internally homogenous. The opposite inference is also common—that when the homogenous groups are comprised of kin, it is individual, not group, selection that is at work.

Although the story of the tall and the short (Section 7.6) may have seemed like a pointless piece of philosophical inventiveness, it is in fact quite relevant to these biological problems. In that thought experiment, I imagined that the populations are internally homogenous for height and that they were constituted in such a way that the group selection hypothesis (selection for groups of greater average height) and the individual selection hypothesis (selection for tall individuals) would generate identical predictions. The way to decide which hypothesis is true, I argued, is to see which property is selected for.

The two hypotheses differ in their claims about which *property* is causally efficacious. They do not disagree over which *objects* will be selected. According to both conjectures, tall individuals will do better than short individuals, and tall populations will do better than short ones. There is no disagreement as to which properties will increase and which will decrease. The dissension has to do with *why*.

Notice that this question has basically nothing to do with the relatedness of the individuals in a group. The six-footers may be the offspring of a single fertilized female. They may be asexual clones of each other. Or they may be entirely unrelated. This has nothing to do with whether there is selection for one property rather than the other.

The significance of relatedness has, I believe, been seriously misunderstood in the units of selection controversy. It is, without question, a property of great moment in the evolution of a system. If groups have their productivities boosted by high frequencies of altruists in conformity with the fitness function shown in Figure 11 (p. 186), it will matter a great deal whether altruists tend to interact mainly with altruists or mainly with selfish individuals. If individuals with a gene for altruism live in kin groups, this tells us something about the probability that those they interact with will be altruists as well. Here we are taking account of relatedness to gauge the *effectiveness* of the selection regimen considered. We are seeing whether selection for groups that are made of altruists is apt to result in the evolution of altruism. But this very important question must not be confused with the question of which properties are selected for and against. Confusing these two issues is, I have argued, similar to conflating the issue of whether a trait is beneficial with the issue of whether it is heritable.

Even though I have discussed at length how group selection connects with the problem of altruism, it is important to stress the conclusion reached in Section 7.7—that group selection need not work in a direction opposite to that of individual selection. The *t* allele example and Wade's (1978) experimental work on *Tribolium* were discussed as examples in which group and individual selection favor the same characteristics. I now want to consider a third possibility, wherein group selection increases a character that is neither favored nor disfavored at the individual level.

As I have already stressed, within-group individual selection will modify the frequency of only those traits that make for *differences* in fitness within the group. This means that a trait that affects group productivity without causing differences in fitness within a group will be invisible to individual selection within the group, although it will be grist for the mill of group selection. Such a trait is not altruistic, since there is no difference within the group between organisms that have it and organisms that do not. Wilson (1980) emphasizes that this so-called neutral pathway provides numerous possible avenues for the operation of group selection. The sort of fitness function we are imagining here is shown in Figure 16. G (whose fitness function is represented by a solid line) is the trait that benefits the group at no cost to itself, and F (whose fitness function is represented by a broken line) is a "free rider."[36] Individual selection will fail to modify the initial frequencies

36. F is a "free rider," not in the sense that it increases in frequency within the group because of its pleiotropic association with some individually advantageous trait (Section 1.2), but because individuals with that trait who happen to inhabit groups in which G predominates will have their fitnesses boosted by the association.

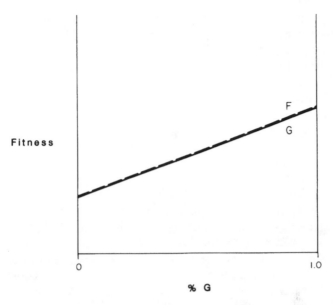

% G

Figure 16. The fitness function for Wilson's (1980) "neutral pathway." Traits F and G are equally fit in each group, but groups with higher concentrations of G show higher productivities. This allows for group selection without any within-group, individual, selection.

of F and G within the group. But group selection, among groups having different concentrations of the group-beneficial trait, should lead to an increase in G in the ensemble of populations.

One class of examples that Wilson develops concerns phenotypic differences in fitness that do not correlate within the group with genotypic differences. Suppose, for instance, that the mature males in a monkey troop drive the juvenile males to the periphery, where they are more vulnerable to predators. The phenotypic trait of being a juvenile male is thereby selected against; possessors of the trait run an increased risk of mortality. But if there is no *genetic* difference between juvenile males and the rest of the population (i.e., if the trait is not *heritable*), then this form of selection will fail to modify trait frequencies in the population. As far as this single selection process is concerned, all *genetic* traits are equal in fitness.[37] Yet the trait that causes juvenile males to run this increased risk may be favored by group selection, in that groups with the trait may enjoy enhanced productivities.

37. To get the point of the example, ignore the fact that the male sex chromosome is at a disadvantage in this process.

A somewhat similar application of this idea is to the dominance/ subordination hierarchy in the Harris sparrow (*Zonotrichia querula*) (Wilson 1980, pp. 76–78). Dominant birds will chase subordinate ones out of the flock when food supplies are low but will be more tolerant of their presence when food supplies are high. This fact, by itself, is of course no challenge whatever to the idea of individual selection. A standard individualist explanation will be that dominant sparrows chase out subordinate ones because it is advantageous for them to do so. The fact that the population will then have fewer mouths to feed in conditions of scarcity is seen as a fortuitous group benefit, not a group adaptation (Section 7.6).

However, Wilson focuses on another question. He asks not how the dominant birds benefit from interfering with subordinates but how it is that interfering birds gain an advantage over the other birds that remain in the flock but do not exert themselves. All the remaining birds benefit from the reduction in flock numbers, but the bird that performs the expulsion incurs an energy cost that the "free riders" avoid paying.

An individualistic account of this phenomenon will look for some special benefit that accrues to the interfering bird alone. If the expelling bird simply excluded the subordinate from its own habitat, without driving it entirely out of the flock, this would count as the sort of differential benefit the individual selection hypothesis requires. But dominant sparrows, apparently, do not stop short.

For both the monkeys and the sparrows, it is probably implausible to claim that individuals who engage in interfering behavior are *exactly* as fit as individuals who benefit from having that behavior performed without engaging in it themselves. The energy costs cannot be ignored. Still, those costs may be quite modest, compared with the benefit obtained by the group. Driving a subordinate bird out of the flock may be only slightly more taxing than driving it out of one's own territory. Wilson believes that the group selection issue has focused too single-mindedly on characteristics that represent severe costs to their possessors. What interests him is behaviors that benefit the group at little or no cost to the individual.

This perspective applies to selection among multispecies communities in the same way it applies to selection among groups from the same species. The standard way that ecologists have understood *mutualism*, for example, is as a one-to-one relationship between *organisms*. A damselfish feeds its own anemone and thereby improves its own protection.[38] Here, the "altruism" of the donor is reciprocated in a way perfectly

---

38. See, for example, the cost-benefit model of this symbiotic interaction developed in Roughgarden (1975).

consistent with the idea of individual selection (Trivers 1971).[39] But what if the receiver of the benefit reciprocates by augmenting the absolute fitness of the *entire population* to which the donor belongs? Here we have "population mutualism" (Wilson 1980, p. 98) via the neutral pathway.

Wilson (1980, pp. 98–106) constructs a hypothetical example to show how this idea works. It is known that earthworms improve the quality of the soil, and hence the life prospects of the plants that live in it, by the way they process food. Similarly, earthworms are affected by the nutrient value of the plant detritus on which they feed and the plant secretions and pathogenic bacteria they encounter. Wilson then considers two types of plant—one beneficial, the other detrimental, to earthworms. Imagine a set of demes that all include earthworms but that differ in the relative representation of the two plant types.

What happens within each deme? The earthworm population increases as a function of the relative abundance of the two plants. But even though the two plant types have different effects on earthworms, all the plants in a locale experience the activity of earthworms in the same way. Wilson's model stipulates that an earthworm does not discriminate between the soil below a beneficial plant and the soil below a harmful one, for example. So within each deme, there is no difference in fitness between a plant that helps earthworms and a plant that harms them. Yet, between demes, groups with higher proportions of plants that benefit earthworms may have higher productivities. Wilson designs his model so that the two plant types may be from the same or from different species. The key to his idea of population mutualism is that the benefits circle back not to the individual organism who was the donor but to the group of which the donor is a member.

Williams (1966, p. 18) had already used the earthworm as an intuitive example of how a benefit to the community should not be credited automatically to selection acting at a level higher than that of the individual organism:

> As the earthworm feeds, it improves the physical and chemical properties of the soil through which it moves. The contribution from each individual is negligible, but the collective contribution, cumulative over decades and centuries, gradually improves the soil as a medium for worm burrows and for the plant growth on which the earthworm's feeding ultimately depends. Should we therefore call the causal activities of the earthworm a soil-improvement mechanism? . . . [I]f we were to examine the digestive

39. As mentioned before, a trait that leads to "reciprocal altruism" is not really altruistic at all, even in the weak sense developed earlier in this section.

system and feeding behavior of an earthworm, I assume that we would find it adequately explained on the assumption of design for individual nutrition. The additional assumption of design for soil improvement would explain nothing that is not also explainable as a nutritional adaptation. It would be a violation of parsimony to assume both explanations when one suffices. Only if one denied that some benefits can arise by chance instead of by design, would there be a reason for postulating an adaptation behind every benefit.

On the other hand, suppose we did find some features of the feeding activities of earthworms that were inexplicable as trophic adaptations but were exactly what we should expect of a system designed for soil improvement. We would then be forced to recognize the system as a soil-modification mechanism, a conclusion that implies a quite different level of adaptive organization from that implied by the nutritional function. As a digestive system, the gut of a worm plays a role in the adaptive organization of that worm and nothing else, but as a soil-modification system it would play a role in the adaptive organization of the whole community.

It is instructive to place the approaches of Wilson and Williams side by side. We need to modify Wilson's discussion just a bit to make them comparable. In Wilson's example, it wasn't the earthworm's gut that evolved but the frequency of beneficial and deleterious plants. The earthworm was treated as an unchanging part of the environment that generates a selection pressure. But we could reverse the items in the model and think of two sorts of earthworm guts and a single plant species. The two gut forms would benefit the plants unequally, and the plants would then reciprocate whatever impact the local deme of earthworms had on them. Earthworm demes containing a higher frequency of the gut that is more beneficial to plants would have their productivities boosted by the neutral pathway. In nature, says Wilson, we should expect influences to flow in both directions.

Wilson's argument shows that the following line of reasoning is mistaken:

> Each earthworm benefits from its association with a plant of a type that helps it. Each plant benefits from its association with an earthworm that helps it. Hence, it is to be expected, *on the basis of individual selection alone,* that we should observe ecological communities in which benefits are traded back and forth.

This inference glosses over the difference between benefits that are distributed differentially within a group and benefits that all the members of a group gain in virtue of their common group membership.

The difference between individual mutualism and population mutualism makes all the difference when it comes to discerning the unit of selection.

Wynne-Edwards' defense of group selection and Williams' critique both focused on traits that benefit groups but impose substantial costs on individuals. The possibility of traits of this sort evolving may be quite limited, especially if the group benefits can be gained by alternative traits that represent little or no cost to individuals (Wilson 1980). Biologists' interest in group selection began as an interest in strong altruism, but the notion of group selection is interesting in its own right. Mention of group selection frequently conjures up the image of a sentinel crow issuing a warning cry and perishing for the good of the group. Perhaps a more useful example for orienting the problem is one suggested by Wilson (1980, p. 106):

> The crab *Pugettia producta* (Randall) concentrates certain trace elements in its feces. This is likely to mean very little directly to the crab, but it means more to a large set of other species in the community that require those nutrients for growth. Among these species one might expect variation in their effect on the crab. The variation can act along multiple pathways, including the inhibition of species that themselves inhibit the crab. . . .
>
> Now the crab is embedded in a community that has a definite relationship with it. It too will evolve in a direction that is beneficial to itself through indirect effects. But how? One way is by increasing the concentration of nutrients in its feces. In this way the community directs the evolution of the species, in much the way as human communities direct the professions of its members. Notice that it is not necessary for the community to have total control over its members, only enough to make it advantageous to evolve in the proper directions.

It is not for a philosopher to say whether this line of thinking will prove to be a significant source of insights about adaptation. My point here is to identify the contours of a concept. Group selection may be defined so as to require altruism. Altruism may be defined so that it cannot evolve by any sort of selection. But neither of these pieces of semantic legislation promises enlightenment. The strategy of averaging reduces to a parsimonious uniformity the variety of causal structures that a selection process may exhibit. But this uniformity arises from glossing over just the sorts of differences that the idea of a unit of selection ought to mark.

9.4 Species Selection

In explaining genic selection, I began with a formulation in which genic properties are causally efficacious in the survival and reproduction of organisms (Section 9.1). This approach works quite well in a range of examples (e.g., in single-locus models of viability selection in which the heterozygote is intermediate in fitness). But examples like meiotic drive and selfish DNA required that this interpretation be expanded. In those cases, it isn't organisms that are helped or hurt by genic properties but objects at lower levels of organization. I concluded that the organism is one, but not the only possible benchmark of genic selection.[40]

It now is time to expand the concept of group selection in a similar way. Examples such as Wade's experiments, myxoma, and the dynamics of the t allele fell nicely into place when interpreted in terms of the organismic approach. In those cases, membership in certain kinds of groups affects the fitness values of *organisms*. Logic suggests that there must be a second notion of group selection—one in which it is groups, not organisms, that serve as benchmarks.

Fortunately, it is not necessary to invent this alternative concept out of philosophical whole cloth. It is already available in the hypothesis of *species selection* that has recently excited controversy in the study of macro-evolution.[41] Theorists such as Eldredge, Gould (Eldredge and Gould 1972, Gould and Eldredge 1977, Gould 1980b, 1980e), Stanley (1975, 1979), and Vrba (1980) have advanced a pair of hypotheses about the pattern and process of extinction and speciation that is meant to challenge the received view of these matters and also to establish macro-evolutionary theory as a domain of inquiry that is, in a sense to be discussed, autonomous from micro-evolutionary thought.

Their hypothesis about pattern is well described by the label used by Eldredge and Gould (1972)—*punctuated equilibrium*. The idea is that most species change very little and that most of the evolutionary change they experience occurs during the first 5 percent of their existence. If the average species lasts about 10 million years, then evolution is supposed to be concentrated in the first 500,000 years, and stasis in

40. A word of clarification about benchmarks: To say that organisms are not the benchmarks for the selection process of meiotic drive is not to say that organisms are never helped or hurt. After all, the t allele is a driving gene that hurts organisms when it occurs in double dose. The point is that meiotic drive as a kind of process *need* not affect the reproductive chances of organisms. So the question of what the benchmark is in a selection process and the question of what objects benefit or suffer are distinct.

41. I treat species selection as a variety of group selection even though species are not groups, in the sense of being local demes. But they can be thought of as more inclusive groups; typically, organisms belong to local demes, which are nested in populations, which belong to species.

the remaining 9,500,000. Although earlier paleobiologists often claimed that the absence of intermediate forms in the fossil record was due to the record's incompleteness, advocates of the thesis of punctuated equilibrium contend that the record is in this respect remarkably complete. The fossil record fails to exhibit gradual modifications in species morphology because for most species there was no such thing.

Even if one accepts this claim about the unevenness of evolutionary rates—and this is still a matter of great controversy—there remains the task of explaining this pattern of evolution and stasis. Two hypotheses have been suggested—one to explain stasis, the other to explain the existence of large-scale macro-evolutionary trends.

Species are thought to change so little in their lifetimes because they are locked into their phenotypes and genotypes. Architectural constraints prevent nature from doing much in the way of ad hoc tinkering. Or, as the idea is sometimes expressed, once a new species is established (i.e., when the process of reproductive isolation from other species is complete),[42] the species is at an adaptive peak, and individual selection is a preventer, not a cause, of change.[43]

If species change very little, how to explain macro-evolutionary patterns and trends—for example, the prevalence of sexual reproduction or the pattern of increase in size found so frequently in vertebrate lineages (Cope's rule)? The second hypothesis put forward is *species selection*. We will attend to this idea in a moment. For now, I want to note that the authors mentioned think of this cluster of ideas as challenging the standard Modern Synthesis view of evolution. They think of the received view as committed to a doctrine of "extrapolationism," according to which macro-evolutionary events—the branching of a phylogenetic tree by speciation and extinction—are simply the cumulative effects of micro-evolutionary events—preeminently intraspecific selection. The received position they call "phyletic gradualism," according to which most evolutionary changes are the result of the gradual modification of populations within lineages.

Advocates of these ideas frequently call them the "punctuated equilibrium" view, although that label literally characterizes only the doc-

42. Although macro-evolutionists both favoring and opposing the punctuated equilibrium hypothesis often use Mayr's (1963) "biological species concept"—i.e., one stated in terms of reproductive isolation—both sides want to consider asexual species. It isn't that an adequate species concept is unproblematic; rather, it's just that this problem is not the one at issue in the controversy discussed here.
43. Recall the conclusion reached in Section 9.3 about the status of the "adaptive peak" idea when selection is frequency-dependent. This language is often used simply to indicate that the population has evolved to a selective equilibrium, without any indication of what happens to $\bar{w}$.

trine about pattern, not the set of ideas about process. Someone might criticize the idea of species selection while granting that evolution exhibits vastly uneven rates, for example. I will follow the practice of using "punctuated equilibrium" to denote the combined views about pattern and process; however, when necessary, we must be prepared to split these conjectures apart.

Eldredge, Gould, Stanley, and Vrba describe their theory as "decoupling" macro-evolution from micro-evolution; they see antireductionist consequences flowing from the punctuated equilibrium idea. Critics (e.g., Lande 1980, Ayala 1983) have responded that even if the punctuated equilibrium view is correct, there is still an important sense in which micro- and macro-evolution remain unified. Ayala (1983, p. 284) argues that macro- and micro-evolution are not decoupled in two senses. They exhibit both

> identity at the level of events and compatibility of theories. First, the populations in which macro-evolutionary patterns are observed are the same populations that evolve at the micro-evolutionary level. Second, macro-evolutionary phenomena can be accounted for as the result of known micro-evolutionary processes. That is, the theory of punctuated equilibrium (as well as the theory of phyletic gradualism) is consistent with the theory of population genetics. Indeed, any theory of macro-evolution that is correct must be compatible with the theory of population genetics, to the extent that this is a well established theory.

Ayala goes on to observe that since population genetics is compatible with both punctualism and gradualism, it entails neither of them. And since we cannot at present deduce the truth about macro-evolution from what we know about micro-evolution, the study of macro-evolution may pursue its own course, asking its own questions and investigating macro-evolutionary phenomena for their own sake.

Ayala's point is explicitly modeled on Ernest Nagel's (1961) general views about reduction. It is possible to maintain that every biological process is physical and still deny the methodological usefulness of investigating speciation from the point of view of the protons, electrons, and neutrons that species are made of. The ontological point—that biological events *are* physical events—seems to hold even if we now don't know how to state a physicalistic characterization of the biological phenomena, and even if it would be fruitless to investigate biological issues as physicists.

The position involved here—that reductionism is correct as an ontological thesis—sounds quite reasonable. The only alternative appears to be vitalism, which holds that biological entities are really not physical at all, because they contain an immaterial fluid—an *élan vital*—that animates them with life. This is a position that very few people would want to take seriously now.

I do not doubt that anything that happens in evolution must be "representable" in the language of population genetics. This is the idea that Dawkins and Williams exploited in one argument for genic selectionism. I argued in Sections 7.4 and 9.1 that the truism has no such implications favoring reductionism. Precisely the same point applies here. Since speciation involves change in gene frequencies, the trajectory of a speciating lineage may be plotted with the equations of population genetics. Yet this should not obscure the fact that species selection is a very different sort of evolutionary mechanism from individual selection. This is the important sense in which the ideas collected under the banner of punctuated equilibrium effectively "decouple" macro- from micro-evolution.[44]

This point goes beyond the fairly obvious idea that macro-phenomena may be studied "at their own level," rather than from a micro point of view. That is a relatively uncontroversial epistemological or methodological thesis. Rather, we have here an ontological claim to the effect that an item at the macro-level is not identical with anything at the micro-level. It is not to be doubted that species are *composed* of organisms. Rather, the idea is that a certain causal mechanism—species selection—is distinct from the array of mechanisms acknowledged by the individual-level science.

Critics of the punctuated equilibrium idea have noted that a pattern of evolution and stasis is by itself no problem for traditional evolutionary models. Directional *individual* selection, followed by stabilizing *individual* selection, may generate the pattern of evolution and stasis that is said to be so important (Lande 1980, Ayala 1983). Five percent of a species' lifetime may provide a huge number of generations and may therefore allow plenty of time for "gradual" Darwinian selection to do its work.

Just as uneven rates of morphological change can be accommodated by possible mechanisms acknowledged by standard population genetics, so too is the picture of speciation itself claimed to be no threat to traditional thinking. The supposed alternative to punctuated equilibrium—phyletic gradualism—is often alleged to be a straw man. Tem-

---

44. Supervenience provides a second reason that ontological reductionism isn't vindicated by the simple fact that species are made of organisms and organisms are made of physical particles (Section 1.5). Although each organism's fitness has a physical basis, fitness is not a physicl property. The same relationship, presumably, obtains between macro- and micro-evolutionary concepts. A given kind of macro-evolutionary event arguably can have many different possible micro-evolutionary realizations.

pleton (1980), for example, has pointed out that virtually no one now thinks that sympatric speciation (i.e., speciation in which geographical isolation occurs after the speciation process has been completed) at a constant evolutionary rate will be the rule.[45] Gould and Eldredge (1977) have acknowledged their debt to Mayr's (1963) ideas about the role of genetic revolutions and founder effects in speciation. Macro-evolution is not decoupled from micro-evolution by the fact that speciation often occurs via the splitting of lineages rather than the slow accumulation of adaptations within lineages.

To see why the punctuated equilibrium idea has antireductionist consequences, we must attend to the claims made about *process*, not to the ideas about *pattern*. Although the claim about rates of stasis and change is interesting and important, it does not in itself require a novel mechanism. The parallel with Wynne-Edwards (1962) on group selection may be helpful. It is not controversial that groups often benefit from the properties of their member organisms. What is controversial is the claim that organisms have these group-beneficial properties because of group selection. It is only the latter claim about mechanism that implies that there is more to evolution than is acknowledged by individual-level theory.

Stanley (1975) and Gould and Eldredge (1977) describe species selection by an analogy. Individual selection occurs when *organisms* survive and reproduce differentially. If species are to be the objects involved in species selection, then species must die and reproduce differentially. But what is "death" and "reproduction" for an object of this sort? The analog of organismic death is easy to foresee: it is species *extinction*. But how is the idea of species reproduction to be interpreted? For an organism, reproduction differs from growth in a fairly obvious way. An organism grows by increasing the number of cells that are part of its body; it reproduces by creating a new functioning entity. The parallel is this: Species reproduce not when the organisms in them grow in number but when they found numerically distinct offspring species. Species selection thus must involve the idea of differential *speciation*.[46]

This way of developing the analogy between organismic and species selection is supplemented by another. The speciation process generates new entities. In this, it is like reproduction. Additionally, speciation generates daughter species that do not resemble their parents. In this,

45. Polyploidy is an uncontroversial example of sympatric speciation.
46. Notice that one of the population parameters that may diminish the effectiveness of intraspecific group selection should not do much to dampen the effect of species selection. Migration among groups may reduce the intergroup variance on which group selection depends. But, if species are reproductively isolated from each other, little mixing of genes will occur.

it resembles mutation; it creates the variation on which a selection process may subsequently operate. This way of understanding the idea of species selection is obviously at work in the formulation used by Gould and Eldredge (1977, pp. 139–140). They say that the following two conditions suffice for species selection to occur:

1. the model of punctuated equilibrium itself;
2. the proposition that a set of morphologies produced by speciation events is essentially random with respect to the direction of evolutionary trends within a clade.

They credit this second idea to Sewall Wright (1967) and suggest that it be called "Wright's rule," although they note that Mayr (1963, p. 621) also asserted that speciation events might be viewed as random, mutation-like inputs into the macro-evolutionary process.

Figure 17 shows the sort of phylogenetic tree that might result if conditions 1 and 2 were satisfied. Branching represents speciation, and the termination of a line represents species extinction. The first condition—punctuated equilibrium—demands that morphological change be concentrated in branching events, rather than in the subsequent lifetime of a species. The tree satisfies this requirement, in that the lines are all perfectly vertical and horizontal. The tree also conforms to the second condition—Wright's rule—since there are as many left branchings as right branchings, and the left and right branchings are of equal magnitude. Notice that some species are more reproductively successful than others (i.e., they speciate more). This is what produces the evolutionary trend of increase in the morphological character in question.

It is clear that differential speciation of the kind illustrated can produce this sort of evolutionary trend, without the punctual pattern holding. Tilting the entire tree to the right provides a phylogeny in which tendencies toward morphological increase *within species* are reinforced by differential reproduction *of species*. This is shown in Figure 18. This possibility should come as no surprise, since we have already seen that group and individual selection can work in the same direction (Section 7.7). The *pattern* of punctuated equilibrium is not a necessary condition for the *process* of species selection.

Similar remarks apply to Wright's rule: The idea of species selection allows but does not require that speciation be at random. One of Williams' (1966, pp. 98–99) comments on the logic of higher-level selection processes is relevant here:

It should be clear that, in general, the fossil record can be a direct source of information on organic evolution only when changes in

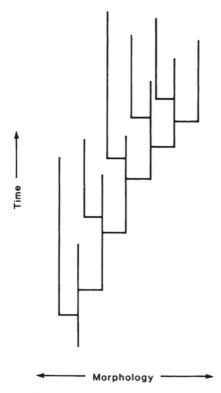

Time →

← Morphology →

Figure 17. A punctuated equilibrium pattern of speciation and extinction. Morphological change occurs at the time of speciation; species remain static thereafter. Speciation is just as likely to decrease as it is to increase the morphological value in question. Yet a bias in speciation rates gives rise to a trend of increase in the characteristic.

single populations can be followed through a continuous sequence of strata. Ordinarily the record tells us only that the biota at time t' was different from that at time t and that it must have changed from one state to the other during the interval. An unfortunate tendency is to forget this and to assume that the biotic change must be ascribed to appropriate organic change. The horse-fauna of the Eocene, for instance, was composed of smaller animals than that of the Pliocene. From this observation, it is tempting to conclude that, at least most of the time and on the average, a larger than mean size was an advantage to an individual horse in its reproductive competition with the rest of its population. So the component populations of the Tertiary horse-fauna are presumed to have been evolving larger size most of the time and on the average.

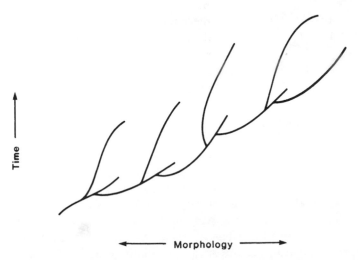

Figure 18. Increase in the morphological character occurs during the lifetime of a species as well as during speciation events. However, as in Figure 17, differential speciation contributes to the trend of morphological increase.

It is conceivable, however, that precisely the opposite is true. It may be that at any given moment during the Tertiary, most of the horse populations were evolving a smaller size. To account for the trend towards larger size it is merely necessary to make the additional assumption that group selection favored such a tendency. Thus, while a minority of the populations may have been evolving a larger size, it could have been this minority that gave rise to most of the populations of a million years later.

Williams (pp. 99–100) traces "the unwarranted assumption of organic evolution as an explanation of biotic evolution" back to Darwin. He describes how Darwin tried to explain "advance in organization" in *The Origin of Species*:

He interpreted the fossil record as indicating that the biota has evolved progressively "higher" forms from the Cambrian to Recent, clearly a change in the biota. His explanation, however, is put largely in terms of the advantage that an individual might have over his neighbors by virtue of a larger brain, greater histological complexity, etc. Darwin's reasoning here is analogous to that of someone who would expect that if the organic evolution of horses proceeded toward larger size during the Tertiary, most equine mutations during this interval must have caused larger size in the affected individuals. I suspect that most biologists would tend toward the opposite view, and expect that random changes in the germ plasm would be more likely to curtail growth than augment it.

The point here is that the variation presented by groups or species serves as raw material for a higher-level selection process. Such variation *may* be produced by chance. In this sense speciation events *may* be to species selection as individual mutations are to individual selection. *But they need not be.*

Suppose that species selection had produced the increase in average horse size during the period mentioned. Species with a greater average height speciated more. This *may* mean that a species of a given average height is no more likely to have a taller daughter species than a shorter one. But it is consistent with the idea of species selection that the speciation process should have been biased. Imagine that the tallest organisms within a species are the ones most likely to be the founders of daughter species. Speciation may thereby show a bias toward increased height; the speciation process will not, in this sense, be like mutation. However, this is perfectly consistent with the operation of species selection.

These arguments—that neither punctuated equilibrium nor Wright's rule is necessary for species selection—may not undermine the characterization of species selection presented by Gould and Eldredge (1977). In the passage cited above, they do not say that the two conditions are *necessary* for species selection but only that the conditions are sufficient.[47] In any event, the raw materials required for showing that the two conditions are not sufficient are available from the discussion before of group selection. They need only be assembled and applied.

We have already seen that differences in *actual* survival and reproduction are no sure sign of selection at any level; drift may change gene frequencies within a single population, and, as Raup and Gould (1974) have argued, chance may also produce interesting patterns in a branching phylogenetic tree. So species selection cannot be said to exist just because there is differential speciation and extinction.

Nor is it sufficient for species selection that the differential speciation and extinction not be due to chance. A species goes extinct when every organism in it is dead. Death may be a purely individual affair. Stanley (1975, p. 648; 1979, p. 197) mentions predation, competition, and habitat alteration as sources of differential extinction. But when the organisms in one prey species are all eaten, while the organisms in a second prey

47. However, they go on to remark that "punctuated equilibrium + Wright's rule = species selection" (p. 140). This perhaps indicates that Gould and Eldredge thought the conditions are necessary as well.

species escape, why think of this as *species* selection? True, all the variation in fitness may be *between* species, if each species is internally homogeneous. But we have already seen that group selection cannot be defined in terms of the pattern of within- and between-group variance in fitness. The same holds for species selection.

When one species is eaten and the other is spared, there is selection *of* one species rather than the other. Think of the predators as so many grim reapers. The first species is selected (to be supper); the second is not. However, it is equally true that the *organisms* in the first species are the ones that are selected. Selection *of* a species does not imply species selection, anymore than the selection *of* green balls implied that there was selection for greenness (Section 3.2). Genic selection (Section 2.10) requires *selection for the property* of possessing a particular gene; group selection (Section 9.2) requires *selection for the property* of belonging to a group of a certain sort. The crucial ingredient in species selection, properly so-called, is not the *selection of* species but *selection for* species-wide properties.[48]

Species selection, as Stanley and others have noted, may proceed via differential extinction (mortality) and differential speciation (reproduction). It is instructive to compare some of the examples that have been offered of these two modes of species selection. Those that proceed by differential extinction frequently seem not to be cases of species selection. However, some of the examples of species selection by differential speciation appear to fill the bill.

Stanley (1979, pp. 200–203) discusses the impact of predation on bivalve mollusks as a case of species selection. He argues that increasingly sophisticated predation during the Mesozoic era resulted in a variety of defensive adaptations among the prey. The responses included thicker and spinier shells, the occupation of crevices in or beneath rocks, and, in the case of scallops (*Pectinidae*), the ability to swim. Stanley attributes these changes to *species* selection presumably because he thinks of there being little or no variation in fitness *within* the various species. So species persisted or went extinct in virtue of the uniform structure exemplified by member organisms.

Note that there is no suggestion here of a species-wide property that affects the fitness of member organisms. Presumably, an *organism* survives because it has a thick shell; the fact that it belongs to a species with some average shell thickness is not invoked in the explanation. The survival of the species may simply be a consequence (a "statistical summation" as Williams says of the fast deer) of the survival of the

---

48. This point should be familiar from the discussion in Section 7.6 of Williams' (1966, p. 16) example of the fast and slow herds of deer.

member organisms. It is not inconceivable that a higher-level selection process should be involved here, but Stanley's account does not suggest that this is so.

Matters change when we turn to some of Stanley's examples of species selection via differential speciation (reproduction). He considers (1979, p. 205) how changes in climate may influence the development of xeric (those adapted to dry conditions) and mesic (those adapted to moderately moist conditions) plants:

> At first consideration, it might seem that a severe shrinkage of tropical forest area and complementary spread of arid territory would favor phylogenetic trends toward xeric adaptations. During the Pleistocene in South America, however, the result of such changes has apparently been quite the reverse. . . . Catastrophic selection seems to have occurred, with new species evolving from small populations isolated in residual islands of forest. . . . The composite result has been a considerable enrichment of the avian and lizard faunas of tropical forests. Whether there has been complementary enrichment of xeric habitats during forest expansion is unknown.

Stanley's hypothesis is that the decline in forest acreage increased the patchiness of the remaining forest territory. This makes it probable that small colonies of isolates formed more frequently in mesic areas. Hence mesic species had a greater chance of speciating.

Another example that illustrates the structure of the species selection idea is Stanley's explanation of why wingless grasshoppers in the genus (*Melanoplus*) tend to be found in numerous, narrowly ranging species, whereas fully winged forms are distributed in a small number of widespread species (Stanley 1979, pp. 261–262). The hypothesis is that the formation of a small isolate population is harder to achieve with winged dispersers. Stanley mentions that this idea also may help explain why speciation is more frequent on land than in the sea.

We might visualize how this hypothesis works by a simple hypothetical example. Let's begin with two species, one winged and the other nonwinged, which initially have the same census sizes. Suppose that winged organisms survive and reproduce exactly as well as their wingless counterparts. But there is a difference: small colonies of wingless individuals become separated from the main population rather frequently; winged organisms, on the other hand, in virtue of their greater mobility, very rarely form isolated subpopulations. If this system is allowed to evolve, we may later find a large number of rather small wingless species and a small number of rather large winged species. Wingless species speciate, whereas winged species grow. Notice that although wingless *species* have had more daughter species, wingless *organisms* are no more reproductively successful than winged ones.

My point here is not to defend Stanley's hypotheses about the influence of climate or wings on the speciation process but to bring out the conceptual structure of the species selection hypothesis. We have here selection for a species property, but without an organismic benchmark. Mesic plants and wingless grasshoppers are not said to have their fitnesses enhanced by belonging to species with certain properties. If the organisms were thus affected by their group membership, species selection would have a conceptual format already familiar from traditional hypotheses of group selection. Rather, the hypothesis of species selection bypasses the organismic level in much the same way that ideas of meiotic drive and selfish DNA represent genic selection without an organismic benchmark. It is *species* (or clades)—not *organisms*—that serve as the benchmark in these hypotheses.

Eldredge (1979) has remarked that the idea of the transformation of *characters* within a lineage has preoccupied evolutionary theory at the expense of the quite different idea of the origin of new *taxa* by cladogenesis. Indeed, the standard definition of evolution as change in gene frequency enshrines the idea that evolution is fundamentally a process in which *properties* shift in their representation; it is quite a different matter to conceive of evolution as occurring when new *objects* are brought into being. Often, there will be no clash between these two points of view, but at times a preoccupation with one can render invisible phenomena that would be striking if we could only see them differently.[49]

In the grasshopper example we examined two properties. There are a large number of small species of wingless grasshoppers and a small number of large species of winged grasshoppers. Let us suppose, for the sake of argument, that there are equal numbers of winged and wingless grasshoppers. This configuration may have resulted from an earlier situation in which there was a single winged species and a single wingless one. The wingless species speciated while the winged species grew large.

Was there selection for winglessness here? Was there evolution with respect to that property? The answer depends on how you count. Winged and wingless *organisms* may not have differed in their (chances of) survival and reproductive success; and the frequency of the two traits

49. I should point out that the idea that species are individuals rather than natural kinds, defended by Ghiselin (1974b) and Hull (1976b, 1978) does not entail the species selection hypothesis or its negation. Whether or not you think species selection occurs, you still may hold that species are individuals—i.e., spatiotemporally localized chunks of the phylogenetic nexus.

in the entire ensemble of species may not have changed. So from an organismic point of view, one would conclude that there was no selection for winglessness and no evolution of winglessness. However, to adhere to this monolithic point of view is to miss something. The traits *did* differ in their fitness values; selection *did* favor one as opposed to the other; the traits *did* change their frequencies. It is the benchmark of species, not organisms, that makes these conclusions available.

Darwin, as we saw in Section 7.1, thought of evolution and natural selection in terms of organisms. Organisms die, reproduce, and thereby effect changes in the frequencies of traits. Although Darwin considered the possibility of group selection (and rejected it in all cases but, perhaps, one), his idea of group selection remained wedded to an organismic paradigm. That is, he considered group selection in which it is organisms that are helped and hurt by group properties.

The conceptual possibilities made available in evolutionary theory have subsequently expanded in two directions. There is the group above and the gene below. Biologists have considered the impact that group and genic *properties* may have on organismic fitness. In addition, they have deprived the organismic benchmark of its exclusive grip on the way we look at natural selection; the properties that are selected for and against may be causal factors in the survival and reproduction of genes and species, as well as of organisms. Evolutionary theory now deploys a striking hierarchy of possible selection mechanisms. Indeed, it is a double hierarchy—of both objects and properties—that contemporary theory has had to consider. Objects at different levels may be selected; and there may be selection for and against properties at different levels as well. The adequacy of the various hyotheses made available by this hierarchy is a subject of continuing empirical controversy, which philosophical reflection cannot resolve. In this book, I have tried to lay out the conceptual structure and lines of reasoning that allow evolutionary theory to pose a set of intricate and multilevel questions for nature to answer.

References

Alexander, R. D., and G. Borgia (1978): Group selection, altruism, and the levels of organization of life. *Annual Review of Ecology and Systematics* 9: 449–474.

Allee, W., A. Emerson, O. Park, T. Park, and K. Schmidt (1949): *Principles of Animal Ecology*. Philadelphia: W. B. Saunders.

Anscombe, G. E. M. (1971): Causality and determination. In E. Sosa (ed.), *Causation and Conditionals*. Oxford: Oxford University Press, 1975.

Arnold, A., and K. Fristrup (1982): The theory of evolution by natural selection: a hierarchical expansion. *Paleobiology* 8: 113–129.

Ayala, F. (1983): Beyond Darwinism? The challenge of macroevolution to the synthetic theory of evolution. In P. Asquith and T. Nickles (eds.), *PSA 1982*, vol. 2, East Lansing: Philosophy of Science Association, pp. 275–292.

Balme, D. (1962): Development of biology in Aristotle and Theophrastus: theory of spontaneous generation. *Phronesis* 2: 91–104.

Bethell, T. (1976): Darwin's mistake. *Harper's*, February, pp. 70–75.

Blalock, H. M. (1961): *Causal Inferences in Nonexperimental Research*. Chapel Hill: University of North Carolina Press.

Block, N., and G. Dworkin (1974): IQ, heritability, and inequality. *Philosophy and Public Affairs* 3: 331–409; 4: 40–99; reprinted in N. Block and G. Dworkin (eds.), *The IQ Controversy*. New York: Random House, 1976.

Block, N., and J. Fodor (1972): What psychological states are not. *Philosophical Review* 81: 159–181.

Boring, E. (1950): *A History of Experimental Psychology*. New York: Appleton-Century-Crofts.

Bowler, P. (1974): Darwinism and the argument from design: suggestions for a reevaluation. *Journal for the History of Biology* 10: 29–43.

Brandon, R. (1978): Adaptation and evolutionary theory. *Studies in the History and Philosophy of Science* 9: 181–206; reprinted in Sober (1984b).

Brandon, R. (1982): The levels of selection. In P. Asquith and T. Nickles (eds.), *PSA 1982*, vol. 1, East Lansing: The Philosophy of Science Association, pp. 315–324.

Bromberger, S. (1966): Why-questions. In *Mind and Cosmos: Essays in Contemporary Science and Philosophy*. Pittsburgh: University of Pittsburgh Press.

Brower, Lincoln (1969): Ecological chemistry. *Scientific American* 229: 22–29.

Brower, Lincoln, F. Harvey Pough, and H. R. Meck (1970): Theoretical investigations in automimicry I. *Proceedings of the National Academy of Sciences, USA* 66: 1059–1066.

Bull, J. (1983): *The Evolution of Sex Determination Mechanisms*. Menlo Park: Benjamin/Cummings.

Byerly, H. (1979): Substantial causes and nomic determination. *Philosophy of Science* 46: 57–81.

Carr-Saunders, A. (1922): *The Population Problem: A Study in Human Evolution*. Oxford: Clarendon Press.

Cartwright, N. (1979): Causal laws and effective strategies. *Nous* 13: 419–437.

Causey, R. (1972): Attribute identities and microreductions. *Journal of Philosophy* 69: 407–422.

Cavalli-Sforza, L., and M. Feldman (1981): *Cultural Transmission and Evolution*. Princeton: Princeton University Press.

Charnov, E. (1982): *The Theory of Sex Allocation*. Princeton: Princeton University Press.

Chomsky, N. (1975): *Reflections on Language*. New York: Random House.

Cohen, G. A. (1978): *Karl Marx's Theory of History: A Defense*. Princeton: Princeton University Press.

Colwell, R. (1981): Group selection is implicated in the evolution of female-biased sex ratios. *Nature* 290: 401–404.

Crow, J. F. (1979): Genes that violate Mendel's rules. *Scientific American* 240: 134–146.

Crow, J. F., and M. Kimura (1970): *An Introduction to Population Genetics Theory*. Minneapolis: Burgess.

Cummins, R. (1975): Functional analysis. *Journal of Philosophy* 72: 741–764; reprinted in Sober (1984b).

Darwin, C. (1859): *On the Origin of Species*. London: J. Murray; Harvard: Harvard University Press, 1964.

Darwin, C. (1871): *The Descent of Man, and Selection in Relation to Sex*. First edition, London: J. Murray; Princeton University Press, 1981.

Darwin, C. (1886): *The Descent of Man, and Selection in Relation to Sex*. Second edition. New York: D. Appleton.

Darwin, C. (1876): *Autobiography*. N. Barlow (ed.). New York: W. W. Norton, 1969.

Darwin, F., and Seward, A. (1903): *More Letters of Charles Darwin*. New York: D. Appleton.

Davidson, D. (1963): Actions, reasons, and causes. In *Essays on Actions and Events*. Oxford: Oxford University Press, 1980, pp. 3–20.

Davies, P. C. (1979): *The Forces of Nature*. Cambridge: Cambridge University Press.

Dawkins, R. (1976): *The Selfish Gene*. Oxford: Oxford University Press.

Dawkins, R. (1978): Replicator selection and the extended phenotype. *Zeitschrift für Tierpsychologie* 47: 61–76; reprinted in Sober (1984b).

Dawkins, R. (1982): *The Extended Phenotype*. Oxford: W. H. Freeman.

Delbrück, M. (1971): Aristotle-totle-totle. In J. Monod and J. Borek (eds.), *Microbes and Life*. New York: Columbia University Press, pp. 50–55.

Dennett, D. (1971): Intentional systems. *Journal of Philosophy* 68: 87–106.

Dennett, D. (1980): True believers: the intentional stance and why it works. *Herbert Spencer Lectures on Scientific Explanation*. Oxford: Oxford University Press.

Doolittle, W., and C. Sapienza (1980): Selfish genes, the phenotypic paradigm, and genome evolution. *Nature* 284: 601–603.

Dretske, F. (1973): Contrastive statements. *Philosophical Review* 81: 411–437.

Dretske, F. (1982): *Knowledge and the Flow of Information*. Cambridge: Bradford/MIT Press.

Dretske, F. (1984): Misrepresentation. In R. Bogdan (ed.), *Belief*. Oxford: Oxford University Press.

Dretske, F., and A. Snyder (1972): Causal irregularity. *Philosophy of Science* 39: 69–71.

Duhem, P. (1914): *The Aim and Structure of Physical Theory*. Princeton: Princeton University Press, 1954.

Earman, J. (1984): Determinism. University of Minnesota: unpublished manuscript.

Earman, J., and M. Friedman (1973): The meaning and status of Newton's law of inertia and the nature of gravitational forces. *Philosophy of Science* 40: 329–359.

Eddington, A. S. (1928): *The Nature of the Physical World*. Cambridge: Cambridge University Press.

Edwards, A. W. F. (1971): *Likelihood*. Cambridge: Cambridge University Press.

Eells, E. (1984): Objective probability theory theory. *Synthese*: forthcoming.

Eells, E., and E. Sober (1983): Probabilistic causality and the question of transitivity. *Philosophy of Science* 50: 35–57.

Eldredge, N. (1979): Alternative approaches to evolutionary theory. *Bulletin of the Carnegie Museum of Natural History* 13: 7–19.

Eldredge, N., and J. Cracraft (1980): *Phylogenetic Patterns and the Evolutionary Process*. New York: Columbia University Press.

Eldredge, N., and S. J. Gould (1972): Punctuated equilibria: an alternative to phyletic gradualism. In T. Schopf (ed.), *Models in Paleobiology*. San Francisco: Freeman Cooper.

Enc, B. (1975): Necessary properties and Linnean essentialism. *Canadian Journal of Philosophy* 5: 83–102.

Fetzer, J. (1981): *Scientific Knowledge*. Dordrecht: Reidel.

Feyerabend, P. (1970): Against method. In *Minnesota Studies in the Philosophy of Science*, vol. 4.

Fisher, R. A. (1922): On the dominance ratio. *Proceedings of the Royal Society of Edinburgh* 42: 321–341.

Fisher, R. A. (1930): *The Genetical Theory of Natural Selection*. Oxford: Oxford University Press; New York: Dover, 1958.

Fisher, R. A. (1959): *Smoking*. London: Oliver and Boyd.

Fodor, J. (1975): *The Language of Thought*. New York: Thomas Crowell.

Fodor, J., and N. Block (1972): What psychological states are not. *Philosophical Review* 81: 159–181.

Frege, G. (1884): *The Foundations of Arithmetic*. Oxford: Basil Blackwell, 1968.

Furth, M. (1975): Essence and individual: reconstruction of an Aristotelian metaphysics, chap. 11; duplicated for the Society for Ancient Greek Philosophy, unpublished.

Futuyama, D. J. (1979): *Evolutionary Biology*. Sunderland, Mass.: Sinauer.

Garfinkel, A. (1981): *Forms of Explanation: Rethinking the Questions of Social Theory*. New Haven: Yale University Press.

Ghiselin, M. (1969): *The Triumph of the Darwinian Method*. Berkeley: University of California Press.

Ghiselin, M. (1974a): *The Economy of Nature and the Evolution of Sex*. Berkeley: University of California Press.

Ghiselin, M. (1974b): A radical solution to the species problem. *Systematic Zoology* 23: 536–544.

Giere, R. (1979): *Understanding Scientific Reasoning*. New York: Holt, Rinehart & Winston.

Giere, R. (1980): Causal systems and statistical hypotheses. In L. J. Cohen (ed.), *Applications of Inductive Logic*. Oxford: Oxford University Press.

Gilpin, M. (1975): *Group Selection in Predator-Prey Communities*. Princeton: Princeton University Press.

Gish, D. (1979): *Evolution? The Fossils Say No!* San Diego: Creation Life Publishers.

Goldschmidt, R. (1940): *The Material Basis for Evolution*. New Haven: Yale University Press.

Good, I. J. (1961–1962): A causal calculus I and II. *British Journal for the Philosophy of Science* 11: 305–318; 12: 43–51; 13: 88.

Gould, S. J. (1977a): Darwin's untimely burial. In *Ever Since Darwin*. New York: W. W. Norton, pp. 39–48; reprinted in Sober (1984b).

Gould, S. J. (1977b): The misnamed, mistreated, and misunderstood Irish elk. In *Ever Since Darwin*. New York: W. W. Norton, pp. 79–90.

Gould, S. J. (1980a): Caring groups and selfish genes. In *The Panda's Thumb*. New York: W. W. Norton, pp. 85–92; reprinted in Sober (1984b).

Gould, S. J. (1980b): Is a new and general theory of evolution emerging? *Paleobiology* 6: 119–130.

Gould, S. J. (1980c): Natural selection and the human brain: Darwin vs. Wallace. In *The Panda's Thumb*. New York: W. W. Norton, pp. 47–58.

Gould, S. J. (1980d): The panda's thumb. In *The Panda's Thumb*. New York: W. W. Norton.

Gould, S. J. (1980e): The promise of paleobiology as a nomothetic, evolutionary discipline. *Paleobiology* 6: 96–118.

Gould, S. J. (1982): Darwinism and the expansion of evolutionary theory. *Science* 216: 380–387.

Gould, S. J. (1983): A hearing for Vavilov. In *Hen's Teeth and Horse's Toes*. New York: W. W. Norton, pp. 134–146.

Gould, S. J., and N. Eldredge (1977): Punctuated equilibria: the tempo and mode of evolution reconsidered. *Paleobiology* 3: 115–151.

Gould, S. J., and R. Lewontin (1979): The spandrels of San Marco and the Panglossian paradigm: a critique of the adaptationist programme. *Proceedings of the Royal Society of London* 205: 581–598; reprinted in Sober (1984b).

Gould, S. J., and E. Vrba (1982): Exaptation—a missing term in the science of form. *Paleobiology* 8: 4–15.

Greeno, J. G. (1971): Explanation and information. In W. Salmon (ed.), *Statistical Explanation and Statistical Relevance*. Pittsburgh: University of Pittsburgh Press, 89–104.

Gruber, H. (1974): *Darwin on Man: A Psychological Study of Scientific Creativity*. London: Wildwood House.

Guhl, A., and W. Allee (1944): Some measurable effects of social organization in flocks of hens. *Physiological Zoology* 17: 320–347.

Hacking, I. (1965): *The Logic of Statistical Inference*. Cambridge: Cambridge University Press.

Hacking, I. (1975): The autonomy of statistical law. Talk delivered at the American Philosophical Association, unpublished.

Haldane, J. B. S. (1932): *The Causes of Evolution*. New York: Cornell University Press, 1966.

Haldane, J. B. S. (1964): A defence of beanbag genetics. *Perspectives in Biology and Medicine* 7: 343–359.

Hamilton, W. D. (1964): The genetical theory of social behavior: I and II. *Journal of Theoretical Biology* 7: 1–52.

Hamilton, W. D. (1967): Extraordinary sex ratios. *Science* 156: 477–488.

Hanson, N. (1958): *Patterns of Discovery*. Cambridge: Cambridge University Press.

Harper, J. (1977): *Population Biology of Plants*. London: Academic Press.

Hempel, C. G. (1945): On the nature of mathematical truth. *American Mathematical Monthly* 52: 543–556. Reprinted in R. Sleigh (ed.), *Necessary Truth*. Englewood Cliffs, N.J.: Prentice-Hall, 34–51.

Hempel, C. G. (1965): *Aspects of Scientific Explanation and Other Essays in the Philosophy of Science*. New York: Free Press.

Hempel, C. G., and P. Oppenheim (1948): Studies in the logic of explanation. *Philosophy of Science* 15: 135–175. Reprinted in Hempel (1965).

Herbert, S. (1971): Darwin, Malthus, and selection. *Journal of the History of Biology* 4: 209–217.

Hesse, M. (1969): Simplicity. In P. Edwards (ed.), *The Encyclopedia of Philosophy*. New York: Macmillan.

Hesslow, G. (1976): Two notes on the probabilistic approach to causality. *Philosophy of Science* 43: 290–292.

Hilts, V. (1973): Statistics and social science. In R. Giere and R. Westfall (eds.), *Foundations of Scientific Method in the Nineteenth Century*. Bloomington: Indiana University Press, pp. 206–33.

Hirshliefer, J. (1977): Economics from a biological viewpoint. *Journal of Law and Economics* 1: 1–52.

Hirshliefer, J. (1982): Evolutionary models in economics and law: cooperation versus conflict strategies. *Research in Law and Economics* 4: 1–60.

Hodge, M. J. (1983): The development of Darwin's general biological theorizing. In D. S. Bendall (ed.), *Evolution from Molecules to Men*. Cambridge: Cambridge University Press, pp. 43–60.

Horn, H. S. (1978): Optimal tactics of reproduction and life history. In J. R. Krebs and N. B. Davies (eds.), *Behavioral Ecology: An Evolutionary Approach*. Oxford: Blackwell, pp. 411–429.

Hull, D. (1965): The effect of essentialism on taxonomy: 2000 years of stasis. *British Journal for the Philosophy of Science* 15: 314–326; 16: 1–18.

Hull, D. (1976a): Central subjects and historical narratives. *History and Theory* 14: 253–274.

Hull, D. (1976b): Are species really individuals? *Systematic Zoology* 25: 174–191; reprinted in Sober (1984b).

Hull, D. (1978): A matter of individuality. *Philosophy of Science* 45: 335–360.

Hume, D. (1779): *Dialogues Concerning Natural Religion*. Indianapolis: Bobbs-Merrill, 1947.

Jacob, F. (1977): Evolution and tinkering. *Science* 196: 1161–1166.

Jammer, M. (1957): *Concepts of Force*. Cambridge: Harvard University Press.

Janzen, D. H. (1977): What are dandelions and aphids? *American Naturalist* 111: 586–589.

Jeffrey, R. (1971): Statistical explanation vs. statistical inference. In W. Salmon (ed.), *Statistical Explanation and Statistical Relevance*. Pittsburgh: University of Pittsburgh Press, pp. 19–28.

Kessler, G. (1980): Frege, Mill and the foundations of arithmetic. *Journal of Philosophy* 77: 65–79.

Kettlewell, H. B. (1973): *The Evolution of Melanism*. Oxford: Oxford University Press.

Kim, J. (1978): Supervenience and nomological incommensurables. *American Philosophical Quarterly* 15: 149–156.

Kimura, M. (1968): Evolutionary rate at the molecular level. *Nature* 217: 624–626.

Kimura, M. (1979): The neutral theory of molecular evolution. *Scientific American* 241:94–104.

Kimura, M. (1983): *The Neutral Theory of Molecular Evolution*. Cambridge: Cambridge University Press.

King, J. L., and T. H. Jukes (1969): Non-Darwinian evolution. *Science* 164: 788–798.

Kitcher, P. (1980): A priori knowledge. *The Philosophical Review* 89: 3–23.

Kitcher, P. (1982a): *Abusing Science: The Case against Creationism*. Cambridge: MIT Press.

Kitcher, P. (1982b): Genes. *British Journal for the Philosophy of Science* 33: 337–359.

Kitcher, P. (1984a): Darwin's achievement. N. Rescher (ed.), *Pittsburgh Studies in the Philosophy of Science*, forthcoming.

Kitcher, P. (1984b): Species. *Philosophy of Science*, forthcoming.

Kohn, D. (1980): Theories to work by: rejected theories, reproduction, and Darwin's path to natural selection. *Studies in the History of Biology* 4: 67–170.

Kojima, K. (1971): Is there a constant fitness value for a given genotype? No! *Evolution* 25: 281–285.

Kripke, S. (1972): Naming and necessity. In D. Davidson and G. Harman (eds.), *Semantics of Natural Language*. Dordrecht: Reidel.

Kripke, S. (1978): Lectures on time and identity, unpublished.

Kuhn, T. (1970): *The Structure of Scientific Revolutions.* Second edition. Chicago: University of Chicago Press.

Lamarck, J. B. (1809): *Zoological Philosophy.* London: Macmillan, 1914.

Lande, R. (1980): Microevolution in relation to macroevolution. *Paleobiology* 6: 233–238.

Laplace, P. (1814): *A Philosophical Essay on Probabilities.* New York: Dover, 1951.

Levene, H., O. Pavlovsky, and T. Dobzhansky (1954): Interaction of the adaptive values in polymorphic experimental populations of *Drosophila pseudoobscura. Evolution* 8: 335–349.

Levins, R. (1966): The strategy of model building in population biology. *American Scientist* 54: 421–431; reprinted in Sober (1984b).

Lewis, D. (1973): Causation. *Journal of Philosophy* 70: 556–567; reprinted in E. Sosa (ed.), *Causation and Conditionals.* Oxford: Oxford University Press, 1975.

Lewis, D. (1984): Causal explanation. In *Collected Papers.* Oxford: Oxford University Press.

Lewontin, R. (1970): The units of selection. *Annual Review of Ecology and Systematics* 1: 1–14.

Lewontin, R. (1974a). The analysis of variance and the analysis of cause. *American Journal of Human Genetics* 26: 400–411; reprinted in N. Block and G. Dworkin (eds.), *The IQ Controversy.* New York: Random House, 1976.

Lewontin, R. (1974b): *The Genetic Basis of Evolutionary Change.* New York: Columbia University Press; Introduction reprinted in Sober (1984b).

Lewontin, R. (1977): Biological determinism as a social weapon. In Ann Arbor Science for the People Editorial Collective, *Biology as a Social Weapon.* Minneapolis: Burgess, pp. 6–20.

Lewontin, R. (1978): Adaptation. *Scientific American* 239: 156–169.

Lewontin, R. (1980): Adattamento. *The Encyclopedia Einaudi.* Milan: Einaudi; reprinted in Sober (1984b).

Lewontin, R. (1983): Darwin's revolution. *New York Review of Books* 30: 21–27.

Lewontin, R., and L. C. Dunn (1960): The evolutionary dynamics of a polymorphism in the house mouse. *Genetics* 45: 705–722.

Lewontin, R., and M. White (1960): Interaction between inversion polymorphisms of two chromosome pairs in the grasshopper *Moraba scurra. Evolution* 14: 116–129.

Limoges, C. (1970): *La Selection Naturelle.* Paris: Presses Universitaires de France.

Lloyd, G. (1968): *Aristotle: The Growth and Structure of His Thought.* Cambridge: Cambridge University Press.

Lorenz, K. (1966): *On Aggression.* New York: Harcourt, Brace, and World.

Lovejoy, A. (1936): *The Great Chain of Being.* Cambridge: Harvard University Press.

Lucas, J. R. (1970): *The Concept of Probability.* Oxford: Clarendon Press.

Malthus, T. (1798): *An Essay on the Principle of Population.* New York: A. M. Kelley, 1965.

Martin, R. D. (1981): Relative brain size and basal metabolic rate in terrestrial vertebrates. *Nature* 293: 57–60.

Maynard Smith, J. (1964): Group selection and kin selection. *Nature* 201: 1145–1147.

Maynard Smith, J. (1974): The theory of games and the evolution of animal conflicts. *Journal of Theoretical Biology* 47: 209–221.

Maynard Smith, J. (1976): Group selection. *Quarterly Review of Biology* 51: 277–283.

Maynard Smith, J. (1978): Optimization theory in evolution. *Annual Review of Ecology and Systematics* 9: 31–56; reprinted in Sober (1984b).

Maynard Smith, J. (1978): *The Evolution of Sex.* Cambridge: Cambridge University Press.

Maynard Smith, J. (1983): *Evolution and the Theory of Games.* Cambridge: Cambridge University Press.

Mayr, E. (1961): Cause and effect in biology. *Science* 134: 1501–1506.

Mayr, E. (1963): *Animal Species and Evolution.* Cambridge: Harvard University Press.

Mayr, E. (1976a): Lamarck revisited. In *Evolution and the Diversity of Life.* Cambridge: Harvard University Press, pp. 222–250.

Mayr, E. (1976b): Typological versus population thinking. In *Evolution and the Diversity of Life.* Cambridge: Harvard University Press, pp. 26–29; reprinted in Sober (1984b).

Mayr, E. (1977): Darwin and natural selection. *American Scientist* 65: 321–327.

Mayr, E. (1982): *The Growth of Biological Thought.* Cambridge: Harvard University Press.

McColloch, J. R. (1830): *The Principles of Political Economy with a Sketch of the Rise and Progress of the Science.* Second edition. London: Longman's.

Michod, R. (1982): The theory of kin selection. *Annual Review of Ecology and Systematics* 13: 23–55.

Mills, S., and J. Beatty (1979): The propensity interpretation of fitness. *Philosophy of Science* 46: 263–288; reprinted in Sober (1984b).

Nagel, E. (1961): *The Structure of Science.* New York: Hackett.

Odum, E. P. (1963): *Ecology.* New York: Holt, Rinehart & Winston.

Orgel, L., and F. Crick (1980): Selfish DNA: the ultimate parasite. *Nature* 284: 604–607.

Ospovat, D. (1981): *The Development of Darwin's Theory.* Cambridge: Cambridge University Press.

Otte, R. (1981): A critique of Suppes' theory of probabilistic causality. *Synthese* 48: 167–181.

Owen, D. (1982): *Camouflage and Mimicry.* Chicago: University of Chicago Press.

Paley, W. (1819): *Natural Theology.* In *Collected Works.* London: Rivington.

Peters, R. (1976): Tautology in evolution and ecology. *American Naturalist* 110: 1–12.

Piaget, J. (1959): *The Language and Thought of the Child.* London: Routledge & Kegan Paul.

Popper, K. (1963): *Conjectures and Refutations.* London: Hutchinson.

Preuss, A. (1975): *Science and Philosophy in Aristotle's Biological Works.* New York: Georg Olms.

Price, G. (1970): Selection and covariance. *Nature* 227: 520–521.

Provine, W. (1971): *The Origins of Theoretical Population Genetics.* Chicago: University of Chicago Press.

Provine, W. (1984): Adaptation and mechanisms of evolution after Darwin: a study in persistent controversies. In D. Kohn (ed.), *The Darwinian Heritage.* Princeton: Princeton University Press.

Putnam, H. (1962): The analytic and the synthetic. In *Mind, Language, and Reality.* Cambridge: Cambridge University Press, 1975, pp. 33–69.

Putnam, H. (1967): The nature of mental states. In *Mind, Language, and Reality.* Cambridge: Cambridge University Press, 1975, pp. 429–440.

Putnam, H. (1970): Is semantics possible? In *Mind, Language, and Reality.* Cambridge: Cambridge University Press, 1975, pp. 139–152.

Putnam, H. (1975): The refutation of conventionalism. In *Mind, Language, and Reality.* Cambridge: Cambridge University Press, 1975, pp. 153–191.

Putnam, H. (1978): There is at least one a priori truth. *Erkenntnis* 13: 153–170.

Quine, W. V. O. (1953): Two dogmas of empiricism. In *From a Logical Point of View.* Cambridge: Harvard University Press, 1980, pp. 20–46.

Quine, W. V. O. (1960): *Word and Object.* Cambridge: MIT Press.

Raup, D., and S. J. Gould (1974): Stochastic simulation and evolution of morphology— towards a nomothetic paleontology. *Systematic Zoology* 23: 305–322.

Reichenbach, H. (1956): *The Direction of Time.* Berkeley: University of California Press.

Rendel, J. (1967): *Canalization and Gene Control.* London: Logos Press.

Richerson, P. J., and R. Boyd (1978): A dual inheritance model of the human evolutionary process I: basic postulates and a simple model. *Journal of Social and Biological Structure* 1: 127–154.

Rosen, D. (1978): In defence of a probabilistic theory of causality. *Philosophy of Science* 45: 604–613.

Rosenberg. A. (1978): The supervenience of biological concepts. *Philosophy of Science* 45: 368–386; reprinted in Sober (1984b).

Roughgarden, J. (1975): Evolution of marine symbiosis—a simple cost benefit model. *Ecology* 56: 1201–1208.

Roughgarden, J. (1978): *Theory of Population Genetics and Evolutionary Ecology*. New York: Macmillan.

Ruse, M. (1975): Charles Darwin and artificial selection. *Journal of the History of Ideas* 36: 339–350.

Ruse, M. (1979): *The Darwinian Revolution: Science Red in Tooth and Claw*. Chicago: University of Chicago Press.

Ruse, M. (1980): Charles Darwin and group selection. *Annals of Science* 37: 615–630.

Russell, B. (1913): On the notion of cause. *Proceedings of the Aristotelian Society* 13: 1–26.

Salmon, W. (1971): Statistical explanation. In W. Salmon (ed.), *Statistical Explanation and Statistical Relevance*. Pittsburgh: University of Pittsburgh Press, pp. 29–88.

Salmon, W. (1975): Theoretical explanation. In S. Korner (ed.), *Explanation*. Oxford: Blackwell.

Salmon, W. (1978): Why ask "Why?" An inquiry concerning scientific explanation. In *Hans Reichenbach: Logical Empiricist*. Dordrecht: Reidel.

Salmon, W. (1980): Probabilistic causality. *Pacific Philosophical Quarterly* 61: 59–74.

Schweber, S. (1977): The origin of the *Origin* revisited. *Journal of the History of Biology* 10: 229–316.

Schweber, S. (1980): Darwin and the political economists. *Journal of the History of Biology* 13: 195–289.

Schweber, S. (1982): Demons, angels, and probability: some aspects of British science in the nineteenth century. In A. Shimony and H. Feshback (eds.), *Physics and Natural Philosophy*. Cambridge: MIT Press.

Schweber, S. (1983): Factors ideologique et intellectuels dans la genese de la theorie de la selection naturelle. In Y. Conry (ed.), *De Darwin au Darwinisme: Science et Ideologie*. Paris: Libraire Philosophique J. Vrin.

Scriven, M. (1959): Explanation and prediction in evolutionary theory. *Science* 130: 477–482.

Simon, H. (1957): *Models of Man*. New York: Wiley.

Simpson, E. H. (1951): The interpretation of interaction in contingency tables. *Journal of the Royal Statistical Society B* 13: 238–241.

Simpson, G. G. (1953): *The Major Features of Evolution*. Chicago: University of Chicago Press.

Sklar, L. (1970): Is probability a dispositional property? *Journal of Philosophy* 67: 355–366.

Skyrms, B. (1980): *Causal Necessity*. New Haven: Yale University Press.

Smith, A. (1776): *An Inquiry into the Nature and Causes of the Wealth of Nations*. New York: The Modern Library, 1937.

Sober, E. (1975): *Simplicity*. Oxford: Oxford University Press.

Sober, E. (1980): Evolution, population thinking, and essentialism. *Philosophy of Science* 47: 350–383.

Sober, E. (1981a): Evolutionary theory and the ontological status of properties. *Philosophical Studies* 40: 147–176.

Sober, E. (1981b): Holism, individualism, and units of selection. In R. Giere and P. Asquith (eds.), *PSA 1980*: volume 2, East Lansing: Philosophy of Science Association. 93–121; reprinted in Sober (1984b).

Sober, E. (1981c): The principle of parsimony. *British Journal for the Philosophy of Science* 32: 145–156.

Sober, E. (1981d): Revisability, a priori truth, and evolution. *Australasian Journal of Philosophy* 59: 68–85.

Sober, E. (1982a): Frequency-dependent causation. *Journal of Philosophy* 79: 247–253.

Sober, E. (1982b): Why must homunculi be so stupid? *Mind* 91: 420–422.

Sober, E. (1983a): Dispositions and subjunctive conditionals; or, dormative virtues are no laughing matter. *Philosophical Review* 91: 591–596.

Sober, E. (1983b): Equilibrium explanation. *Philosophical Studies* 43: 201–210.

Sober, E. (1983c): Parsimony in systematics: philosophical issues. *Annual Review of Ecology and Systematics* 14: 335–358.

Sober E. (1983d): Parsimony methods in systematics. In N. Platnick and V. Funk (eds.), *Advances in Cladistics*, vol. 2, New York: Columbia University Press, pp. 37–48.

Sober, E. (1984a): Common cause explanation. *Philosophy of Science*, forthcoming.

Sober, E. (1984b): *Conceptual Issues in Evolutionary Biology: An Anthology.* Cambridge: Bradford/MIT Press.

Sober, E. (1984c): Fact, fiction, and fitness: A reply to Rosenberg. *Journal of Philosophy*, forthcoming.

Sober, E. (1984d): Sets, species, and natural kinds: Comments on Philip Kitcher's paper. *Philosophy of Science*, forthcoming.

Sober, E., and R. Lewontin (1982): Artifact, cause, and genic selection. *Philosophy of Science* 47: 157–180; reprinted in Sober (1984b).

Sokal, R., and F. Rohlf (1969): *Biometry.* San Francisco: W. H. Freeman.

Stampe, D. (1977): Towards a causal theory of linguistic representation. In P. French (ed.), *Midwest Studies in Philosophy* 2: 42–63.

Stanley, S. (1975): A theory of evolution above the species level. *Proceedings of the National Academy of Sciences, USA* 72: 646–650.

Stanley, S. (1979): *Macroevolution: Pattern and Process.* San Francisco: W. H. Freeman.

Suppes, P. (1970): *A Probabilistic Theory of Causality.* Amsterdam: North Holland Publishing Co.

Suppes, P., and M. Zinotti (1976): On the determinism of hidden variable theories with strict correlation and conditional statistical independence of observables. In P. Suppes, *Logic and Probability in Quantum Mechanics.* Dordrecht: Reidel.

Templeton, A. (1980): Macroevolution. *Evolution* 34: 1224–1227.

Templeton, A. (1982): Adaptation and the integration of evolutionary forces. In R. Milkman (ed.), *Perspectives on Evolution.* Sunderland: Sinauer, 15–31.

Trivers, R. (1971): The evolution of reciprocal altruism. *Quarterly Review of Biology* 46: 35–57.

Van Fraassen, B. C. (1977): Relative frequencies. In W. Salmon (eds.), *Hans Reichenbach: Logical Empiricist.* Dordrecht: Reidel, pp. 129–167.

Van Fraassen, B. C. (1980): *The Scientific Image.* Oxford: Oxford University Press.

Van Valen, L. (1973): A new evolutionary law. *Evolutionary Theory* 1: 1–30.

Vorzimmer, P. (1970): *Charles Darwin: The Years of Controversy.* Philadelphia: Temple University Press.

Vrba, E. (1980): Evolution, species, and fossils: how does life evolve? *South African Journal of Science* 76: 61–84.

Waddington, C. (1957): *The Strategy of the Genes.* London: Allen and Unwin.

Wade, M. (1976): Group selection among laboratory populations of *Tribolium. Proceedings of the National Academy of Sciences, USA* 73: 4604–4607.

Wade, M. (1978): A critical review of the models of group selection. *Quarterly Review of Biology* 53: 101–114.

Wallace, A. (1905): *Darwinism*. London: Macmillan.

Watson, J. D. (1976): *The Molecular Biology of the Gene*. Third edition. Menlo Park: W. A. Benjamin.

Weismann, A. (1889): The continuity of germ-plasm as the foundation of a theory of heredity. In *Essays upon Heredity Kindred and Biological Problems*. Oxford: Clarendon Press, pp. 161–248.

Wickler, W. (1968): *Mimicry in Plants and Animals*. London: Weidenfeld and Nicolson.

Wiggins, D. (1980): *Sameness and Substance*. Oxford: Oxford University Press.

Williams, G. C. (1966): *Adaptation and Natural Selection*. Princeton: Princeton University Press.

Williams, G. C. (1975): *Sex and Evolution*. Princeton: Princeton University Press.

Williams, G. C., and D. C. Williams (1957): Natural selection of individually harmful social adaptations among sibs with special reference to social insects. *Evolution* 11: 32–39.

Williams, M. B. (1970): Deducing the consequences of evolution: a mathematical model. *Journal of Theoretical Biology* 29: 343–385.

Williams, M. B. (1973): The logical status of the theory of natural selection and other evolutionary controversies. In M. Bunge (ed.), *The Methodological Unity of Science*. Dordrecht: Reidel; reprined in Sober (1984b).

Wilson, D. S. (1975): A theory of group selection. *Proceedings of the National Academy of Sciences, USA* 72: 143–146.

Wilson, D. S. (1980): *The Natural Selection of Populations and Communities*. Menlo Park: Benjamin/Cummings.

Wilson, D. S. (1983): The group selection controversy: history and current status. *Annual Review of Ecology and Systematics* 14: 159–188.

Wilson, D. S., and R. Colwell (1981): Evolution of sex ratio in structured demes. *Evolution* 35: 882–897.

Wilson, E. O. (1975): *Sociobiology: The New Synthesis*. Cambridge: Harvard University Press; abridged edition, 1980.

Wimsatt, W. (1980): Reductionistic research strategies and their biases in the units of selection controversy. In T. Nickles (ed.), *Scientific Discovery*, vol. 2, Dordrecht: Reidel.

Wimsatt, W. (1981): The units of selection and the structure of the multi-level genome. In P. Asquith and R. Giere (eds.), *PSA 1980*, vol. 2, East Lansing: Philosophy of Science Association, pp. 122–186.

Wright, L. (1973): Functions. *Philosophical Review* 85: 70–86; reprinted in Sober (1984b).

Wright, S. (1931): Evolution in Mendelian populations. *Genetics* 16: 97–159.

Wright, S. (1945): Tempo and mode in evolution: a critical review. *Ecology* 26: 415–419.

Wright, S. (1967): Comments on the preliminary working papers of Eden and Waddington. In P. Moorehead and M. Kaplan (eds.), *Mathematical Challenges to the Neo-Darwinian Theory of Evolution*. Wistar Institute Symposium 5: 117–120.

Wright, S. (1978): *Evolution and the Genetics of Populations*. Vol. 4: *Variability within and among Natural Populations*. Chicago: University of Chicago Press.

Wright, S. (1980): Genic and organismic selection. *Evolution* 34: 825–843.

Wynne-Edwards, V. C. (1962): *Animal Dispersion in Relation to Social Behavior*. Edinburgh: Oliver and Boyd.

Young, R. (1971): Darwin's metaphor: does nature select? *The Monist* 55: 442–503.

Index

Adaptation. *See also* Fitness; Natural
  selection; Genic selectionism
  adaptive peaks and, 26, 320 ff., 357
  definition of, 208
  evolutionary versus ontogenetic, 203 ff.
  group versus individual, 2–5, 255 ff.
  response and, 202
  selection and, 196 ff.
Adaptationism, 24–27
Alexander, R., 332, 334 ff.
Allee, W., 222
Altruism, 26, 184, 216, 239–240, 262 ff.,
  326 ff.
ANOVA, 266 ff., 307–308, 317, 333
A priori truth
  causation and, 77–78
  empirical revisability and, 65, 73
  explanation and, 62, 74–75, 79–80
  natural selection and, 27–28, 63 ff.
  negative and positive characterizations
  of, 64 ff.
  prediction and, 62
  Quine's critique of, 62
Aristotle, 32, 72, 157 ff., 161–163, 270
Arnold, T., 275
Artificial selection, 17–20, 201
Assimilation, genetic, 199 ff.
Ayala, F., 358, 359

Beatty, J., 48, 64, 74, 138
Behaviorism, 154
Block, N., 49, 84, 108, 128
Borgia, G., 332, 334
Brandon, R., 48, 64, 74, 229
Byerly, H., 298

Canalization, 200
Carr-Saunders, A., 224

Cartwright, N., 32, 286, 291, 296
Causality. *See also* A priori truth;
  Correlation; Determinism
  artifacts and, 2 ff., 256 ff., 311–312, 365
  averaging and, 89 ff., 133–134,
  303–304, 311
  context-dependence of, 230–233, 263,
  266 ff., 273, 297, 313–314
  determinism and, 143–144
  disjunctive properties and, 92 ff., 102,
  209–210
  events versus standing conditions,
  77–78
  explanation and, 139–142
  frequency dependent, 287–288
  individual- versus population-level,
  295–296
  natural selection and, 100–101, 255,
  278
  Pareto-formulation, 293–294, 301, 331
  principle of common cause and,
  292–293
  probabilistic criterion, 285 ff., 293, 294,
  301
  Simpson's Paradox and, 285, 286,
  326 ff.
  status as scientific, 7–8
  transitivity of, 227–230, 233, 298–299,
  313
Chance. *See also* Determinism; Fitness;
  Probability
  explanation and, 143–144
  random genetic drift and, 25–26, 34,
  103, 110 ff., 124, 130, 153
  survival/reproduction and, 44, 46, 103,
  261
Colwell, R., 342 ff.
Competition, 17, 186–187